T0210752

Graduate Texts in Physics

Graduate Texts in Physics publishes core learning/teaching material for graduate- and advanced-level undergraduate courses on topics of current and emerging fields within physics, both pure and applied. These textbooks serve students at the MS- or PhD-level and their instructors as comprehensive sources of principles, definitions, derivations, experiments and applications (as relevant) for their mastery and teaching, respectively. International in scope and relevance, the textbooks correspond to course syllabi sufficiently to serve as required reading. Their didactic style, comprehensiveness and coverage of fundamental material also make them suitable as introductions or references for scientists entering, or requiring timely knowledge of, a research field.

Rainer Oloff

The Geometry
of Spacetime

A Mathematical Introduction to
Relativity Theory

 Springer

Rainer Oloff
Fakultät für Mathematik und Informatik
Universität Jena
Jena, Thüringen, Germany

ISSN 1868-4513   ISSN 1868-4521   (electronic)
Graduate Texts in Physics
ISBN 978-3-031-16141-4   ISBN 978-3-031-16139-1   (eBook)
https://doi.org/10.1007/978-3-031-16139-1

Translation from the German language edition: "Geometrie der Raumzeit" by Rainer Oloff, © Springer Spektrum 2018. Published by Springer Spektrum. All Rights Reserved.

This Springer imprint is published by the registered company Springer Nature Switzerland AG
The registered company address is: Gewerbestrasse 11, 6330 Cham, Switzerland

Preface

This textbook on the Theory of Relativity is a translation of my textbook 'Geometrie der Raumzeit' which was originally published by Vieweg-Verlag in 1999. Since then, several editions have followed, most recently the sixth edition by Springer in 2018. The Theory of Relativity was developed by Albert Einstein at the beginning of the twentieth century. Still today, it is regarded to be a complicated and difficult to understand area of theoretical physics. One approach to understanding this complex theory is the application of mathematics and more specifically the use of differential geometry on manifolds. This is what I set out to do in my original work. Tensors are used for describing curvature. They are introduced as multilinear forms. Rotating and non-rotating black holes are also discussed. I wish to thank Prof. Amol Sasane for providing the initial translation. I thank my grandson Tom Zierbock and his wife Lea for digitising and typesetting the LaTex sources. Furthermore, I thank Angela Lahee, Chandra Sekaran Arjunan and everyone at Springer for their constructive support of this project.

Jena, Germany   Rainer Oloff
March 2022

Introduction

In Maxwell's electrodynamics theory completed in 1860, it was concluded that light is an electromagnetic wave. Since usually wave phenomena require a medium, it was proposed that there is a fictitious medium called **ether**, which carries the electric and magnetic field strengths. In a frame which is at rest relative to this ether, the speed of light is expected the same in all directions. In a second frame that is moving with respect to the first one, the direction of motion is expected to influence the speed of light. However, scientists have failed to confirm this experimentally and in all experiments done since 1881, the speed of light was observed to be independent of the motion of the reference frame.

Newtonian mechanics is based on the basic law that *force is mass times acceleration*, in other words that *force is the time derivative of momentum*. This applies to the particle path $x(t)$, $y(t)$, $z(t)$ in inertial frames: coordinate systems in which particles are either at rest or move in a straight line at constant velocity given no forces are acting on them. A different coordinate system x', y', z' which is related to x, y, z by the Galileo-Transformation $x' = x - vt$, $y' = y$, $z' = z$ is also inertial. Also part of the Galileo-Transformation is the identity $t' = t$ which is self-evident when considering the synchronisation of the clocks of both reference frames.

The validity of Newton's laws of inertial systems gives rise to the Galilean relativity principle that velocity of intertial reference frames cannot be determined by mechanical experiments. In his 1905 published special theory of relativity, Einstein formulated his special relativity principal stating that all inertial reference frames are equivalent, that is, in all inertial reference frames all physical laws are the same. Together with the constancy of the speed of light c, he showed that in a coordinate system x', y', z' in which the origin of an inertial frame x, y, z moves with a velocity of $(v, 0, 0)$, one has

$$x' = \frac{x - vt}{\sqrt{1 - (v/c)^2}}$$

$$y' = y$$

$$z' = z$$

$$t' = \frac{t - vx/c^2}{\sqrt{1 - (v/c)^2}}$$

([R] page 49). This mapping of x, y, z, t to x', y', z', t' is called a *Lorentz trans-formation*. It was derived by H. A. LORENTZ in 1895 within another context. What is remarkable about this transformation is that it mixes space and time. There are two noteworthy conclusions we can draw from this ([Schr] pages 36, 38): *Length contraction* (*a rod along the x'-axis of length l has a length of only $l\sqrt{1 - (v/c)^2}$ in the x-coordinate system*) and **time dilation** (*a time interval Δt in one coordinate system is measured as $\Delta t / \sqrt{1 - (v/c)^2}$ in the other one*).

In Newtonian mechanics, the speeds in an unprimed coordinate system are obtained by subtracting $(v, 0, 0)$ from the speeds in the primed system. In special relativity, this is only valid if v is negligible in comparison to c. A particle has the velocity (v_x, v_y, v_z) in the unprimed system, and the velocity $(v_{x'}, v_{y'}, v_{z'})$ in the primed system, then

$$v_{x'} = \frac{v_x - v}{1 - vv_x/c^2}$$

$$v_{y'} = \frac{v_y\sqrt{1 - (v/c)^2}}{1 - vv_x/c^2}$$

$$v_{z'} = \frac{v_z\sqrt{1 - (v/c)^2}}{1 - vv_x/c^2}$$

([Schr] page 43).

While in Newtonian physics velocities are added as vectors, this needs to be revised in relativity physics, since otherwise speeds larger than c are possible. We now formulate the special relativity law by setting $v = -v_1$ and $(v_x, v_y, v_z) = (v_2, 0, 0)$ in the above formulae. Then the x-component of the sum of $(v_1, 0, 0)$ and $(v_2, 0, 0)$ is given by $(v_1 + v_2)/(1 + v_1 v_2/c^2)$. The argument of this number $|v_i| < c$ is again less than c since from

$$0 < (c - v_1)(c - v_2) = c(c + v_1 v_2/c - (v_1 + v_2))$$

follows

$$v_1 + v_2 < c(1 + v_1 v_2/c^2).$$

The basic concepts of special relativity theory are expressed most simply in Minkowski space $\mathbb{R}^4$, equipped with the bilinear form

$$g((\xi^0, \xi^1, \xi^2, \xi^3), (\eta^0, \eta^1, \eta^2, \eta^3)) = c^2\xi^0\eta^0 - \xi^1\eta^1 - \xi^2\eta^2 - \xi^3\eta^3,$$

The canonical basis

$$e_0 = (1, 0, 0, 0)$$

$$e_1 = (0, 1, 0, 0)$$

$$e_2 = (0, 0, 1, 0)$$

$$e_3 = (0, 0, 0, 1)$$

has the property $g(e_0, e_0) = c^2$, $g(e_i, e_i) = -1$ for $i = 1, 2, 3$ and $g(e_j, e_k) = 0$ for $j \neq k$. Numerous other bases also have this property, and we call such as Basis a *Lorentz* basis. In particular for every β, the vectors

$$e_0' = \frac{e_0 + \beta c e_1}{\sqrt{1 - \beta^2}}$$

$$e_1' = \frac{e_1 + (\beta/c)e_0}{\sqrt{1 - \beta^2}}$$

$$e_2' = e_2$$

$$e_3' = e_3$$

formulate a *Lorentz* basis. The coordinates t', x', y', z' for the primed basis are related to the coordinates t, x, y, z of the canonical basis by

$$t' = \frac{t - (\beta/c)x}{\sqrt{1 - \beta^2}}$$

$$x' = \frac{x - \beta ct}{\sqrt{1 - \beta^2}}$$

$$y' = y$$

$$z' = z.$$

This is the Lorentz transformation for $v = \beta c$. As suggested by the notation, the 0th coordinate is interpreted as time, the other terms being spacial coordinates.

The 4-velocity of a particle, with speed (v_x, v_y, v_z) in a given inertial frame, is the following vector in Minkowski space, with respect to the corresponding Lorentz basis:

$$\left(\frac{c}{\sqrt{1-\beta^2}}, \frac{v_x}{\sqrt{1-\beta^2}}, \frac{v_y}{\sqrt{1-\beta^2}}, \frac{v_z}{\sqrt{1-\beta^2}} \right)$$

where $\beta = \sqrt{(v_x)^2 + (v_y)^2 + (v_z)^2}/c$ ([R] page 106]). The 4-momentum ([R] page 111) is

$$(p_0, p_x, p_y, p_z) = \left(\frac{cm_0}{\sqrt{1-\beta^2}}, \frac{m_0 v_x}{\sqrt{1-\beta^2}}, \frac{m_0 v_y}{\sqrt{1-\beta^2}}, \frac{m_0 v_z}{\sqrt{1-\beta^2}} \right)$$

obtained by multiplying the 4-velocity by the number m_0, interpreted as the *rest* mass. The expression

$$m = \frac{m_0}{\sqrt{1-\beta^2}}$$

is the *relativistic* mass and the triple

$$(p_x, p_y, p_z) = (mv_x, mv_y, mv_z)$$

is the relativistic momentum. The 0th component $p_0 = mc$ is up to the factor c, the *Energy* E, $E = mc^2$ ([R] page 110).

The electromagnetic field in Maxwell's electrodynamics is described by the electric field (E_x, E_y, E_z) and the magnetic field (B_x, B_y, B_z). A particle with charge e and velocity (v_x, v_y, v_z) experiences the Lorentz force

$$e[(E_x, E_y, E_z) + (v_x, v_y, v_z) \times (B_x, B_y, B_z)].$$

In relativity theory, the components of the field strength are described by the antisymmetric matrix

$$(F_{ik}) = \begin{pmatrix} 0 & \frac{1}{c}E_x & \frac{1}{c}E_y & \frac{1}{c}E_z \\ -\frac{1}{c}E_x & 0 & -B_z & B_y \\ -\frac{1}{c}E_y & B_z & 0 & -B_x \\ -\frac{1}{c}E_z & -B_y & B_x & 0 \end{pmatrix}$$

([R] page 143). Applying this matrix to the 4-velocity of the particle gives the time derivation of the 4-momentum. For conversion of the field strength components to another frame of reference moving in the x-direction with a velocity v, one has the following formulae ([R] page 144):

$$
\begin{aligned}
E_{x'} &= E_x & B_{x'} &= B_x \\
E_{y'} &= \frac{E_y - v B_z}{\sqrt{1-(v/c)^2}} & B_{y'} &= \frac{B_y + (v/c^2) E_z}{\sqrt{1-(v/c)^2}} \\
E_{z'} &= \frac{E_z + v B_y}{\sqrt{1-(v/c)^2}} & B_{z'} &= \frac{B_z - (v/c^2) E_y}{\sqrt{1-(v/c)^2}}.
\end{aligned}
$$

Thus ends our brief discussion of the basic concepts of the special theory of relativity, at least in general to those notions which will play a role in the following chapters. For further motivations, more detailed explanations and applications, we refer to [R] and [Sch]. Following on from here, no knowledge of the special theory of relativity is assumed, and all necessary terms are systematically introduced from a mathematical point of view.

In the general theory of relativity, published in summary in 1916, A. Einstein gave the concept of *gravitation* a completely new content. The starting point of his considerations were apparently the following two principles. One is the point of view formulated by E. Mach in 1893 that a body in an otherwise empty universe would have no inertial properties. Since motion can only ever be described in relation to other masses, the notion of acceleration only makes sense with respect to a certain distribution of matter, i.e., the distribution of matter determines the geometry of space. The second principal is the *equivalency principal*, which is the observation that the inertial mass is equal to the gravitational mass. This means body B is twice as inertially massive as body A in terms of acceleration produced for the same applied force on each of them, and the body B is also twice as heavy as body A in a gravitational field (passive gravity), and also generates a gravitational field twice as strong as body A in the sense of Newton's laws of gravitation (active gravity).

The fundamental message of the general theory of relativity is that the distribution of matter determines the curvature of spacetime and that particles and photons move along geodesics. In other words [MTW]: *M*atter tells space how to curve, and space tells matter how to move. The relationship between matter and curvature is given in the famous *E*instein field equations. The components of two symmetric, doubly covariant tensors are equated up to a factor. On one side is the *E*instein tensor determined by the curvature, and on the other side the *e*nergy-momentum tensor describing the distribution of matter.

A standard task within general relativity is to construct spacetime with its metric for a given distribution of matter. For the simplest case of a non-rotating fixed star, this was carried out as early as 1915 by K. Schwarzschild. By an application of Einstein's theory, he determined the line element as

$$ds^2 = (1 - \frac{2G\mu}{c^2 r})c^2 dt^2 - \frac{1}{1 - \frac{2G\mu}{c^2 r}} dr^2 - r^2 d\vartheta^2 - r^2 \sin^2\vartheta\, d\varphi^2.$$

Here, μ is the mass of the fixed star, G is the gravitational constant and r, ϑ, φ are the spherical coordinates.

The rotation of our Sun can be neglected. The solution to the geodesic problem in Schwarzshild spacetime provides results that are in even better agreement with astronomical data than classical Newtonian celestrial mechanics. As is well known, according to Kepler's laws, the planets move in elliptical orbits, with the Sun at one focus. Since the planets also influence each other through gravitation, the nearest point to the Sun along the ellipse (perihelion) can shift. It has been known since the mid-nineteenth century that Mercury's perihelion revolves around the Sun at an angular velocity of 43 arc s per year. The theory of relativity exactly accounts for this value. The perihelion rotation effect is of course also present in the other planets far from the Sun, but it is not so pronounced there.

The theory of relativity received another brilliant confirmation by observing the deflection of light. In the Schwarzschild spacetime, you can precisely calculate the angle of deflection of a light beam that passes the fixed star. As is well known, this effect is not included in classical radiation optics and was not initially noticed by astronomers. In 1919, this phenomenon, predicted by A. EINSTEIN, was targeted and verified during a solar eclipse. It was actually found that stars that were visible close to the edge of the darkened Sun disk were slightly shifted outwards from their position on the fixed star sky. This result contributed significantly to the recognition of the theory of relativity in the professional world. Nowadays, astronomers can use powerful telescopes to convince themselves that distant star systems act as **gravitational lenses**.

The (outer) Schwarzschild metric applies only to the area outside the fixed star, and there the radial coordinate r is greater than its radius. The fact that the metric coefficient $r_\mu = 2G\mu/c^2$ becomes singular, and that particles and photons cannot escape in the range $0 < r < r_\mu$, initially received little attention, since in the case of our Sun, this number r_μ is just under 3km, while the radius of the Sun is almost 696,000 km. A fixed star having a huge density so that its radius is smaller than r_μ was considered impossible for a long time. Finally in the 1960s, the realisation prevailed that, under certain circumstances, stars can collapse into practically a point at the end of their life. J. A. WHEELER coined the term **black hole** for this. Today, it is believed that black holes are quite commonplace in the universe. Further details on black holes and spectacular consequences described in popular science language can be found in [L].

We now give a brief overview of the contents. The following chapters deal with the mathematical treatment of relativity. The mathematical preliminaries are introduced systematically. The spacetime model is formulated axiomatically, and the most important relativistic effects are derived from it. In this regard, this textbook is based on the following quote from A. EINSTEIN: *The greatest task of physicists*

*is to seek out those most general laws from which the world view can be obtained by
pure deduction. There is no logical path to these elementary laws, only the intuition
based on experience* ([Schr], p. 9). Of course, this textbook is only an exposition
of known things. The main tool is those areas of mathematics that are called *new
differential-geometric methods* in the table of contents of [SU] two decades ago
and were also several decades old at that time. The fact that local coordinates can
be used everywhere requires the concept of an n-dimensional manifold, introduced
systematically in Chap. 1. In Chap. 2, tangent vectors are introduced as functionals
on a function space. This gives one the n-dimensional tangent space at every point,
on which tensor algebra (Chap. 3) can then be developed which is also of interest
from the point of view of physical applications. Basically, tensors are multilinear
forms, which, depending on their type, operate on a certain number of linear forms
and vectors. The selection of a coordinate basis on the tangent spaces gives rise
to a basis in each tensor space. Accordingly, every change of the coordinate basis
induces a change of tensor components. One thus reaches the standpoint of classical theoretical physics where one understands a tensor as a system of components
with rules for their transformation under coordinate system changes.

In the theory of relativity, tensors are usually referred to as tensor fields
(Chap. 4). The metric tensor in particular is a doubly covariant symmetric tensor field. It is important for the formulation of the special theory of relativity that
this bilinear form is indefinite at each point of the four-dimensional manifold; if
the component matrix is diagonal, then there must be one positive and three negative numbers in the diagonal (one negative and three positive numbers can also be
required, but this is a question of convention). At the end of Chap. 4, the concept
of spacetime is then clarified.

In Chap. 5, the most important concepts of special relativity are introduced.
The observer takes the place of the inertial system, according to the logic of having a coordinate-free representation. Without using the Lorentz transformation,
how geometric-physical objects are transformed from one observer to another is
determined. In order to avoid index pulling as much as possible and to make the
formulae as simple as possible, momentum is interpreted as a linear form. Over
80 years ago, H. WEYL noted that the force should be better understood as a linear
form ([We], p. 34). Roughly speaking, the force linearly assigns the work to the
path. According to Newton's law, *force is the rate of change of momentum*, and so
momentum is also a linear form.

A differential form is a skew-symmetric covariant tensor field. Differentiation
and integration also do not depend on a metric. Upon differentiating (Chap. 6) the
order increases by one. Integration (Chap. 13) only makes sense for a differential
form whose order is the dimension of the manifold. When the support of the
differential form is contained in a chart, one inserts its coordinate vector fields
into the differential form and calculates the integral of the resulting real-valued
function. If the support is not contained in a chart, the differential form must be
decomposed into appropriate summands. In no case the unpleasant but widespread
concept of tensor density is required.

The covariant derivative of vector fields (Chap. 7) essentially depends on the metric and is used decisively in the definition of the curvature tensor (Chap. 8). Examples are dealt with in detail: the curved surface in three-dimensional space (for methodological reasons) and the Schwarzschild spacetime (for later usage). It turns out that the curvature scalar of a curved surface is twice the Gaussian curvature. As a by-product, one obtains an elegant proof of the famous theorema egregium of C. F. GAUSS, which says that the Gaussian curvature (so-called today) is bending-invariant.

In Chap. 9, Einstein's field equation is formulated. The main difficulty is the explanation of the relativistic energy-momentum tensor (stress-energy tensor) and the comparison with the Newtonian concepts of energy density, energy current density, momentum density and momentum current density. The energy-momentum tensor is determined for ideal flow (perfect fluid) and the electromagnetic field.

The axiomatic meaning of the concept of a geodesic (Chap. 10) has already been mentioned. The relativistic effects in our solar system can now be recalculated here. Covariant derivation and Lie derivation of tensor fields are deliberately dealt with in later chapters (Chaps. 11 and 12). While the special cases of the covariant derivation of vector fields and Lie brackets are used in important situations, the more general concepts have no such fundamental meaning for a first understanding of the theory of relativity and are also laboriously exhaustive to explain. Black holes (Chap. 14) have already been mentioned. Fortunately, the effects discussed in popular science media can be understood mathematically here. In Chap. 15, the Robertson-Walker metric is derived from the cosmological principle and thus cosmological models are constructed. In the middle of Chap. 16, the Kerr metric is discussed, which describes the gravitation field of a collapsing rotating star.

In contrast to this introductory chapter, the main text uses **relativistic (geometric) units**. These are those in which the speed of light c and the gravitational constant G are equal to 1. The standard unit of length is 1m. Simultaneously, this is also a unit of time, namely the time in which light covers this distance. As a result, one second is the same as $299, 793 \cdot 10^8$m. The gravitational constant G, in classical units, is $6, 673 \cdot 10^{-11}$m^3/(kgs^2). When one converts seconds into meters, one has

$$\frac{6, 673}{(2, 99793)^2} \cdot 10^{-27} \frac{m}{kg} = 7, 425 \cdot 10^{-28} \ m/kg.$$

The requirement that $G = 1$ implies that $1 \, kg = 7, 425 \cdot 10^{-28}$ m. For example, our Sun has a mass of

$$1, 989 \cdot 10^{30} \, kg = 1, 989 \cdot 7, 425 \cdot 10^2 \ m = 1, 477 \cdot 10^3 \ m.$$

This text provides enough material for a one-semester course (with two double hours plus one double hour of exercises per week). Even if less time is available, it is possible to skip Chaps. 6, 11, 12, 13 and Sects. 1.2, 1.3, 5.3, 7.4, 8.2, 9.2, 9.4, 9.5 and 9.8 to get to black holes.

This textbook is deliberately not titled *Theory of Relativity for Mathematicians*. All that is needed is linear algebra and differential and integral calculus of functions of several variables, and the required level of abstraction does not go beyond this level. The reader who is only interested in physics can ignore the proofs, but the basic mathematical concepts should still be understandable.

Contents

Differentiable Manifolds

1.1 Charts and Atlases

Surfaces in three-dimensional space are often described by a parametrisation. A map φ assigns a pair of parameter values u and v to each point P of the surface M. The map φ from M to the parameter set $\Gamma \subseteq \mathbb{R}^2$ is bijective, i.e., it is invertible. A parametrisation is usually specified by giving three real-valued functions which are the three Cartesian coordinates x, y, z as functions of u and v for the inverse map φ^{-1}. Let us consider the usual parametrisation of a cylinder by the three equations $x = r \cos u, y = r \sin u, z = v$ for $(u, v) \in \Gamma = (0, 2\pi) \times \mathbb{R}$, and of a sphere by $x = r \sin u \cos v, y = r \sin u \sin v, z = r \cos u$ for $(u, v) \in \Gamma = (0, \pi) \times (0, 2\pi)$ with the geometric interpretations as shown in Fig. 1.1.

These two examples already illustrate the difficulties that occur with closed surfaces at the points that correspon to the edge of the parameter set. There the continuity of φ is violated or the definition of φ is not meaningful, as φ^{-1} is not injective.

These difficulties can be avoided by using different parametrisations for different parts of M, with at least one parametrisation possible in the neighborhood of each point. Figure 1.2 shows two overlapping subsets U and V of M, where parametrisations φ and ψ are available. There is a natural bijection between the images of the shaded intersection $U \cap V$ which is often continuously differentiable. If we now relax the fact that the parametrisations of our surfaces consist of two parameters, we essentially arrive at the notion of an n-dimensional manifold.

The above motivation and discussion is now summarised with a precise definition.

Definition 1.1 Let M be a set, and n a natural number. An n-**dimensional chart** of M is a pair (U, φ), where U is a subset of M and φ is a bijection from U onto an open subset of $\mathbb{R}^n$. Let k we a natural number or ∞. A C^k-**atlas** $\mathcal{A}$ of M is a family of charts (U_i, φ_i), $(i \in I)$ with the properties

© The Author(s), under exclusive license to Springer Nature Switzerland AG 2023
R. Oloff, *The Geometry of Spacetime*, Graduate Texts in Physics,
https://doi.org/10.1007/978-3-031-16139-1_1

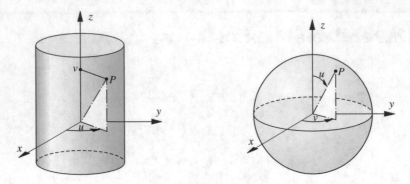

Fig. 1.1 Parametrisation of a cylinder and a sphere

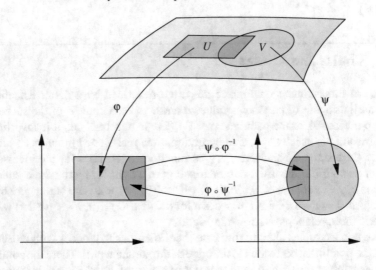

Fig. 1.2 Compatible charts (U, φ) and (V, ψ)

(MA1)
$$M = \bigcup_{i \in I} U_i$$

(MA2) The charts are pairwise compatible in the following sense: For i, $j \in I$ and
$U_i \cap U_j \neq \emptyset$, $\varphi_j \circ \varphi_i^{-1}$ is a C^k-diffeomorphism, that is, it is bijective and in
both directions k-times continuously differentiable.

♦

The first of the following examples reviews the surface considered at the outset.
Examples 2 and 3 are intended to illustrate that maps need not be compatible at all.
We will repeatedly refer to Example 4 because of its special importance in the theory
of relativity.

Example 1 For the sphere

$$M = \{ (x, y, z) \in \mathbb{R}^3 : x^2 + y^2 + z^2 = 1 \}$$

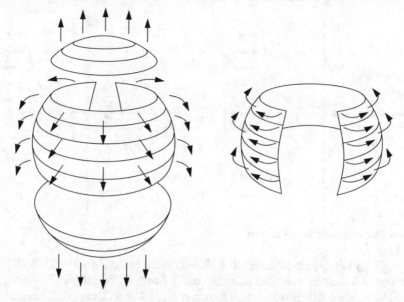

Fig. 1.3 An atlas for the sphere

the atlas could consist of four charts (U_i, φ_i), where the subsets U_i are sketched in Fig. 1.3, and maps φ_i are depicted with arrows there. φ_1 and φ_2 are the orthogonal projections of the two spherical caps U_1 and U_2 to the planes $z = \pm1$. φ_3 first projects U_3 horizontally from the z-axis to the cylindrical surface $x^2 + y^2 = 1$, which is then cut open and rolled out on a flat surface. φ_4 acts analogously on U_4.

Example 2 For the curve

$$M = \{ (x, y) \in \mathbb{R}^2 : y = x^3 \}$$

the chart (M, φ) with $\varphi(x, y) = y$ is an atlas (see Fig. 1.4).

Example 3 The same curve is now equipped with three charts (see Fig. 1.4). Let

$$U_1 = \{ (x, y) \in M : y > 1 \}$$

with $\varphi_1(x, y) = y$, and

$$U_2 = \{ (x, y) \in M : y < -1 \}$$

with $\varphi_2(x, y) = y$. As the third chart, take

$$U_3 = \{ (x, y) \in M : -2 < x < 2 \}$$

and $\varphi_3(x, y) = x$. The overlapping regions $U_1 \cap U_3$ and $U_2 \cap U_3$ give rise to the bijections $\left(\varphi_3 \circ \varphi_1^{-1}\right)(y) = \sqrt[3]{y}$ on the interval $(1, 8)$ to the interval $(1, 2)$ and

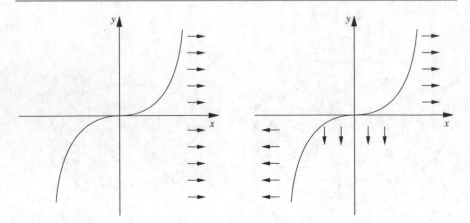

Fig. 1.4 Two atlases for the cubic curve

$(\varphi_3 \circ \varphi_2^{-1})(y) = \sqrt[3]{y}$ on the Intervall $(-8, -1)$ to the interval $(-2, -1)$. Both these functions and their inverses are infinitely many times differentiable. However, the map (U_3, φ_3) is not compatible with the chart (U, φ) from Example 2, because the bijection $(\varphi_3 \circ \varphi^{-1})(y) = \sqrt[3]{y}$ is not differentiable at $y = 0$.

Example 4 For a positive number μ consider the point set

$$M = \mathbb{R} \times (2\mu, +\infty) \times S_2,$$

the Cartesian product of the set of real numbers $\mathbb{R}$, the open interval $(2\mu, +\infty)$ and the surface S_2 which is the unit sphere in $\mathbb{R}^3$. Each chart (U_i, φ_i) of and atlas for S_2 induces a chart $(\mathbb{R} \times (2\mu, +\infty) \times U_i, \psi_i)$, where ψ_i sends a number t, a number $r > 2\mu$ and a point $P \in S_2$ to the quadruple (t, r, p, q) where $(p, q) = \varphi_i(P)$. These charts constitute an atlas for M.

Definition 1.2 Two C^k-atlases $\mathcal{A}$ and $\mathcal{B}$ of a set M are called **equivalent** whenever $\mathcal{A} \cup \mathcal{B}$ is a C^k-atlas of M, i.e., every chart of $\mathcal{A}$ is compatible with that of $\mathcal{B}$ and vice versa. ◆

The atlases considered in Examples 2 and 3 are not equivalent.

Definition 1.3 A C^k-atlas $\mathcal{A}$ is **maximal** if $\mathcal{A}$ contains every atlas compatible with $\mathcal{A}$. ◆

Clearly each atlas can be extended to a amaximal atlas by adding all conceivable charts that are compatible with every chart of the atlas.

A manifold is understood to mean a set equipped with an atlas, with equivalent atlases leading to the same manifold. Since two atlases on the same set are equivalent

if and only if they are contained in the same maximal atlas, this prompts the following:
A C^k-manifold $[M, \mathcal{A}]$ is a set M together with a maximal C^k-atlas $\mathcal{A}$.

But this is not yet a suitable definition, because we will later need the notion of continuity of a real-valued function defined on a manifold.

1.2 Topologisation

As is well known, the continuity of a map can be characterised by the concept of open sets. A map φ is continuous if and only if the preimage $\varphi^{-1}(G)$ is open for each open subset G. A map φ on a manifold maps to $\mathbb{R}^n$. There the usual notion of distance defines what the open subsets are. In this section we will choose the open subsets of a manifold so that the maps are continuous.

Definition 1.4 A **topological space** $[M, \mathcal{G}]$ is a set M together with a collection $\mathcal{G}$ of open subsets of M such that

(G1) $M, \emptyset \in \mathcal{G}$

(G2) For $G_i \in \mathcal{G}$ also $\bigcup_i G_i \in \mathcal{G}$

(G3) For $G_1, \dots, G_m \in \mathcal{G}$ also $G_1 \cap \dots \cap G_m \in \mathcal{G}$.

The collection $\mathcal{G}$ is called a **topology**, and its members are called open sets. ◆

As it is known, the concept of topology enables one to define the notion of convergence: A sequence of elements $P_1, P_2, \dots$ in a topological space M converges to an element $P \in M$ if for each $G \in \mathcal{G}$ such that $P \in G$, there exists an index m_0 such that for every natural number $m > m_0$ $P_m \in G$. The uniqueness of the limit is enforced by a so-called separation axiom:

Definition 1.5 A topological space $[M, \mathcal{G}]$ is called a **Hausdorff space** if for every two distinct $P, Q \in M$, there exist open sets $G, H \in \mathcal{G}$ such that $P \in G$ and $Q \in H$. ◆

For two different limiting values P and Q of the same sequence in a Hausdorff space, we could choose disjoint open sets G and H with $P \in G$ and $Q \in H$, and then for large m, all P_m would have to lie in both G and H, a contradiction, since $G \cap H = \emptyset$.

An atlas creates a topology on the base set M with the goal of making the maps homeomorphisms, that is, they must be continuous in both directions.

Definition 1.6 A subset G of a set M equipped with an atlas is called **open** if for each chart (U, φ), the subset

$$\varphi(U \cap G) = \{ \varphi(P) : P \in U \wedge P \in G \}$$

of $\mathbb{R}^n$ is open. ◆

These distinguished sets form a topology in the sense of Definition 1.4: the total set M and the empty set $\emptyset$ obviously belong to it, and (G2) and (G3) follow from

$$\varphi\left(U \cap \bigcup_i G_i\right) = \bigcup_i \varphi(U \cap G_i)$$

and

$$\varphi(U \cap G_1 \cap \cdots \cap G_m) = \varphi(U \cap G_1) \cap \cdots \cap \varphi(U \cap G_m).$$

In order to show the continuity of the map φ, we examine the preimage $\varphi^{-1}(H)$ of an open subset H of $\mathbb{R}^n$. For a (different) chart (V, ψ), we have

$$\psi\left(V \cap \varphi^{-1}(H)\right) = \left(\varphi \circ \psi^{-1}\right)^{-1}(H).$$

The map $\varphi \circ \psi^{-1}$ from $\psi(U \cap V)$ to $\varphi(U \cap V)$ is k-times continuously differentiable, and in particular continuous. Thus the preimage of the open set H under $\varphi \circ \psi^{-1}$ is open, and so $\varphi^{-1}(H)$ is open in the sense of Definition 1.6.

Unfortunately, it is not possible to prove the separation axiom formulated in Definition 1.5 for the topology on M introduced by Definition 1.6. The following counterexample shows that this topology does not actually have to be Hausdorff.

Let M be the set of real numbers without zero, but extended with the complex numbers i and $-i$, so that $M = \left(\mathbb{R}\backslash\{0\}\right) \cup \{i\} \cup \{-i\}$. This set is equipped with the two charts

$$U_1 = \left(\mathbb{R}/\{0\}\right) \cup \{i\}, \qquad \varphi_1(x) = \begin{cases} x & \text{for } x \in \mathbb{R}\backslash\{0\} \\ 0 & \text{for } x = i \end{cases}$$

and

$$U_2 = \left(\mathbb{R}/\{0\}\right) \cup \{-i\}, \qquad \varphi_2(x) = \begin{cases} x & \text{for } x \in \mathbb{R}\backslash\{0\} \\ 0 & \text{for } x = -i \end{cases}.$$

Every open set, in the sense of Definition 1.6, which contains i, also contains a punctured neighborhood $(-\varepsilon, 0) \cup (0, \varepsilon)$ of 0, and the same holds for one containing $-i$. Thus such sets can never be disjoint.

Since the separation axiom is not automatically fulfilled, we require that the topology generated by the maximal atlas makes M a Hausdorff space. Likewise, a so-called **second countability axiom** must be demanded: *There exists a sequence of open subsets $G_1, G_2, \ldots$ of M, so that every open subset G of M can be written as a union of a finite or infinite number of these G_i.* We will only need this axiom when integrating on manifolds, but it is usually already demanded in the definition of a manifold.

The discussion so far motivates the following definition:

Definition 1.7 An n-**dimensional** C^k-**manifold** $[M, \mathcal{A}]$ is a second countable Hausdorff space M with a C^k-atlas $\mathcal{A}$, consisting of charts (U_i, φ_i), where the φ_i are homeomorphisms of the open subset U_i of M onto open subsets of $\mathbb{R}^n$. The number n is called the **dimension** of the manifold. ♦

1.3   Submanifolds of $\mathbb{R}^m$

To further illustrate the concept of manifolds and to prepare for the axiomatic development of the concept of tangent vectors, to be introduced in the next chapter, we examine in this section manifolds that arise as the intersection of hypersurfaces in $\mathbb{R}^m$.

Suppose that $f_1, \ldots, f_l$ are l $(l < m)$ continuously differentiable real-valued functions on $\mathbb{R}^m$. The independent variables are labelled $\xi^1, \ldots, \xi^m$. The Jacobian matrix

$$\begin{pmatrix} \dfrac{\partial f_1}{\partial \xi^1} & \cdots & \dfrac{\partial f_1}{\partial \xi^m} \\ \vdots & & \vdots \\ \dfrac{\partial f_l}{\partial \xi^1} & \cdots & \dfrac{\partial f_l}{\partial \xi^m} \end{pmatrix}$$

is assumed to have rank l everywhere. If the functions were linear forms on $\mathbb{R}^m$, then this would mean that the coefficient matrix of the corresponding linear system of equations has rank l, and the solution set would be n-dimensional, with $n = m - l$. In a certain sense, this is also valid in the general case. That is,

$$M = \left\{ (\xi^1, \ldots, \xi^m) \in \mathbb{R}^m : f_i(\xi^1, \ldots, \xi^m) = 0, \ i = 1, \ldots l \right\}$$

is an n-dimensional manifold. For the proof, we use the fact that the Jacobian matrix has rank l on M. Let $(\xi_0^1, \ldots, \xi_0^m) \in M$. To avoid indices on indices, let us assume for simplicity that the square matrix

$$\begin{pmatrix} \dfrac{\partial f_1}{\partial \xi^1} & \cdots & \dfrac{\partial f_1}{\partial \xi^l} \\ \vdots & & \vdots \\ \dfrac{\partial f_l}{\partial \xi^1} & \cdots & \dfrac{\partial f_l}{\partial \xi^l} \end{pmatrix}$$

is regular at the point $(\xi_0^1, \ldots, \xi_0^m)$. As is known, the system of equations

$$f_i(\xi^1, \ldots, \xi^l, \xi^{l+1}, \ldots, \xi^m) = 0, \ i = 1, \ldots, l$$

can then be solved locally for $(\xi^1, \ldots, \xi^l)$, that is, for $(\xi^{l+1}, \ldots, \xi^m)$ from a neighborhood of $(\xi_0^{l+1}, \ldots, \xi_0^m)$, there exists exactly one l-tuple $(\xi^1, \ldots, \xi^l)$ from a neighborhood of $(\xi_0^1, \ldots, \xi_0^l)$, so that $(\xi^1, \ldots, \xi^m) \in M$. Hence the projection

$$\varphi(\xi^1, \ldots, \xi^l, \xi^{l+1}, \ldots, \xi^m) = (\xi^{l+1}, \ldots, \xi^m),$$

restricted to a sufficiently small neighborhood of $(\xi_0^1, \ldots, \xi_0^m)$ is invertible, and its inverse is continuously differentiable at least as many times as the given functions f_i. The compatibility of these maps is beyond question, and the metric and thus a Hausdorff topology is transferred from $\mathbb{R}^m$ to M.

With this discussion in mind, the following definition is well-motivated:

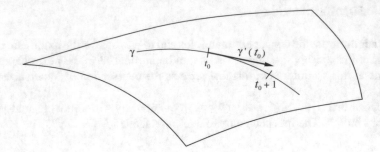

Fig. 1.5 Curve and tangent vector

Definition 1.8 An n-dimensional C^k-**submanifold** M in $\mathbb{R}^m$ $(n < m)$ is a nonempty set of the form

$$M = \left\{ (\xi^1, \ldots, \xi^m) \in \mathbb{R}^m : \; f_i(\xi^1, \ldots, \xi^m) = 0, \; i = 1, \ldots, m - n \right\}$$

where f_i are k times continuously differentiable functions on $\mathbb{R}^m$ whose Jacobian matrix on M has the rank $m - n$.   ◆

For a smooth surface in $\mathbb{R}^3$, what a tangent vector should be is geometrically clear: the base point of such a vector lies at a point P on the surface, and the tip and thus the entire arrow lies in the tangent plane to the surface at the point P. The aim of the following discussion is the formulation of this term for submanifolds in $\mathbb{R}^m$ and the preparation of the axiomatic approach for the general case of a manifold.

Definition 1.9 A C^k-**curve** in a C^k-submanifolds of $\mathbb{R}^m$ is a map γ from an interval I to M, the components γ^i of which defined by

$$\gamma(t) = \left(\gamma^1(t), \ldots, \gamma^m(t) \right),$$

are k-times continuously differentiable.   ◆

Thus a curve is described using a parametric representation. We take the viewpoint that different parameter representations give rise to different curves. Tangent vectors can be described as velocity vectors if we think of a curve as describing the trajectory of a point mass (Fig. 1.5).

Definition 1.10 The vector

$$\gamma'(t) = \lim_{h \to 0} \frac{\gamma(t + h) - \gamma(t)}{h}$$

for a curve γ in the submanifold M is called the **tangent vector** at the point $\gamma(t)$. The set of all tangent vectors at a point P is denoted by M_P and is called the **tangent space** at P.   ◆

Theorem 1.1 *Every tangent space of an n-dimensional submanifold in $\mathbb{R}^m$ is an n-dimensional linear space.*

Proof In order to obtain the whole tangent space M_P, it is sufficient to consider the derivatives of all curves γ with $\gamma(0) = P$. Suppose the given point $P \in M$ lies in the chart (U, φ). By composition with the map, a curve γ gives rise to k-times continuously differentiable real-valued functions $u^1, \ldots, u^n$, defined by

$$\left(u^1(t), \ldots, u^n(t) \right) = \varphi\left(\gamma(t) \right).$$

If we apply the chain rule to the composition $\gamma = \varphi^{-1} \circ (\varphi \circ \gamma)$, then we get

$$\gamma' = \begin{pmatrix} \dfrac{\partial \xi^1}{\partial u^1} & \cdots & \dfrac{\partial \xi^1}{\partial u^n} \\ \vdots & & \vdots \\ \dfrac{\partial \xi^m}{\partial u^1} & \cdots & \dfrac{\partial \xi^m}{\partial u^n} \end{pmatrix} \begin{pmatrix} \dfrac{du^1}{dt} \\ \vdots \\ \dfrac{du^n}{dt} \end{pmatrix}.$$

Every curve γ with $\gamma(0) = P$ creates a vector

$$\left(\frac{du^1}{dt}, \ldots, \frac{du^n}{dt} \right)_{t=0} \in \mathbb{R}^n,$$

and vice versa for each n-tuple $v = (v^1, \ldots, v^n)$, there exists a curve with $\gamma(0) = P$ and

$$\frac{du^k}{dt}(0) = v^k,$$

for example the curve $u = vt$ lifted to the manifold using the map. Summarising, we see that the tangent space is the image space of the matrix $(\frac{\partial \xi^i}{\partial u^k})$, calculated at the n-tuple $(u^1, \ldots, u^m) = \varphi(P)$. This shows the linearity of M_P, and it remains to show that this matrix has rank n.

The tangent space and its dimension if of course independent of the choice of the chart. So we can take φ to be a projection that assigns the m numbers $\xi^1, \ldots, \xi^m$ n amongst them. Possibly after a renumbering we may assume that these are the first n. Thus $u^k = \xi^k$ for $k = 1, \ldots, n$. The first n rows of the matrix $(\frac{\partial \xi^i}{\partial u^k})$ are then the rows of the identity matrix, and so the rank is n.

A tangent vector can be used as a directional derivative. It will be seen that the tangent vector can be identified by the effect the directional derivative has on real-valued functions.

Definition 1.11 For a C^∞-submanifold M on $\mathbb{R}^m$, let $\mathcal{F}(M)$ be the linear space of all real-valued functions f on M, such that for every chart φ, the function $f \circ \varphi^{-1}$ is infinitely many times differentiable. ◆

A function $f \in \mathcal{F}(M)$ will often interest us in its behavior on an open set containing P (open neighborhood of P). Any function defined in an open neighborhood of P for which $f \circ \varphi^{-1}$ is infinitely many times differentiable can be extended to all of M while maintaining this property.

Definition 1.12 Let $P \in M$, $x \in M_P$ and $f \in \mathcal{F}(M)$. The number $xf = (f \circ \gamma)'(0)$ with $\gamma(0) = P$ and $\gamma'(0) = x$ is called the **directional derivative** of f in the direction x at the point P.   ♦

The fact that the directional derivative does not depend on the choice of the curve is evident in the case when the partial derivatives of f can be formed using the coordinates of $\mathbb{R}^m$, using the equation

$$\left(f \circ \gamma\right)'(0) = \operatorname{grad} f(P) \cdot \gamma'(0).$$

If f is only defined on M, we must instead use a suitable map φ and factor $f \circ \gamma$ into $f \circ \gamma = g \circ u$ with $g = f \circ \varphi^{-1}$ and $u = \varphi \circ \gamma$. By the chain rule, we then have

$$(f \circ \gamma)' = \operatorname{grad} g\bigl(\varphi(P)\bigr) \cdot u'(0).$$

Since

$$x = \gamma'(0) = (\varphi^{-1} \circ u)'(0)$$

$u'(0)$ is the preimage of x under the Jacobian matrix of φ^{-1}. But this has rank n, and so $u'(0)$ is uniquely determined by x, and so $\left(f \circ \gamma\right)'(0)$ is also uniquely determined by x.

Example. Let i be one of the number 1 to m, and f_i be the function that assigns to every point of a submanifold M of $\mathbb{R}^m$ its ith coordinate. For the tangent vector $y = (y^1, \ldots, y^m)$ we have

$$yf_i = \operatorname{grad} f \cdot y = y^i.$$

By using a map φ, one gets the same result, as follows: The function $g = f_i \circ \varphi^{-1}$ assigns the n-tuple $(x^1, \ldots, x^n)$ its ith coordinate. We have $yf_i = \operatorname{grad} g \cdot v$ where v is the preimage of y under the matrix $(\frac{\partial \xi^i}{\partial u^k})$. Since $\operatorname{grad} g$ is the ith row of this matrix, the scalar product $\operatorname{grad} g \cdot v$ is the ith component of the tangent vector y.

The following simple arithmetic rules follow immediately from the definition of the directional derivative.

Theorem 1.2 *For $P \in M$, $x \in M_P$, $f, g \in \mathcal{F}(M)$ and real numbers λ and μ, the following hold:*
(L) $x(\lambda f + \mu g) = \lambda(xf) + \mu(xg)$
(P) $x(fg) = (xf)g(P) + f(P)(xg)$

Theorem 1.3 *If there are two tangent vectors x, $y \in M_P$ such that for all $f \in \mathcal{F}(M)$ their directional derivatives xf and yf coincide, then x must be equal to y.*

Proof For the functions f_i discussed in the example above, we have by Definition 1.12 that $xf_i = x^i$. By the equality of the directional derivatives, it follows that $x^i = y^i$ for all $i = 1, \ldots, m$, and so $x = y$.

Tangent Vectors

2.1   The Tangent Space

In this chapter, M is an n-dimensional C^∞-manifold in the sense of Definition 1.7. The function space $\mathcal{F}(M)$ introduced here is as in Definition 1.11. In Sect. 1.3, we introduced the concept of n-dimensional submanifolds of $\mathbb{R}^m$ in the sense of Definition 1.8, and the notion of tangent vectors was introduced. Definition 1.10 cannot be used directly in the general case, but Theorems 1.2 and 1.3 motivate the following:

Definition 2.1 Let $P \in M$. A map $x : \mathcal{F}(M) \longrightarrow \mathbb{R}$ such that
(L) $x(\lambda f + \mu g) = \lambda(xf) + \mu(xg)$
(P) $x(fg) = (xf)g(P) + f(P)(xg)$
is called a **tangent vector** at P (also **tangential vector**, **vector** or **contravariant vector**). The set of all tangent vectors at a point $P \in M$ is called the **tangent space** M_P. ♦

For an n-dimensional submanifold of $\mathbb{R}^m$, each tangent vector in the sense of Definition 1.10 is also a tangent vector in the sense of Definition 2.1. The converse will become clear only in the next section.

Theorem 2.1 *Every tangent space M_P is a linear space.*

Proof For $x, y \in M_P$ we need to show that the map $\alpha x + \beta y : \mathcal{F}(M) \longrightarrow \mathbb{R}$ defined by

$$(\alpha x + \beta y)f = \alpha x f + \beta y f$$

© The Author(s), under exclusive license to Springer Nature Switzerland AG 2023
R. Oloff, *The Geometry of Spacetime*, Graduate Texts in Physics,
https://doi.org/10.1007/978-3-031-16139-1_2

satisfies linearity (L) and the product rule (P). We have

$$(\alpha x + \beta y)(\lambda f + \mu g) = \alpha x(\lambda f + \mu g) + \beta y(\lambda f + \mu g)$$
$$= \alpha\lambda xf + \alpha\mu xg + \beta\lambda yf + \beta\mu yg = \lambda(\alpha x + \beta y)f + \mu(\alpha x + \beta y)g$$

and

$$(\alpha x + \beta y)(fg) = \alpha x(fg) + \beta y(fg)$$
$$= \alpha(xf)g(P) + \alpha f(P)(xg) + \beta(yf)g(P) + \beta f(P)(yg)$$
$$= \big((\alpha x + \beta y)f\big)g(P) + f(P)(\alpha x + \beta y)g.$$

In the following two theorems, properties of tangent vectors are given, which in the special case of a submanifold of $\mathbb{R}^m$, are obvious for the directional derivative.

Theorem 2.2 *Let $P \in M$, $x \in M_P$ and f be a constant. Then $xf = 0$.*

Proof Because of homogeneity, $x(\lambda f) = \lambda xf$, and so it is enough to consider the case $f = 1$. We have

$$x1 = x(1 \cdot 1) = (x1)1 + 1(x1) = 2x1$$

and so $x1 = 0$.

Theorem 2.3 *Let f and g coincide in an open neighborhood of P. Then for all $x \in M_P$ $xf = xg$.*

Proof Let U be an open neighborhood of P where f and g coincide. We choose a function $h \in \mathcal{F}(M)$ such that $h(P) = 1$ and $h(Q) = 0$ for $Q \notin U$. That such a function exists is entirely plausible, but a precise reason will be given in Sect. 13.2. The product rule (P) gives in the present case that

$$x(hf) = xh \cdot f(P) + 1 \cdot xf$$

and

$$x(hg) = xh \cdot g(P) + 1 \cdot xg.$$

Since $hf = hg$ and $f(P) = g(P)$, it follows that $xf = xg$.

This last theorem gives rise to the following concepts.

Definition 2.2 Two functions $f, g \in \mathcal{F}(M)$ are called P-**equivalent**, if they coincide on an open neighborhood of P.   ♦

P-equivalence is an equivalence relation, and thus creates a partition of $\mathcal{F}(M)$ into equivalence classes.

Definition 2.3 $\mathcal{F}(P)$ is the collection of all equivalence classes of P-equivalent functions from $\mathcal{F}(M)$. ◆

The space $\mathcal{F}(P)$ can thus be written as the quotient space $\mathcal{F}(P) = \mathcal{F}(M)/\mathcal{N}(P)$. Here $\mathcal{N}(P)$ is the subspace of all $g \in \mathcal{F}(M)$ for which there exists an open neighborhood U of P such that $g(Q) = 0$ for $Q \in U$. Another description is as follows: $\mathcal{F}(P)$ consists of all restrictions of C^∞-functions defined on an open neighborhood of P. For such a function f defined on an open set U, there exists a C^∞-function g defined on the whole of M which coincides with f on an open subset V of U containing P (see Sect. 13.2). Conversely, one can select a representative from a class of P-equivalent functions and restrict it to an open neighborhood of P. The space $\mathcal{F}(P)$ is important in connection with tangent vectors, since by Theorem 2.3, one has that in Definition 2.1, the function space $\mathcal{F}(M)$ can be replaced by the space $\mathcal{F}(P)$.

2.2 Generation of Tangent Vectors

To define a curve in a manifold, we modify the wording of Definition 1.9.

Definition 2.4 A C^∞**-curve** γ in an n-dimensional C^∞-manifold M is a map γ from an interval I to M, such that for every map φ, the map $\varphi \circ \gamma$ is infinitely many times differentiable. ◆

In contrast to the earlier situation described in Sect. 1.3, the curve itself can no longer be differentiated, but for every C^∞-function f, we have $f \circ \gamma = (f \circ \varphi^{-1}) \circ (\varphi \circ \gamma)$ is differentiable. With this in mind, each C^∞-curve γ with $\gamma(0) = P$ generated a tangent vector $x \in M_P$

$$f \xrightarrow{x} (f \circ \gamma)'(0),$$

and then there holds

$$\big((\lambda f + \mu g) \circ \gamma\big)'(0) = \lambda(f \circ \gamma)'(0) + \mu(g \circ \gamma)'(0)$$

and

$$\big((fg) \circ \gamma\big)'(0) = \big((f \circ \gamma)(g \circ \gamma)\big)'(0) = (f \circ \gamma)'(0)(g \circ \gamma)(0) + (f \circ \gamma)(0)(g \circ \gamma)'(0).$$

Using a map, we can express the directional derivative as

$$xf = (f \circ \gamma)'(0) = \big((f \circ \varphi^{-1}) \circ (\varphi \circ \gamma)\big)'(0) = \sum_{i=1}^{n} \frac{\partial(f \circ \varphi^{-1})}{\partial u^i}(\varphi(P)) \frac{du^i}{dt}(0).$$

This formula shows that each tangent vector generated by a curve is a linear combination of the special tangent vectors introduced in the following definition.

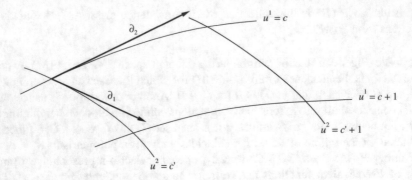

Fig. 2.1 Coordinate induced tangent vectors

Definition 2.5 At a given point P of an n-dimensional C^∞-manifold and a chart map φ, let $\frac{\partial}{\partial u^i}(P)$ or even briefer ∂_i denote the tangent vector, that assigns for a function $f \in \mathcal{F}(P)$ the number

$$\frac{\partial(f \circ \varphi^{-1})}{\partial u^i}(\varphi(P)).$$

◆

Applying the tangent vector ∂_i to f thus means to partially differentiate the chart representative $f \circ \varphi^{-1}$, using the chart map φ, with respect to the ith coordinate. In the case of an n-dimensional submanifold of $\mathbb{R}^m$ with the local representation

$$(\xi^1, \ldots, \xi^m) = \varphi^{-1}(u^1, \ldots, u^n)$$

this procedure coincides with the procedure in the sense of Definition 1.12 with the vector

$$\partial_i = \left(\frac{\partial \xi^1}{\partial u^i}, \ldots, \frac{\partial \xi^m}{\partial u^i} \right).$$

In Fig. 2.1, the special case when $m = 3$ and $n = 2$ is depicted. The rest of this section is devoted to demonstrating that the tangent vectors $\partial_1, \ldots, \partial_n$ form a basis for the tangent space M_P.

Theorem 2.4 *The tangent vectors* $\frac{\partial}{\partial u^1}(P), \ldots, \frac{\partial}{\partial u^n}(P)$ *are linearly independent.*

Proof If

$$\sum_{i=1}^n \lambda_i \frac{\partial}{\partial u^i}(P) = 0,$$

then for $f \in \mathcal{F}(P)$ we have

$$\sum_{i=1}^n \lambda_i \frac{\partial(f \circ \varphi^{-1})}{\partial u^i}(\varphi(P)) = 0.$$

Applied to the function f_k, which maps the kth coordinate of the points in the chart φ, we obtain $\lambda_k = 0$.

Theorem 2.5 *Let B be an open ball with center 0 in $\mathbb{R}^n$, and let g be a C^∞-function defined on B. Then there exist C^∞-functions $g_1, \ldots, g_n$ such that*

$$g_i(0) = \frac{\partial g}{\partial u^i}(0),$$

so that for $u \in B$ there holds

$$g(u) = g(0) + \sum_{i=1}^n g_i(u)u^i.$$

Proof For $u = (u^1, \ldots, u^n) \in B$, we apply to the function $h(t) = g(tu)$ the identity

$$h(1) = h(0) + \int_0^1 h'(t)\,dt$$

giving

$$g(u) = g(0) + \int_0^1 \sum_{i=1}^n \frac{\partial g}{\partial u^i}(tu) \cdot u^i\,dt.$$

The sought functions are given by the integrals

$$g_i(u) = \int_0^1 \frac{\partial g}{\partial u^i}(tu)\,dt.$$

Theorem 2.6 *The tangent vectors $\dfrac{\partial}{\partial u^1}(P), \ldots, \dfrac{\partial}{\partial u^n}(P)$ form a basis for the tangent space M_P.*

Proof We have to show that each tangent vector $x \in M_P$ is a linear combintation of $\frac{\partial}{\partial u^i}(P)$. We can assume that $\varphi(P) = 0$. Again, let f_i be the function that assigns to each point its i-th coordinate under the map φ. For $f \in \mathcal{F}(P)$, we will show that

$$xf = \sum_{i=1}^n xf_i \frac{\partial(f \circ \varphi^{-1})}{\partial u^i}(0),$$

that is, the directional derivatives xf_i will be the sought coefficients in the linear combination. To this end, we apply Theorem 2.5 to the function $g = f \circ \varphi^{-1}$ and obtain

$$f\left(\varphi^{-1}(u)\right) = f\left(\varphi^{-1}(0)\right) + \sum_{i=1}^{n} g_i(u)u^i \,,$$

and so

$$f(Q) = f(P) + \sum_{i=1}^{n} (g_i \circ \varphi)(Q)f_i(Q) \,.$$

Thus the function f acquires the form

$$f = f(P) + \sum_{i=1}^{n} (g_i \circ \varphi)f_i \,.$$

Hence

$$xf = \sum_{i=1}^{n} x(g_i \circ \varphi)f_i(P) + \sum_{i=1}^{n} (g_i \circ \varphi)(P)xf_i \,.$$

By the definition of f_i, we have $f_i(P) = 0$, and thanks to the property possessed by the g_i stipulated in Theorem 2.5, we have

$$(g_i \circ \varphi)(P) = g_i(0) = \frac{\partial g}{\partial u^i}(0) = \frac{\partial (f \circ \varphi^{-1})}{\partial u^i}(0) \,.$$

This completes the proof of the theorem.

2.3   Vector Fields

Definition 2.6 A **vector field** X on a C^∞-manifold M is a map that for every $P \in M$ assigns a tangent vector $X(P) \in M_P$. X is a C^∞-vector field if for all $f \in \mathcal{F}(M)$, the real-valued function Xf, defined by $Xf(P) = X(P)f$, is infinitely many times differentiable. The linear space of all C^∞-vector fields is denoted by $\mathcal{X}(M)$. $\mathcal{X}(P)$ denotes the space $\mathcal{X}(M)$ quotiented by the subspace of all vector fields that are zero in a neighborhood of P.   ◆

Example. Let P be a point of an n-dimensional C^∞-manifold. Each chart (U, φ) with $P \in U$ gives rise to vector fields $\frac{\partial}{\partial u^i} \in \mathcal{X}(P)$, defined by

$$\frac{\partial}{\partial u^i} f(P) = \frac{\partial (f \circ \varphi^{-1})}{\partial u^i}(\varphi(P))$$

Fig. 2.2 A vector field

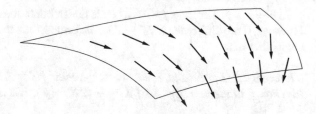

(see Definition 2.5). These are the n **coordinate vector fields** induced by the chart (U, φ). As mentioned earlier, we will sometimes denote them by ∂_i.

To further illustrate the concept of a vector field, we now consider curved surfaces in $\mathbb{R}^3$. A tangent vector is then an arrow pinned at a point P in the tangential plane. With this in mind, a vector field X attaches an arrow at each point P (Fig. 2.2), while a C^∞-vector field arranges for the arrows to differ only slightly if their base points are slightly apart. This system of arrows can be interpreted as a flow, where the tangent vector prescribes a velocity at each point on the surface. For a function $f \in \mathcal{F}(P)$, the directional derivative $Xf(P)$ then describes the change in the function value of f registered by an observer moving with the flow at the moment when it passes the point P.

The n coordinate vector fields $\frac{\partial}{\partial u^1}, \dots, \frac{\partial}{\partial u^n}$, give for each point Q in the chart, the tangent vectors $\frac{\partial}{\partial u^i}(Q)$, which form a basis for the tangent space M_Q. However, the converse is in general not true. If there are n vector fields $X_1, \dots, X_n$, such that for each point Q in U, the tangent vectors $X_1(Q), \dots, X_n(Q)$ are linearly independent, there in general is no map φ on U, such that $X_1, \dots, X_n$ are the corresponding coordinate vector fields. This is illustrated by the following counterexample.

Let $M = \mathbb{R}^2 \backslash (-\infty, 0]$, equipped with the polar coordinates ϱ and φ. Define the vector fields X and Y by

$$X(P) = \frac{\partial}{\partial \varrho}(P) \quad \text{and} \quad Y(P) = \frac{1}{\varrho}\frac{\partial}{\partial \varphi}(P).$$

For every P, the vectors $X(P)$ and $Y(P)$ are orthogonal, and in particular, also linearly independent. If X and Y were coordinate vector fields, then as with the polar coordinates, concentric circles and rays from the origin would be the coordinate lines, and $||Y(P)||$ should be proportional to the distance from the origin. This contradicts $||Y(P)|| = 1$.

The concept of the Lie bracket (named after the Norwegian mathematician SOPHUS LIE 1842–1899) will be dealt with, among other things, in the next section, in order to determine which vector fields are coordinate vector fields.

2.4   The Lie Bracket

A vector field X can be applied point by point to a scalar field f (i.e., to a real-valued function), and thereby generate a number $X(P)f$ at each point, hence giving a new scalar field $g = Xf$. If we now apply another vector field Y on g, we get a third scalar field $h = YXf$. The map $f \longrightarrow h(P)$ is linear, but it does not obey the product rule. However, if the difference $XYf - YXf$ is formed, then the troublesome terms cancel each other out.

Theorem 2.7 *Let X and Y be two C^∞-vector fields. The map, sending the C^∞-function f to the number $(XYf)(P) - (YXf)(P)$, is a tangent vector in M_P.*

Proof The needed linearity is satisfied by each of the summands. For the first one,

$$XY(\lambda f + \mu g)(P) = X(P)(\lambda Yf + \mu Yg)$$
$$= \lambda X(P)Yf + \mu X(P)Yg = \lambda XYf(P) + \mu XYg(P)$$

and analogously for the second one:

$$YX(\lambda f + \mu g)(P) = \lambda YXf(P) + \mu YXg(P).$$

The product rule for vector fields gives

$$XY(fg)(P) - YX(fg)(P)$$
$$= X(P)\big((Yf)g\big) + X(P)\big(f(Yg)\big) - Y(P)\big((Xf)g\big) - Y(P)\big(f(Xg)\big).$$

For the first summand, we have using the product rule for the vector field X:

$$X(P)\big((Yf)g\big) = X(P)(Yf)g(P) + (Y(P)f)(X(P)g).$$

Adding and subtracting analogous results for the other three summands, we obtain

$$\big(X(P)Y - Y(P)X\big)(fg) =$$
$$= \big((X(P)Y - Y(P)X)f\big)g(P) + f(P)\big(X(P)Y - Y(P)X\big)g.$$

Definition 2.7 The **Lie bracket** is the operation in $\mathcal{X}(M)$, that maps the vector fields X and Y to the vector field $[X, Y]$, defined by

$$[X, Y](P)f = XYf(P) - YXf(P).$$

$\blacklozenge$

Thus

$$[X, Y]f = XYf - YXf,$$

and even more briefly,

$$[X, Y] = XY - YX.$$

Theorem 2.8 *The Lie bracket of any two coordinate vector fields is the zero vector field.*

Proof We have

$$\left[\frac{\partial}{\partial u^i}, \frac{\partial}{\partial u^k}\right] f = \frac{\partial}{\partial u^i}\frac{\partial}{\partial u^k}(f \circ \varphi^{-1}) - \frac{\partial}{\partial u^k}\frac{\partial}{\partial u^i}(f \circ \varphi^{-1}) = 0,$$

because, as is well known, the order does not matter in partial differentiation.

From Definition 2.7, it can be seen immediately that the Lie bracket is a skew-symmetric bilinear operation on $\mathcal{X}(M)$ or $\mathcal{X}(P)$. We have

$$[X, Y] = -[Y, X]$$
$$[X + Y, Z] = [X, Z] + [Y, Z]$$
$$[X, Y + Z] = [X, Y] + [X, Z]$$
$$[\lambda X, \mu Y] = \lambda\mu[X, Y].$$

This last equality does not always hold if the numbers λ and μ are replaced by functions.

Theorem 2.9 *For $X, Y \in \mathcal{X}(P)$ and $f, g \in \mathcal{F}(P)$, there holds*

$$[fX, gY] = fg[X, Y] + f(Xg)Y - g(Yf)X.$$

Proof The product rule for vector fields implies

$$[fX, gY]h = fX(gYh) - gY(fXh)$$
$$= f(Xg)(Yh) + fgXYh - g(Yf)(Xh) - gfYXh.$$

When a chart is chosen, vector fields can be represented using the n coordinate vector fields $\partial_1, \ldots, \partial_n$. The coefficients are then scalar fields.

Theorem 2.10 *For $X = \sum_{i=1}^n X^i \partial_i$ and $Y = \sum_{k=1}^n Y^k \partial_k$, there holds*

$$[X, Y] = \sum_{j=1}^n \left(\sum_{i=1}^n (X^i \partial_i Y^j - Y^i \partial_i X^j)\right)\partial_j.$$

Proof By Theorem 2.9,

$$[X, Y] = \sum_{i,k=1}^n [X^i \partial_i, Y^k \partial_k] = \sum_{i,k=1}^n X^i (\partial_i Y^k)\partial_k - \sum_{i,k=1}^n Y^k (\partial_k X^i)\partial_i.$$

Theorem 2.11 *For vector fields* X, Y, Z, *there holds*

$$[[X, Y], Z] + [[Y, Z], X] + [[Z, X], Y] = 0 \,.$$

Proof This so-called **Jaboci identity** follows from the fact that all twelve summands cancel each other out in pairs.

Tensors

3

3.1 Introduction

The representation of elements of a finite-dimensional space E, of linear forms on E, and of linear maps on E, as n-tuples or as matrices, depends essentially on a choice of basis in E. The concept of tensor introduced in the next section generalises these objects, and thus facilitates, among other things, a unified theory of coordinate transformation. In later applications, the underlying space E will always be the tangent space of a manifold, but this fact will not play a role in this chapter. In this respect, the present chapter is completely independent of the considerations of the first two chapters.

In tensor algebra, there are always sums with often times several summation indices. To simplify the notations, the following so-called **Einstein notation** is commonly used: *When an index occurs twice in one term, once as a subserscript and once as a subscript, then a summation is implied over the range of this index.* Thus instead of writing $\sum_{i=1}^{n} \lambda^i x_i$, we shorten this to $\lambda^i x_i$. An equation of the form $\eta^k = \sum_{i=1}^{n} \alpha_i^k \xi^i$ is replaced by $\eta^k = \alpha_i^k \xi^i$. The summation index range is no longer specified, but is usually clear from the context.

The use of the summation convention requires uniform regulations determining which indices have to be at the top and which at the bottom. We will now index vectors below when numbering them, and so the coefficients in linear combinations of vectors must then of course be indexed above. Since the coordinates of a vector in such a representation are coefficients, their running index must also be at the top. Matrices, which are to represent linear maps, will be applied on coordinates, and produce coordinates. Thus the column index is at the bottom and the row index is at the top. If a matrix is to be used as the coefficient matrix of a bilinear form, then both indices must occur below. Linear forms are indexed at the top, and their coordinates at the bottom.

© The Author(s), under exclusive license to Springer Nature Switzerland AG 2023
R. Oloff, *The Geometry of Spacetime*, Graduate Texts in Physics,
https://doi.org/10.1007/978-3-031-16139-1_3

To demonstrate the use of the summation convention, we will now discuss coordinate transformations for vectors and linear maps, using the new notation. The basis is always for the n-dimensional space E.

Let two bases $x_1, \ldots, x_n$ and $y_1, \ldots, y_n$, related by $y_k = \alpha_k^i x_i$ and $x_i = \beta_i^k y_k$, be given. The representation $z = \xi^i x_i$ of a vector $z \in E$ should be converted to the other basis. To this end, we express the x_i in terms of the other basis vectors using the conversion formula given above, and obtain $z = \xi^i \beta_i^k y_k$. We can then read off the coordinates as $\eta^k = \xi^i \beta_i^k$.

The matrix (τ_k^i) describes a map $A \in L(E, E)$ with respect to a basis $x_1, \ldots, x_n$, that is, $Ax_k = \tau_k^i x_i$. The matrix (σ_k^j) of A with respect to the other basis should be determined. By using both conversion formulae, we obtain

$$Ay_k = A\alpha_k^i x_i = \alpha_k^i Ax_i = \alpha_k^i \tau_i^l x_l = \alpha_k^i \tau_i^l \beta_l^j y_j$$

and so we read off that $\sigma_k^j = \alpha_k^i \tau_i^l \beta_l^j$.

3.2   Multilinear Forms

We are familiar with the notion of multilinear forms via our acquaintance with determinants: the rows of a square matrix are assigned a number, and this assignment is known to be linear with respect to each individual row (the others being kept fixed).

Let E be an n-dimensional vector space, and E^* denote its dual space, consisting of all linear forms a on E. We denote the action of a on an element x by $\langle x, a \rangle$.

Definition 3.1 Let p and q be nonnegative integers. A map $f : E^{*p} \times E^q \longrightarrow \mathbb{R}$ such that

$$f(a^1, \ldots, a^{i-1}, \lambda a + \mu b, a^{i+1}, \ldots, a^p, x_1, \ldots, x_q)$$
$$= \lambda f(a^1, \ldots, a, \ldots, a^p, x_1, \ldots, x_q) + \mu f(a^1, \ldots, b, \ldots, a^p, x_1, \ldots, x_q)$$

and

$$f(a^1, \ldots, a^p, x_1, \ldots, x_{k-1}, \lambda x + \mu y, x_{k+1}, \ldots, x_q)$$
$$= \lambda f(a^1, \ldots, a^p, x_1, \ldots, x, \ldots, x_q) + \mu f(a^1, \ldots, a^p, x_1, \ldots, y, \ldots, x_q)$$

is called a p-**fold contravariant and** q-**fold covariant tensor** on E, or even more briefly, a (p, q)-**tensor**.

Example 1 The function f defined on $E^* \times E$, given by $f(a, x) = \langle x, a \rangle$ is a $(1, 1)$-tensor and is called the **unit tensor**.

Example 2 Let $E = \mathbb{R}^3$. The function f on $E^* \times E \times E$ defined by $f(a, x_1, x_2) = \langle x_1 \times x_2, a \rangle$ is a $(1, 2)$-tensor on $\mathbb{R}^3$.

Example 3 For given vectors $x_1, \ldots, x_p \in E$ and linear forms $a^1, \ldots, a^q$, the function f on $E^{*p} \times E^q$ defined by

$$f(b^1, \ldots, b^p, y_1, \ldots, y_q) = \langle x_1, b^1 \rangle \cdots \langle x_p, b^p \rangle \langle y_1, a^1 \rangle \cdots \langle y_q, a^q \rangle$$

is a (p, q)-tensor. This is denoted by

$$f = x_1 \otimes \cdots \otimes x_p \otimes a^1 \otimes \cdots \otimes a^q .$$

Such tensors are called **simple**.

For a given real vector space E and non-negative integers p and q, the set of p-fold contravariant and q-fold covariant tensors form a linear space with respect to pointwise addition and pointwise multiplication. This is denoted by E_q^p or by $E \otimes \cdots \otimes E \otimes E^* \otimes \cdots \otimes E^*$. For example E_2^1 or $E \otimes E^* \otimes E^*$. By E_0^0 we mean of course $\mathbb{R}$.

The following theorem gives a wealth of other examples and shows the great generality of the concept of tensors.

Theorem 3.1 *The following equalities hold in the sense of isomorphisms:*

(1) $E = E_0^1$
(2) $E^* = E_1^0$
(3) $L(E, E) = E_1^1$
(4) $L(E, L(E, E)) = E_2^1$
(5) $L(L(E, E), L(E, E)) = E_2^2$

The isomorphisms do not depend on the choice of a basis.

Proof A vector x generates a real function Jx on E^* by setting $(Jx)(a) = \langle x, a \rangle$. The linearity of the mapping $a \longrightarrow \langle x, a \rangle$ follows from the definition of addition in E^*, and the linearity of $x \longrightarrow Jx$ is exactly the linearity of linear forms. The map $J : E \longrightarrow E_0^1$ is injective, since it is known that if x_1 and x_2 are distinct vectors, then there exists a linear form a which distinguished them, that is $\langle x_1, a \rangle \neq \langle x_2, a \rangle$. The agreement of the dimensions of E and E_0^1, which we will show in a more general context in the next section, finally ends the proof of (1). The equality (2) is trivial. To prove the other equalities, we only give the isomorphisms. Given a linear map A on E, we assign to it the bilinear form $f(a, x) = \langle Ax, a \rangle$. For $T \in L(E, L(E, E))$ we assign the trilinear form $f(a, x_1, x_2) = \langle (Tx_1)x_2, a \rangle$, which is the corresponding $(1, 2)$-tensor. $T \in L(L(E, E), L(E, E))$ is interpreted as the $(2, 2)$-tensor given by

$$f(a^1, a^2, x_1, x_2) = \langle (T(x_1 \otimes a^1))x_2, a^2 \rangle .$$

In light of the identifications in (1) and (2), we call vectors as **contravariant vectors** and linear forms as **covariant vectors**.

The following two recursion formulae generalise the equalities (3), (4) and (5) of Theorem 3.1, and allow the construction of tensors of arbitrary 'heights'.

Theorem 3.2 *Let E be a finite-dimensional space, and p, q, r, s, t, u be nonnegative integers. Then in the sense of isomorphisms, we have*

$$L(E_s^r, E_t^u) = E_{r+t}^{s+u}$$

and in particular,

$$L(E, E_q^p) = E_{q+1}^p .$$

Proof As the isomorphism, we use the map J, which sends $T \in L(E_s^r, E_t^u)$ to the tensor

$$f(a^1, \ldots, a^s, a^{s+1}, \ldots, a^{s+u}, x_1, \ldots, x_r, x_{r+1}, \ldots, x_{r+t})$$
$$= (T(x_1 \otimes \cdots \otimes x_r \otimes a^1 \otimes \cdots \otimes a^s))(a^{s+1}, \ldots, a^{s+u}, x_{r+1}, \ldots, x_{r+t}).$$

Obviously, f is multilinear and J is linear. If one already uses here that one can form for E_s^r a basis using simple tensors (Theorem 3.3. in the next section), then one also sees that $f = 0$ implies $T = 0$. Thus J is injective. As it will be shown that both spaces have the same dimensions, J is also surjective. The second equality in Theorem 3.2 is the special case of the first one with $r = 1, s = 0, t = q, u = p$.

3.3   Components

Every basis $x_1, \ldots, x_n$ for a linear space is known to give rise to a basis $a^1, \ldots, a^n$ for the dual space E^*, defined by

$$\langle \lambda^i x_i, a^k \rangle = \lambda^k .$$

This basis $a^1, \ldots, a^n$ is called the **dual basis** to the given basis $x_1, \ldots, x_n$.

Theorem 3.3 *Let $x_1, \ldots, x_n$ be a basis for E and let $a^1, \ldots, a^n$ be the dual basis for E^*. The n^{p+q}-tensors*

$$x_{i_1} \otimes \cdots \otimes x_{i_p} \otimes a^{j_1} \otimes \cdots \otimes a^{j_q} ,$$

where the indices $i_1, \ldots, i_p, j_1, \ldots, j_q$ all belong to $\{1, ..n\}$, form a basis for E_q^p. Thus it follows that

$$dim E_q^p = n^{p+q} .$$

Proof By the definition of simple tensors, and the dual basis,

$$x_{i_1} \otimes \cdots \otimes x_{i_p} \otimes a^{j_1} \otimes \cdots \otimes a^{j_q}(a^{i'_1}, \ldots, a^{i'_p}, x_{j'_1}, \ldots, x_{j'_q}) = 1$$

if and only if all the corresponding indices match in pairs, and otherwise the function is zero. With this knowledge, linear independence is easily demonstrated. Let

$$\lambda^{i_1 \cdots i_p}_{j_1 \cdots j_q} x_{i_1} \otimes \cdots \otimes x_{i_p} \otimes a^{j_1} \otimes \cdots \otimes a^{j_q} = 0$$

(n^{p+q} summations).

Applying this on $a^{i'_1}, \ldots, a^{i'_p}, x_{j'_1}, \ldots, x_{j'_q}$ makes all the summands disappear except for one, and we obtain

$$\lambda^{i'_1 \cdots i'_p}_{j'_1 \cdots j'_q} = 0 \,,$$

and as the indices $i'_1, \ldots, i'_q$ were chosen arbitrarily the linear independence follows. It remains to show that these simple tensors span E^p_q. Let

$$f = \lambda^{i_1 \cdots i_p}_{j_1 \cdots j_q} x_{i_1} \otimes \cdots \otimes x_{i_p} \otimes a^{j_1} \otimes \cdots \otimes a^{j_q} \,.$$

Applying this to the variables $a^{i'_1}, \ldots, x_{j'_q}$ gives

$$f(a^{i'_1}, \ldots, a^{i'_p}, x_{j'_1}, \ldots, x_{j'_q}) = \lambda^{i'_1 \cdots i'_p}_{j'_1 \cdots i'_q} \,.$$

Given $f \in E^p_q$, we show that

$$f = f(a^{i_1}, \ldots, a^{i_p}, x_{j_1}, \ldots, x_{j_q}) x_{i_1} \otimes \cdots \otimes x_{i_p} \otimes a^{j_1} \otimes \cdots \otimes a^{j_q} \,.$$

This equation is correct when the variables are taken from the basis of E and the corresponding dual base. Due to multi-linearity it is also correct for the general case. This also settles Theorem 3.1 and 3.2, for which we had used Theorem 3.3 in their proofs.

From the above proof, we see that the coordinates of a tensor can be conveniently calculated with respect to a basis.

Theorem 3.4 *Let $x_1, \ldots, x_n$ be a basis for E. The coordinates of $f \in E^p_q$ with respect to the basis of tensors $x_{i_1} \otimes \cdots \otimes x_{i_p} \otimes a^{j_1} \otimes \cdots \otimes a^{j_q}$ are given by*

$$f^{i_1 \cdots i_p}_{j_1 \cdots j_q} = f(a^{i_1}, \ldots, a^{i_p}, x_{j_1}, \ldots, x_{j_q}) \,.$$

The numbers are called the **components** *of f with respect to the basis $x_1, \ldots, x_n$.*

Example. A linear mapping $T \in L(E, E)$ is given by a matrix (τ_k^i) with respect to the basis $x_1, \ldots, x_n$ for E, that is,

$$T(\xi^k x_k) = \tau_k^l \xi^k x_l.$$

The linear form a^i of the dual basis sends an element $x \in E$ to its i-th coordinate with respect to the given basis. The tensor $f \in E_1^1$ corresponding to the mapping $T \in L(E, E)$ defined by $f(a, x) = \langle Tx, a \rangle$ and has the components

$$f_j^i = f(a^i, x_j) = \langle Tx_j, a^i \rangle = \langle \tau_j^l x_l, a^i \rangle = \tau_j^i.$$

The elements of the matrix are thus the components of the tensor with respect to the same basis.

Next we investigate the effect of a change of basis in E on the components of a tensor $f \in E_q^p$. In addition to a basis $X = \{x_1, \ldots, x_n\}$ for E, suppose there is given another basis $\overline{X} = \{\overline{x}_1, \ldots, \overline{x}_n\}$. The two bases are related to each other by the conversion formulae $\overline{x}_i = \alpha_i^j x_j$ and $x_j = \beta_j^i \overline{x}_i$. Of course the two matrices (α_i^j) and (β_j^i) are inverses of each other. We now determine the relation between the dual bases $A = \{a^1, \ldots, a^n\}$ for X and the dual basis $\overline{A} = \{\overline{a}^1, \ldots, \overline{a}^n\}$ for $\overline{X}$. We set $\overline{a}^k = \gamma_l^k a^l$ and apply this linear form on the basis elements $x_1, \ldots, x_n$. Then as

$$\langle x_j, \overline{a}^k \rangle = \langle \beta_j^i \overline{x}_i, \overline{a}^k \rangle = \beta_j^i \langle \overline{x}_i, \overline{a}^k \rangle = \beta_j^k$$

and

$$\langle x_j, \gamma_l^k a^l \rangle = \gamma_l^k \langle x_j, a^l \rangle = \gamma_j^k$$

it follows that $\overline{a}^k = \beta_l^k a^l$, and since (α_i^j) is the inverse of (β_j^i), also $a^l = \alpha_k^l \overline{a}^k$.

Now the components $f_{j_1 \cdots j_q}^{i_1 \cdots i_p}$ of a tensor $f \in E_q^p$ with respect to the basis X can easily be related to the components $\overline{f}_{j_1 \cdots j_q}^{i_1 \cdots i_p}$, with respect to $\overline{X}$. We have

$$\overline{f}_{j_1 \cdots j_q}^{i_1 \cdots i_p} = f(\overline{a}^{i_1}, \ldots, \overline{a}^{i_p}, \overline{x}_{j_1}, \ldots, \overline{x}_{j_q})$$

$$= f(\beta_{l_1}^{i_1} a^{l_1}, \ldots, \beta_{l_p}^{i_p} a^{l_p}, \alpha_{j_1}^{k_1} x_{k_1}, \ldots, \alpha_{j_q}^{k_q} x_{k_q}) = \beta_{l_1}^{i_1} \cdots \beta_{l_p}^{i_p} f_{k_1 \cdots k_q}^{l_1 \cdots l_p} \alpha_{j_1}^{k_1} \cdots \alpha_{j_q}^{k_q}.$$

We summarise this in the following result.

Theorem 3.5 *Suppose the bases $X = \{x_1, \ldots, x_n\}$ and $\overline{X} = \{\overline{x}_1, \ldots, \overline{x}_n\}$ for E are related by $\overline{x}_i = \alpha_i^j x_j$ and $x_j = \beta_j^i \overline{x}_i$. Then the components of $f \in E_q^p$ with respect to X and $\overline{X}$ are related by*

$$\overline{f}_{j_1 \cdots j_q}^{i_1 \cdots i_p} = \beta_{l_1}^{i_1} \cdots \beta_{l_p}^{i_p} f_{k_1 \cdots k_q}^{l_1 \cdots l_p} \alpha_{j_1}^{k_1} \cdots \alpha_{j_q}^{k_q}$$

.

Example. Let $f \in E_1^1$ be the tensor corresponding to the linear map $T \in L(E, E)$ with matrix (τ_k^i) with respect to the basis X for E. We had already seen that the elements τ_k^i are the components f_k^i of the tensor f with respect to X. Wtih respect to another basis $\overline{X}$, the components are $\overline{f}_j^i = \beta_l^i f_k^l \alpha_j^k$. We had already met this formula at the end of Sect. 3.1.

3.4 Operations with Tensors

First of all E_q^p is a vector space, and so two tensors f and g are of the same type (also said to have the same index structure) can be added

$$(f + g)(a^1, \ldots, a^p, x_1, \ldots, x_q) = f(a^1, \ldots, a^p, x_1, \ldots, x_q)$$
$$+ g(a^1, \ldots, a^p, x_1, \ldots, x_q).$$

For components, one has then

$$(f + g)_{j_1 \cdots j_q}^{i_1 \cdots i_p} = f_{j_1 \cdots j_q}^{i_1 \cdots i_p} + g_{j_1 \cdots j_q}^{i_1 \cdots i_p}.$$

The multiplication by scalars is given by

$$(\lambda f)(a^1, \ldots, a^p, x_1, \ldots, x_q) = \lambda f(a^1, \ldots, a^p, x_1, \ldots, x_q)$$

and for component one then has

$$(\lambda f)_{j_1 \cdots j_q}^{i_1 \cdots i_p} = \lambda f_{j_1 \cdots j_q}^{i_1 \cdots i_p}.$$

Definition 3.2 The **tensor product** $f \otimes g$ of two tensors $f \in E_q^p$ and $g \in E_s^r$ is a tensor in E_{q+s}^{p+r} defined by

$$f \otimes g(a^1, \ldots, a^p, a^{p+1}, \ldots, a^{p+r}, x_1, \ldots, x_q, x_{q+1}, \ldots, x_{q+s})$$
$$= f(a^1, \ldots, a^p, x_1, \ldots, x_q) \, g(a^{p+1}, \ldots, a^{p+r}, x_{q+1}, \ldots, x_{q+s}).$$

$\blacklozenge$

The tensor product of two tensors can be formed whenever the same vector space E is underlying both the tensors. The components of the tensor product are given in terms of the components of the two factors by the relation

$$(f \otimes g)_{j_1 \cdots j_q j_{q+1} \cdots j_{q+s}}^{i_1 \cdots i_p i_{p+1} \cdots i_{p+r}} = f_{j_1 \cdots j_q}^{i_1 \cdots i_p} g_{j_{q+1} \cdots j_{q+s}}^{i_{p+1} \cdots i_{p+r}}.$$

The following calculation rules can be easily established:

$$(f \otimes g) \otimes h = f \otimes (g \otimes h)$$
$$(f + g) \otimes h = (f \otimes h) + (g \otimes h)$$
$$f \otimes (g + h) = (f \otimes g) + (f \otimes h)$$
$$(\lambda f) \otimes g = \lambda (f \otimes g) = f \otimes (\lambda g)$$

The notation for simple tensors established in Example 3 of Sect. 3.2 is a special case of the one for the more general tensor products. The simple tensor

$$x_1 \otimes \cdots \otimes x_p \otimes a^1 \otimes \cdots \otimes a^q \in E_q^p$$

is the tensor product of the tensors $x_1, \ldots, x_p, a^1, \ldots, a^q$ of E_0^1 and E_1^0.

Definition 3.3 For $r \in \{1, \ldots, p\}$, $s \in \{1, \ldots, q\}$ and $f \in E_q^p$, let $C_s^r f \in E_{q-1}^{p-1}$ be the tensor defined by the sum (summation over k)

$$C_s^r f(a^1, \ldots, a^{r-1}, a^{r+1}, \ldots, a^p, x_1, \ldots, x_{s-1}, x_{s+1}, \ldots, x_q)$$
$$= f(a^1, \ldots, a^{r-1}, b^k, a^{r+1}, \ldots, a^p, x_1, \ldots, x_{s-1}, y_k, x_{s+1}, \ldots, x_q),$$

where $y_1, \ldots, y_n$ is a basis for E and $b^1, \ldots, b^n$ is the dual basis for E^*. The transition of the tensor f to the tensor $C_s^r f$ is called **contraction**.

It needs to be clarified why the definition of $C_s^r f$ is independent of the choice of the basis $y_1, \ldots, y_n$ for E. So we consider a second basis $z_i = \alpha_i^j y_j$ with the dual basis $c^1, \ldots, c^n$. As we had found out earlier in the proof of Theorem 3.5, there holds $c^j = \beta_i^j b^i$, where (β_i^j) is the inverse of the matrix (α_i^j). Thus

$$f(a^1, \ldots, c^k, \ldots, a^p, x_1, \ldots, z_k, \ldots, x_q)$$
$$= f(a^1, \ldots, \beta_l^k b^l, \ldots, a^p, x_1, \ldots, \alpha_k^m y_m, \ldots, x_q)$$

$$= \beta_l^k \alpha_k^m f(a^1, \ldots, b^l, \ldots, a^p, x_1, \ldots, y_m, \ldots, x_q)$$
$$= f(a^1, \ldots, b^l, \ldots, a^p, x_1, \ldots, y_l, \ldots, x_q).$$

The determination of the components of the contracted tensor is much easier than the abstract definition of contraction. Applying the definition of the contraction of a tensor and of the components of a tensor, we see that

$$C_s^r f \, {}^{i_1 \cdots i_{r-1} i_{r+1} \cdots i_p}_{j_1 \cdots j_{s-1} j_{s+1} \cdots j_q} = f \, {}^{i_1 \cdots i_{r-1} k \, i_{r+1} \cdots i_p}_{j_1 \cdots j_{s-1} k \, j_{s+1} \cdots j_q}.$$

Example. A linear map T in E gives rise to a tensor $f \in E_1^1$. The elements τ_k^i of the matrix of T with respect to a basis for E are the components f_k^i of the tensor f with respect to that basis. The contracted tensor is the number $f_i^i = \tau_i^i$. This is the trace of the matrix.

Definition 3.4 For $f \in E_q^p$ and $g \in E_s^r$, let $h \in E_{q+s-1}^{p+r-1}$ be the tensor $h = C_{q+u}^t(f \otimes g)$ where $t \in \{1, \ldots, p\}$ and $u \in \{1, \ldots, s\}$ or $h = C_u^{p+t}(f \otimes g)$ where $t \in \{1, \ldots, r\}$ and $u \in \{1, \ldots, q\}$. We call such an h as the **contraction of a pair** of tensors. ◆

Since the contraction of a pair of tensors involves the tensor product and taking a contraction, it is independent of the selection of a basis $y_1, \ldots, y_n$ and its dual basis $b^1, \ldots, b^n$. The components of the tensor h are given by

$$h^{i_1\cdots i_{t-1}i_{t+1}\cdots i_p i_{p+1}\cdots i_{p+r}}_{j_1\cdots j_q j_{q+1}\cdots j_{q+u-1}j_{q+u+1}\cdots j_{q+s}} = f^{i_1\cdots i_{t-1}k i_{t+1}\cdots i_p}_{j_1\cdots j_q} \cdot g^{i_{p+1}\cdots i_{p+r}}_{j_{q+1}\cdots j_{q+u-1}k j_{q+u+1}\cdots j_{q+s}}$$

respectively by

$$h^{i_1\cdots i_p i_{p+1}\cdots i_{p+t-1}i_{p+t+1}\cdots i_{p+r}}_{j_1\cdots j_{u-1}j_{u+1}\cdots j_q j_{q+1}\cdots j_{q+s}} = f^{i_1\cdots i_p}_{j_1\cdots j_{u-1}k j_{u+1}\cdots j_q} \cdot g^{i_{p+1}\cdots i_{p+t-1}k i_{p+t+1}\cdots i_{p+r}}_{j_{q+1}\cdots j_{q+s}}.$$

Example. A linear map $T \in L(E, E)$ gives rise to a tensor $f \in E_1^1$ and a vector $x \in E$ is a tensor $g \in E_0^1$. The only possible contraction of the pair F and g is a tensor $h \in E_0^1$, defined using a basis $x_1, \ldots, x_n$ and its dual basis $a^1, \ldots, a^n$, as

$$h(a) = f(a, x_k)g(a^k) = \langle Tx_k, a\rangle\langle x, a^k\rangle = \langle\langle x, a^k\rangle Tx_k, a\rangle = \langle Tx, a\rangle.$$

We used the fact that $x = \langle x, a^k\rangle x_k$ in order to obtain the last equality. The above shows that the contraction of the pair is the tensor corresponding to the vector Tx which is the image of x under T. The application of a linear map on a vector can thus be viewed as the contraction of a pair of tensors. One can also see this in components

$$h^i = f_k^i g^k = \tau_k^i \xi^k.$$

It can also be seen from the components of the contraction of a tensor with the unit tensor (Example 1 from Sect. 3.2) has no effect, since the components of the unit tensor are given by the Kronecker symbol.

3.5 Tensors on Euclidean Spaces

An inner product is given on a finite dimensional vector space E, that is a positive definite, symmetric bilinear form, i.e. a $(0, 2)$-tensor. We denote it by g and call it a **metric tensor**. Instead of $x \cdot y$, we now write $g(x, y)$.

Since the inner product depends linearly on the second factor, one gets an isomorphism $J : E \longrightarrow E^*$, characterised by

$$\langle y, Jx\rangle = g(x, y) \qquad \text{for all } x, y \in E.$$

For each basis $x_1, \ldots, x_n \in E$ with its dual basis $a^1, \ldots, a^n \in E^*$, and a vector $x = \xi^k x_k$, one can express the components λ_i of the linear form $Jx = \lambda_i a^i$ by

$$\lambda_i = \langle x_i, \lambda_j a^j \rangle = \langle x_i, Jx \rangle = g(x, x_i) = g(\xi^k x_k, x_i) = \xi^k g_{ki} \, ,$$

so that

$$Jx = \xi^k g_{ki} a^i \, ,$$

and in particular

$$Jx_i = g_{ij} a^j \, .$$

If the basis $x_1, \ldots, x_n$ is orthonormal, then the components of g with respect to this basis are given by the Kronecker symbol δ_{ij}:

$$g_{ij} = \delta_{ij} = \begin{cases} 1 & \text{for} \quad i = j \\ 0 & \text{for} \quad i \neq j \end{cases} .$$

The basis for E^* that is dual to $x_1, \ldots, x_n$ is, in this case, $Jx_1, \ldots, Jx_n$. From $x = \xi^k x_k$ it follows that since J is linear,

$$Jx = \xi^k Jx_k \, ,$$

that is, the linear form Jx has the same components with respect to $Jx_1, \ldots, Jx_n$ as the vector x with respect to $x_1, \ldots, x_n$. However, Jx_k should have the index at the top as a linear form and ξ^k should have the index at the bottom as the component of a linear form. Thus the effect of J is called **index pulling**, in this case "from top to bottom", since the k become the indices of a basis for the space of linear forms. More generally, the isomorphism $J : E \longrightarrow E^*$ generates by index pulling from a (p, q)-tensor f with $p \geq 1$, a $(p-1, q+1)$-tensor h, defined by

$$h(b^2, \ldots, b^p, y_0, y_1, \ldots, y_q) = f(Jy_0, b^2, \ldots, b^p, y_1, \ldots, y_q)$$

for $y_0, \ldots, y_q \in E$ and $b^2, \ldots, b^q \in E^*$. The components of the new tensor h with respect to a not necessarily orthonormal basis $x_1, \ldots, x_n$ are given by

$$h^{i_2 \cdots i_p}_{j_0 j_1 \cdots j_q} = h(a^{i_2}, \ldots, a^{i_p}, x_{j_0}, x_{j_1}, \ldots, x_{j_q}) = f(Jx_{j_0}, a^{i_2}, \ldots, a^{i_p}, x_{j_1}, \ldots, x_{j_q})$$

$$= f(g_{j_0 i_1} a^{i_1}, a^{i_2}, \ldots, a^{i_p}, x_{j_1}, \ldots, x_{j_q}) = g_{j_0 i_1} f^{i_1 i_2 \cdots i_p}_{j_1 \cdots j_q} \, .$$

Thus index pulling (from top to bottom) is the contraction of the pair of tensors consisting of the given tensor (with respect to its first contravariant argument) and the metric tensor. One can also see this from the following

$$f(Jy, b^2, \ldots, b^p, y_1, \ldots, y_q)$$
$$= y^k g_{jk} f(a^j, b^2, \ldots, b^p, y_1, \ldots, y_q) = g(y, x_j) f(a^j, b^2, \ldots, b^p, y_1, \ldots, y_q) .$$

The inner product on a finite-dimensional Euclidean space E also gives rise to a natural inner product on the dual space E^*. This $(2, 0)$-tensor is called a **contravariant metric tensor** and is again denoted by g. It is defined by

$$g(a, b) = g(J^{-1}a, J^{-1}b).$$

Which of the two tensors is meant by the symbol g can be deduced from the type of arguments used. One can immediately see that the contravariant metric tensor is also symmetric and positive definite.

By the contraction of the pair of tensors consisting of the contravariant metric tensor with the metric tensor, i.e., by pulling index of the contravariant metric tensor down, a $(1, 1)$-tensor (the unit tensor) is created, which corresponds to the identity map in the sense of Theorem 3.1. This is confirmed by observing that $Jx = \xi^k g_{jk} a^j = g(x, x_j) a^j$ and so

$$g(x, x_j)\, g(a^j, a) = g(Jx, a) = g(a, Jx) = g(J^{-1}a, x) = \langle x, a \rangle.$$

Consequently, with respect to every basis in E, the component matrices (g_{ik}) and (g^{jl}) of the two metric tensors are inverses of each other.

Index pulling can also be done "from the bottom to the top". Here a (p, q)-tensor f with $q \geq 1$ defines a $(p+1, q-1)$-tensor h by

$$h(b^0, b^1, \ldots, b^p, y_2, \ldots, y_q) = f(b^1, \ldots, b^p, J^{-1}b^0, y_2, \ldots, y_q).$$

Since $J^{-1}a^i = g^{ij}x_j$, in terms of components, we have

$$h^{i_0 i_1 \cdots i_p}_{j_2 \cdots j_q} = f(a^{i_1}, \ldots, a^{i_p}, J^{-1}a^{i_0}, x_{j_2}, \ldots, x_{j_q}) = g^{i_0, j_1} f^{i_1 \cdots i_p}_{j_1 j_2 \cdots j_q}.$$

It is again a contradiction of a pair of tensors, but this time with the contravariant metric tensor. Obviously, the two types of index pulling cancel each other out.

In Euclidean spaces, the use of orthonormal bases brings considerable computational advantages. Hence one often just performs the calculation of tensor components only for orthonormal bases. To do this one needs the transformation formula (Theorem 3.5) for the conversion of components with respect to one orthonormal basis $X = \{x_1, \ldots, x_n\}$ to another orthonormal basis $\tilde{X} = \{\tilde{x}_1, \ldots, \tilde{x}_n\}$. The conversion matrix (α_i^j) of the relations $\tilde{x}_i = \alpha_i^j x_j$ is orthogonal, that is its inverse matches its transpose. For the coefficients of the inverse relations $x_j = \beta_j^i \tilde{x}_i$, thus there holds $\beta_j^i = \alpha_i^j$. This isomorphism $J : E \longrightarrow E^*$ has the effect that the components of a vector x with respect to an orthonormal basis $X = \{x_1, \ldots, x_n\}$ are the same as the components of the image linear form Jx with respect to the dual basis $JX = \{Jx_1, \ldots, Jx_n\}$. With this identification of E and E^* by means of J, there is no longer any reason to distinguish between vectors and linear forms, both of which are only n-tuples depending on the choice of the orthonormal basis. The components of a (p, q)-tensor f are generally indexed below, and so

$$f_{i_1 \cdots i_p j_1 \cdots j_q} = f(Jx_{i_1}, \ldots, Jx_{i_p}, x_{j_1}, \ldots, x_{j_q}).$$

The transformation formula for tensor components with respect to orthonormal bases then has the form

$$\tilde{f}_{i_1 \cdots i_p j_1 \cdots j_q} = \alpha_{i_1}^{l_1} \cdots \alpha_{i_p}^{l_p} f_{l_1 \cdots l_p k_1 \cdots k_q} \alpha_{j_1}^{k_1} \cdots \alpha_{j_q}^{k_q}.$$

The difference between covariant and contravariant thus disappears. The number $p+q$ is called the **rank** of the tensor. The transformation formula for the components of a rank r-tensor f on a finite-dimensional Euclidean space is then

$$\tilde{f}_{i_1 \cdots i_r} = \alpha_{i_1}^{k_1} \cdots \alpha_{i_r}^{k_r} f_{k_1 \cdots k_r}$$

when converting from one orthonormal basis $x_1, \ldots, x_n$ to another one with $\tilde{x}_i = \alpha_i^k x_k$.

Example. If we rephrase the $(1, 2)$-tensor f on $\mathbb{R}^3$ treated in Example 2 of Sect. 3.2 in the sense just described by equipping the Euclidean space $\mathbb{R}^3$ with its usual inner product, then we obtain the rank 3-tensor

$$\varepsilon(y_1, y_2, y_3) = f(Jy_1, y_2, y_3) = \langle y_2 \times y_3, Jy_1 \rangle = y_1 \cdot (y_2 \times y_3),$$

called the ε-**tensor**. With respect to the canonical basis, it has the components

$$\varepsilon_{123} = \varepsilon_{231} = \varepsilon_{312} = 1$$

and

$$\varepsilon_{321} = \varepsilon_{213} = \varepsilon_{132} = -1$$

and $\varepsilon_{ijk} = 0$ for the other 21 cases. The components with another orthonormal basis $\{x_1, x_2, x_3\}$ with $x_3 = x_1 \times x_2$ are the same. One can see this both using the definition of this tensor and by using the transformation formula

$$\tilde{\varepsilon}_{ijk} = \alpha_i^s \alpha_j^t \alpha_k^u \varepsilon_{stu}$$

where (α_i^s) is a rotation matrix, and so by Sarrus' rule

$$\tilde{\varepsilon}_{123} = \tilde{\varepsilon}_{231} = \tilde{\varepsilon}_{312} = \det(\alpha_i^s) = 1$$

and

$$\tilde{\varepsilon}_{321} = \tilde{\varepsilon}_{213} = \tilde{\varepsilon}_{132} = -\det(\alpha_i^s) = -1,$$

and for the remaining cases, the determinant of a matrix appears in which certain rows coincide.

In the next chapter, we will deal with symmetric bilinear forms g also, but which are not positive definite. We will see that the linear mapping $J : E \longrightarrow E^*$ defined by $\langle y, Jx \rangle = g(x, y)$ is also then bijective, if g is assumed only to be non-degenerate, that is, the determinant of the components of g is non zero (regular). All the results of this section apply in this more general case.

Semi-Riemannian Manifolds

4

4.1 Tensor Fields

Each tangent space M_P of a C^∞-manifold M is an n-dimensional vector space. Thereby for nonnegative integers p and q, one obtains the tensor spaces $(M_P)_q^p$. In particular, one forms the dual space $M_P^* = (M_P)_1^0$.

Definition 4.1 The dual space $M_P^* = (M_P)_1^0$ to the tangent space M_P is called the **cotangent space**, and its elements are called **cotangent vectors** or **covector** or **covariant vectors**. $\blacklozenge$

Definition 4.2 A **covector field** K on M is a map assigning to each $P \subset M$ a covector $K(P) \in (M_P)_1^0$. K is a C^∞-**covector field**, if for every C^∞-vector field X on M, the real-valued function $P \longrightarrow \langle X(P), K(P) \rangle$ is inifinitely many times differentiable. $\blacklozenge$

As is well-known, a chart gives rise to n-coordinate vector fields $\frac{\partial}{\partial u^i}$ $(i = 1, \ldots, n)$. For each point $P \in U$, the tangent vectors $\frac{\partial}{\partial u^i}(P)$ form a basis for M_P.

Definition 4.3 The n covector fields $du^1, \ldots, du^n$ assign to each point P in the chart with coordinate vector fields $\frac{\partial}{\partial u^i}$ the elements of the basis that is dual to the basis $\frac{\partial}{\partial u^1}(P), \ldots, \frac{\partial}{\partial u^n}(P)$. $\blacklozenge$

Usually, the notation does not distinguish between the covector field and the linear form it assumes at point P. Thus

© The Author(s), under exclusive license to Springer Nature Switzerland AG 2023
R. Oloff, *The Geometry of Spacetime*, Graduate Texts in Physics,
https://doi.org/10.1007/978-3-031-16139-1_4

$$\left\langle \frac{\partial}{\partial u^i} (P), du^k \right\rangle = \begin{cases} 1 & \text{for} \quad i = k \\ 1 & \text{for} \quad i \neq k \end{cases},$$

that is, du^k assigns to the tangent vector $x^i \partial_i$ its k-th component x^k.

One associates with the symbol du^k a small change in the variable u^k when modelling a physical process. This point of view can be met in the interpretation of the following situation: A particle moves along a curve and at time t_0 has the velocity $x = x^i \partial_i$. How fast does the k-th coordinate of its position change? We have

$$\frac{du^k}{dt} = x \, u^k = x^i \partial_i u^k = x^k,$$

which is commonly written in the Physics literature as

$$du^k = x^k dt.$$

Definition 4.4 A (p, q)-**tensor field** T on M is a map that assigns to every $P \in M$ a tensor $T(P) \in (M_P)^p_q$. T is a C^∞-tensor field if for C^∞-vector fields $X_1, \ldots, X_q$ and C^∞-covector fields $K^1, \ldots, K^p$, the real-valued function is infinitely many times differentiable.   ♦

Definition 4.5 $T^p_q(M)$ denotes the vector space of all C^∞-(p, q)-tensor fields on M. $\mathcal{K}(M) = T^0_1(M)$ is the vector space of all C^∞-covector fields. For $P \in M$, $T^p_q(P)$ denotes the quotient space of $T^p_q(M)$ with the subspace of tensor fields that are zeor in a neighborhood of P. There also holds that $\mathcal{K}(P) = T^0_1(P)$ (other interpretation as in the discussion following Definition 2.3).   ♦

A (p, q)-tensor field assigns to a point $P \in M$, q vectors in M_P and p linear forms on M_P, a number. By inserting q vector fields and p covector fields, we obtain a scalar field. In particular a $(0, 2)$-tensor field gives a scalar field if we apply it to two vector fields. More precisely, a $(0, 2)$-tensor field assigns a scalar field to two vector fields in a bilinear manner. But not every bilinear map from $\mathcal{X}(P) \times \mathcal{X}(P)$ to $\mathcal{F}(P)$ is a $(0, 2)$-tensor field. For example, the Lie bracket $[X, Y](P)$ is clearly not determined by $X(P)$ and $Y(P)$ alone, since the partial derivatives of the components of X, Y are also relevant, i.e., the values of X and Y in a neigborhood of P are still required.

Definition 4.6 A multilinear map

$$A : \underbrace{\mathcal{K}(P) \times \cdots \times \mathcal{K}(P)}_{p\text{-times}} \times \underbrace{\mathcal{X}(P) \times \cdots \times \mathcal{X}(P)}_{q\text{-times}} \longrightarrow \mathcal{F}(P)$$

is called $\mathcal{F}$-**homogeneous** if for all covector fields $K^1, \ldots, K^p \in \mathcal{K}(P)$, vector fields $X_1, \ldots, X_q \in \mathcal{X}(P)$ and scalar fields $f_1, \ldots, f_p, g^1, \ldots, g^q \in \mathcal{F}(P)$, we have

$$A(f_1 K^1, \ldots, f_p K^p, g^1 X_1, \ldots, g^q X_q) = f_1 \cdots f_p g^1 \cdots g^q A(K^1, \ldots, K^p, X_1, \ldots, X_q).$$

$\blacklozenge$

Theorem 4.1 *A multilinear map*

$$A : \mathcal{K}(P) \times \cdots \times \mathcal{K}(P) \times \mathcal{X}(P) \times \cdots \times \mathcal{X}(P) \longrightarrow \mathcal{F}(P)$$

is a tensor field if and only if it is $\mathcal{F}$-homogeneous.

Proof A (p, q)-tensor field T generates a multilinear map A via

$$A(K^1, \ldots, K^p, X_1, \ldots, X_q)(P) = T(P)(K^1(P), \ldots, K^p(P), X_1(P), \ldots, X_q(P)),$$

which is clearly $\mathcal{F}$-homogeneous. Conversely, an $\mathcal{F}$-homogeneous multilinear map A generates a tensor field T as follows: If the linear forms $a^1, \ldots, a^p \in M_P^*$ and the vectors $x_1, \ldots, x_q \in M_P$ continue to the covector fields $K^1, \ldots, K^p$ and vector fields $X_1, \ldots, X_q$, then we set

$$T(P)(a^1, \ldots, a^p, x_1, \ldots, x_q) = A(K^1, \ldots, K^p, X_1, \ldots, X_q)(P).$$

It remains to be shown that this number on the left-hand side is independent of the manner of the continuation. To this end we choose a chart and thereby coordinate vector fields $\partial_1, \ldots, \partial_n$ and covector fields $du^1, \ldots, du^n$. As A is multilinear and $\mathcal{F}$-homogeneous, there holds

$$A(K^1, \ldots, K^p, X_1, \ldots, X_q)(P) = A(K_{i_1}^1 du^{i_1}, \ldots, K_{i_p}^p du^{i_p}, X_1^{k_1} \partial_{k_1}, \ldots, X_q^{k_q} \partial_{k_q})(P)$$

$$= K_{i_1}^1(P) \cdots K_{i_p}^p(P) X_1^{k_1}(P) \cdots X_q^{k_q}(P) A(du^{i_1}, \ldots, du^{i_p}, \partial_{k_1}, \ldots, \partial_{k_q})(P),$$

and this shows the independence from the continuation.

4.2 Riemannian Manifolds

Definition 4.7 A **Riemannian manifold** $[M, g]$ is a finite-dimensional C^∞-manifold M on which a doubly covariant C^∞-tensor field g is also defined, such that the bilinear form $g(P)$ on the tangent space M_P is an inner product. A tensor field g is called the **fundamental tensor**, **metric tensor** or **metric** for short. The associated contravariant tensor field (see Sect. 3.5) is called the **contravariant metric tensor** and is also denoted by g.

$\blacklozenge$

Example. A metric g is induced on a curved surface in $\mathbb{R}^3$ by restricting the canonical scalar product to the tangent spaces. A chart (parametrisation) gives rise to two coordinate vector fields ∂_1 and ∂_2. The corresponding components of g are denoted, in the classical Gaussian notation, by

$$E = g_{11} = g\,(\partial_1, \partial_1)$$
$$F = g_{12} = g\,(\partial_1, \partial_2)\,.$$
$$G = g_{22} = g\,(\partial_2, \partial_2)$$

The contravariant metric tensor has the component matrix

$$\begin{pmatrix} E & F \\ F & G \end{pmatrix}^{-1} = \frac{1}{EG - F^2}\begin{pmatrix} G & -F \\ -F & E \end{pmatrix}.$$

A metric gives rise to the notion of **arc length** $s(\gamma)$ of a curve γ as the integral

$$s(\gamma) = \int_{t_1}^{t_2} \sqrt{g(\gamma'(t), \gamma'(t))}\, dt\,.$$

The fact that curves differing only in their parametrisation have the same arc length follows from the substitution rule for Riemann integrals.

If we choose a chart with the coordinates $u^1, ..., u^n$, then the tangent vector is $\frac{du^k}{dt}\partial_k$, and the arc length becomes

$$s(\gamma) = \int_{t_1}^{t_2} \sqrt{g_{ik} \frac{du^i}{dt} \frac{du^k}{dt}}\, dt$$

and for its derivative

$$\frac{ds}{dt} = \sqrt{g_{ik} \frac{du^i}{dt} \frac{du^k}{dt}}\,.$$

In the Physics literature, one often expresses this last equation in the form

$$ds^2 = g_{ik}\, du^i\, du^k\,.$$

This is the so-called **line element**, and one reads off the metric components g_{ik} from it. For example, the line element in $\mathbb{R}^3$ for the Cartesian coordinates is

$$ds^2 = dx^2 + dy^2 + dz^2$$

and for the spherical coordinates is

$$ds^2 = dr^2 + r^2\, d\vartheta^2 + r^2 \sin^2\!\vartheta\, d\varphi^2\,.$$

4.3   Bilinear Forms

An inner product is a positive definite symmetric bilinear form. In the theory of relativity, one is forced to drop the positive definiteness and replace it with another requirement about symmetric bilinear forms.

The components $g_{ik} = g(x_i, x_k)$ of a bilinear form g with respect to the basis $x_1, \ldots, x_n$ can be converted to another basis $\overline{x}_j = \tau^i_j x_i$, according to Theorem 3.5, by the formula

$$\overline{g}_{jl} = g_{ik} \tau^i_j \tau^k_l .$$

In matrix notation, one has $B = T^T A T$ with $A = (g_{ik})$, $B = (\overline{g}_{ik})$ and $T = (\tau^i_k)$. The following so-called **Sylvester law of inertia** gives the extent to which the component matrix can be simplified by using a particularly suitable basis.

Theorem 4.2 *Each symmetrical bilinear form g on an n-dimensional real vector space E has a basis $x_1, \ldots, x_n$, such that the component matrix of g with respect to this basis has the form*

$$
\begin{pmatrix}
1 & \overbrace{0 \cdots}^{p} & \overbrace{\cdots}^{q} & \cdots & 0 \\
0 & 1 & & & \vdots \\
 & & -1 & & \\
 & & & -1 & \\
 & & & & 0 \quad 0 \\
0 & \cdots & \cdots & 0 & 0
\end{pmatrix}
$$

The nonnegative integers p and q are independent of the choice of basis.

Proof We start with an arbitrary basis $y_1, \ldots, y_n$ in E, and let A be the symmetric component matrix of g with respect to this basis. By the well-known spectral theorem, there is an orthogonal matrix D, such that the product $D^T A D$ has the diagonal form

$$
D^T A D = \begin{pmatrix}
\lambda_1 & 0 & \cdots & \cdots & 0 \\
0 & & & & \vdots \\
\vdots & & \ddots & & \\
\vdots & & & & 0 \\
0 & \cdots & \cdots & 0 & \lambda_n
\end{pmatrix}
$$

We may assume that

$$\begin{aligned}
\lambda_i &> 0 \quad \text{for} \quad i = 1, \ldots, p \\
\lambda_i &< 0 \quad \text{for} \quad i = p+1, \ldots, p+q \,. \\
\lambda_i &= 0 \quad \text{for} \quad i = p+q+1, \ldots, n
\end{aligned}$$

We factorise the diagonal matrix as

$$
\begin{pmatrix}
\lambda_1 & 0 & \cdots & 0 \\
0 & \ddots & & \vdots \\
\vdots & & \ddots & 0 \\
0 & \cdots & 0 & \lambda_n
\end{pmatrix}
=
$$

$$
=
\begin{pmatrix}
\gamma_1 & 0 & \cdots & 0 \\
0 & \ddots & & \vdots \\
\vdots & & \ddots & 0 \\
0 & \cdots & 0 & \gamma_n
\end{pmatrix}
\begin{pmatrix}
1 & 0 & \cdots & \cdots & \cdots & 0 \\
0 & 1 & & & & \vdots \\
\vdots & & -1 & & & \\
\vdots & & & -1 & & \vdots \\
\vdots & & & & 0 & 0 \\
0 & \cdots & \cdots & \cdots & 0 & 0
\end{pmatrix}
\begin{pmatrix}
\gamma_1 & 0 & \cdots & 0 \\
0 & \ddots & & \vdots \\
\vdots & & \ddots & 0 \\
0 & \cdots & 0 & \gamma_n
\end{pmatrix}
$$

where

$$
\gamma_i =
\begin{cases}
\sqrt{|\lambda_i|} & \text{for } i = 1, \ldots, p+q \\
1 & \text{otherwise}
\end{cases}
$$

and get the desired result

$$
\begin{pmatrix}
1 & 0 & \cdots & & & & & 0 \\
 & \ddots & & & & & & \vdots \\
0 & & 1 & & & & & \\
\vdots & & & -1 & & & & \vdots \\
\vdots & & & & \ddots & & & \\
\vdots & & & & & -1 & & \\
\vdots & & & & & & 0 & 0 \\
0 & \cdots & & & & & 0 & 0
\end{pmatrix}
= T^T A T
$$

with

$$
T = D
\begin{pmatrix}
\gamma_1^{-1} & 0 & \cdots & \cdots & 0 \\
0 & \ddots & & & \vdots \\
\vdots & & \ddots & & \vdots \\
\vdots & & & \ddots & 0 \\
0 & \cdots & & 0 & \gamma_n^{-1}
\end{pmatrix}
$$

To prove the independence of the numbers p and q from the choice of basis, we give the numbers p and $p + q$ and interpretation that is independent of the basis used. The number $p + q$ is obviously the rank of the component matrix, and the fact that all these matrices have the same rank follows from the general inequalities

$$\text{rank}\,(AB) \leq \text{rank}\,(A) \quad \text{and} \quad \text{rank}\,(AB) \leq \text{rank}\,(B)\,.$$

We now show that p is the maximum dimension of those subspaces M of E such that for all nonzero $x \in M$, $g(x, x)$ is positive. Let $x_1, \ldots, x_n$ be a basis as stated in the theorem, so that

$$g(\xi^i x_i,\ \xi^k x_k) = \sum_{j=1}^{p} (\xi^j)^2 - \sum_{j=p+1}^{p+q} (\xi^j)^2\,.$$

For nonzero elements x belonging to the linear span M_0 of the basis vectors $x_1, \ldots, x_p$, we have $g(x, x) > 0$. We now show that for every subspace M with the property that for all nonzero $x \in M$, $g(x, x) > 0$, we must have $\dim M \leq p$. To this end, let us consider the linear map S, which sends every element

$$x = \sum_{i=1}^{n} \xi^i x_i\ \in M$$

to its projection

$$Sx = \sum_{i=1}^{p} \xi^i x_i\ \in M_0\,.$$

Then S is injective, since if $Sx = 0$, we have

$$x = \sum_{i=p+1}^{n} \xi^i x_i$$

implying

$$g(x, x) = -\sum_{i=p+1}^{p+q} (\xi^i)^2 \leq 0\,.$$

As $x \in M$ and thanks to the assumed property of M, it follows that $x = 0$. For the injection $S : M \longrightarrow M_0$, we have

$$\dim M \leq \dim M_0 = p\,.$$

This proves the Sylvester law of inertia.

Because the component matrices A and B of bilinear symmetric forms in different bases are related by $B = T^T A T$ with an invertible T, the statement that the determinant of the component matrix is different from zero, is independent of the choice of basis.

Definition 4.8 A symmetric bilinear form g is called **nondegenerate** if its component matrix (g_{ik}) has a nonzero determinant.   ◆

An inner product enables linear forms to be viewed as vectors, as described at the
beginning of Sect. 3.5. This also workds in the general case.

Theorem 4.3 *For a nondegenerate symmetric bilinear form g on E, the linear map
$J : E \longrightarrow E^*$ defined by $\langle y, Jx \rangle = g(x, y)$ is an isomorphism.*

Proof We need to show that J is injective. Let $Jx = 0$, i.e., $g(x, y) = 0$ for all
$y \in E$. In terms of components, we have

$$g_{ik}x^i y^k = 0$$

for all n-tuples $(y^1, \ldots, y^n)$. So we obtain the homogeneous system of equations

$$g_{ik}x^i = 0,$$

and since $\det(g_{ik}) \neq 0$, this has the trivial solution $x^i = 0$.

For every orthonormal basis $x_1, \ldots, x_n$ of a Euclidean space, the linear forms
$Jx_1, \ldots, Jx_n$ form the dual basis. If a nondegenerate symmetric bilinear form is
used instead of an inner product, this result holds in a slightly modified form.

Theorem 4.4 *Let g be a nondegenerate symmetric bilinear form and $x_1, \ldots, x_n$ be
a basis such that*

$$g(x_i, x_k) = 0 \quad \text{for } i \neq k$$

and

$$g(x_i, x_i) = \varepsilon_i = \pm 1.$$

Then the linear forms $\varepsilon_i J x_i$ form the dual basis to the basis $x_1, \ldots, x_n$.

Proof This follows from

$$\langle x_i, \varepsilon_k J x_k \rangle = \varepsilon_k g(x_i, x_k).$$

4.4   Orientation

Definition 4.9 Two bases $x_1, \ldots, x_n$ and $y_1, \ldots, y_n$ of a real vector space are said
to be **co-oriented** if the matrix (α_k^i) such that $y_k = \alpha_k^i x_i$ satisfies $\det(\alpha_k^i) > 0$.   ◆

Co-orientation is an equivalence relation that partitions the set of all bases for the
vector space into two equivalence classes.

Definition 4.10 An n-dimensional real vector space is said to be **oriented** if a choice is made for one of the equivalence classes of co-oriented bases. The bases of the chosen equivalence classes are said to **lie in the orientation** or to be **positively oriented**. The dual basis for the dual space, which is dual to a positively oriented basis, is also said to be positively oriented. ◆

Example 1 In the vector space $\mathbb{R}^n$ the canonical basis is usually given the positive orientation. Then a basis of n-tuples $(\xi_1^i, \ldots, \xi_n^i)$ lies in the orientation if and only if $\det(\xi_k^i) > 0$.

Example 2 A curved surface in $\mathbb{R}^3$ is oriented if one of the two sides is distinguished. In the case of a closed surface (here supposed to be the hot surface of a body), this is usually the outside, and in the case of a surface given by $z = f(x, y)$, the top side.

Sometimes it is not possible to distinguish a side. This situation occurs with the **Möbius strip** and the **Klein bottle** (Fig. 4.1). The orientation of a surface means that one of the two unit normal vectors is selected at each point. This normal vector n orients the tangent space by the convention that a basis x_1, x_2 is positively oriented if the triple product $(x_1 \times x_2) \cdot n$ is positive. This is actually an orientation in the sense of Definition 4.10, since another basis of tangent vectors

$$y_1 = \alpha_1^1 x_1 + \alpha_1^2 x_2 \quad \text{and} \quad y_2 = \alpha_2^1 x_+ \alpha_2^2 x_2$$

is positively oriented if it is co-oriented with x_1, x_2 in the sense of Definition 4.9, thanks to the fact that

$$(y_1 \times y_2) \cdot n = \left[\left(\alpha_1^1 x_1 + \alpha_1^2 x_2 \right) \times \left(\alpha_2^1 x_1 + \alpha_2^2 x_2 \right) \right] \cdot n = \left(\alpha_1^1 \alpha_2^2 - \alpha_1^2 \alpha_2^1 \right) (x_1 \times x_2) \cdot n \,.$$

The notion of orientation of a surface can be extended to manifolds. First of all, one demands that the manifold is connected in the sense that any two points can

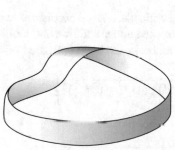

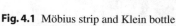
Fig. 4.1 Möbius strip and Klein bottle

always be connected by a path, i.e. for P and Q in M, there exists a continuous map $w : [0, 1] \longrightarrow M$ such that $w(0) = P$ and $w(1) = Q$. Now one wants all tangent spaces to be oriented and that these orientations have been selected in a way that they are compatible with each other. The examples of the Möbius strip and the Klein bottle show that this is not always possible.

Definition 4.11 An atlas is said to be **oriented** if for all charts (U, φ) and (V, ψ) with $U \cap V \neq \emptyset$ the Jacobi-determinant of the diffeomorphism $\varphi \circ \psi^{-1}$ is everywhere positive. An **oriented manifold** is a manifold equipped with a maximal oriented atlas. Each tangent space is then **oriented** by the convention that the basis $\partial_1, \ldots, \partial_n$ is positively oriented.   ◆

In the theory of relativity, the so-called Lorentzian manifolds are used instead of Riemannian manifolds. The common generic term is semi-Riemannian manifold.

Definition 4.12 A **semi-Riemannian manifold** is a finite-dimensional C^∞-manifold with a symmetric doubly covariant C^∞-tensor field g, such that the bilinear form $g(P)$ is everywhere nondegenerate. g is also called the fundamental tensor, metric tensor or metric, and the associated contravariant version, also labelled g, is again called a contravariant metric tensor.   ◆

Definition 4.13 A **Lorentzian manifold** is a semi-Riemannian manifold in which the metric has the property that everywhere its normal form (in the sense of Sylvesters law of inertia, Theorem 4.2) has one $+1$ and the rest -1.   ◆

For Lorentzian manifolds, there is yet another orientation concept, the so-called time-orientation. To explain this, we must first examine the situation in tangent spaces.

Definition 4.14 An n-dimensional real vector space, equipped with a nondegenerate symmetric bilinear form g, for which there exists a basis $x_0, x_1, \ldots, x_{n-1}$ satisfying $g(x_i, x_k) = 0$ for $i \neq k$, $g(x_0, x_0) = 1$ and $g(x_i, x_i) = -1$ for $i = 1, \ldots, n - 1$, is called a **Lorentzian space**, and such a basis is called a **Lorentzian basis**.   ◆

The set of all vectors x in a Lorentzian space with $g(x, x) > 0$ consists of two disjoint convex cones. A cone here means a subset closed under addition and multiplication by positive scalars. When using a Lorentzian basis, the two cones are

$$K_1 = \left\{ \xi^i x_i : \ \xi^0 > \sqrt{(\xi^1)^2 + \cdots + (\xi^{n-1})^2} \right\}$$

and

$$K_2 = \left\{ \xi^i x_i : \ \xi^0 < -\sqrt{(\xi^1)^2 + \cdots + (\xi^{n-1})^2} \right\}.$$

Fig. 4.2 Indefinite metric on a cylinder, that does not admit a time-orientation

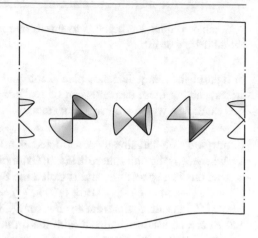

Each of these cones K_j is convex, since if $x, y \in K_j$ and $\lambda, \mu > 0$, then $\lambda x + \mu y$ also belongs to K_j. This follows for K_1 by the Minkowski-inequality

$$\lambda \xi^0 + \mu \eta^0 > \lambda \sqrt{(\xi^1)^2 + \cdots + (\xi^{n-1})^2} + \mu \sqrt{(\eta^1)^2 + \cdots + (\eta^{n-1})2}$$
$$\geq \sqrt{(\lambda \xi^1 + \mu \eta^1)^2 + \cdots + (\lambda \xi^{n-1} + \mu \eta^{n-1})^2}$$

and analogously for K_2 (Fig. 4.2).

Definition 4.15 A connected Lorentzian manifold is said to be **time-oriented**, if for each tangent space, a choice is made of one of the two cones that make up the set $\{x : g(x, x) > 0\}$, in such a way that the following holds for each C^∞-vector field X satisfying $g(X, X) > 0$ everywhere: If $X(P)$ belongs to the chose cone in M_P, then for every Q, $X(Q)$ belongs to the chosen cone in M_Q. ◆

A Lorentzian manifold may not be time-orientable. As a counterexample, consider a cylinder on which a metric is defined in such a way that the cones, which are sectors here, rotate by 180° in the course of traversing a circumference of the cylinder.

4.5 Spacetime

In the theory of relativity the world is understood as a spacetime.

Definition 4.16 A **spacetime** is a four-dimensional oriented and time-oriented Lorentzian manifold. ◆

Definition 4.17 A tangent vector x belonging to a spacetime is
 timelike, if $g(x, x) > 0$,
 spacelike, if $g(x, x) < 0$ or $x = 0$,
 lightlike, if $g(x, x) = 0$ and $x \neq 0$.

A timelike vector that belongs to the light cone chosen by the time-orientation, is called **future pointing**.   ♦

In light of the many demands placed for qualification as a spacetime, the question arises whether these demands can be realised at all. A positive answer is given by two examples, which are important standard models for the theory of relativity.

Example 3 We had already constructed an atlas for the set $M = \mathbb{R} \times (2\mu, +\infty) \times S_2$, where S_2 is the unit sphere in $\mathbb{R}^3$, in Example 4, Sect. 1.1. Equipped with a certain metric, this is a spacetime, and is called the **Schwarzschild spacetime**. The charts $(\mathbb{R} \times (2\mu, +\infty) \times U_i, \psi_i)$ with $\psi_i(t, r, Q) = (t, r, \varphi_i(Q))$ form an atlas for M, where (U_i, φ_i) are charts from an atlas for S_2. The usual topologies of $\mathbb{R}$, $(2\mu, +\infty)$ and S_2 can be used to equip M with the product topology. The required separation axiom is fulfilled. M is also clearly connected. We introduce the metric g as a matrix with respect to the following basis: Given numbers $t \in \mathbb{R}$ and $r \in (2\mu, +\infty)$ and given a point $Q \in S_2$, the vectors $x_1 = \frac{\partial}{\partial t}$ and $x_2 = \frac{\partial}{\partial r}$ act on a function $f \in \mathcal{F}(M)$ as follows

$$x_1 f = \frac{\partial f}{\partial t}(t, r, Q) \quad \text{and} \quad x_2 f = \frac{\partial f}{\partial r}(t, r, Q).$$

The other two basis vectors x_3 and x_4 are induced from a chart for S_2 and a positively oriented basis y_1 and y_2 in $(S_2)_Q$. For $f \in \mathcal{F}(M)$, let $g = f(t, r, .)$, and x_3 and x_4 be defined by

$$x_3 f = y_1 g \quad \text{and} \quad x_4 f = y_2 g.$$

This basis x_1, x_2, x_3, x_4 is considered to be positively oriented. Then only those charts that are compatible with this chart in the sense of Definition 4.11 are used. Then the resulting orientation on M is independent of the choice of the initial chart on S_2 used. The Lorentz metric g now based on this basis is given by the matrix

$$(g_{ik}) = \begin{pmatrix} 1 - \dfrac{2\mu}{r} & 0 & 0 & 0 \\ 0 & -\left(1 - \dfrac{2\mu}{r}\right)^{-1} & 0 & 0 \\ 0 & 0 & -h_{11} & -h_{12} \\ 0 & 0 & -h_{21} & -h_{22} \end{pmatrix},$$

where the submatrix h describes r^2 times the Euclidean metric on S_2. Finally, the time-orientation is given by setting timelike vectors with a positive $\frac{\partial}{\partial t}$-component to be future pointing.

Example 4 The **Einstein-de Sitter spacetime** is the set

$$M = (0, +\infty) \times \mathbb{R}^3 = \{(u^0, u^1, u^2, u^3) \in \mathbb{R}^4 : u^0 > 0\}.$$

With the identity map I, the pair (M, I) is a chart, and in fact an atlas. With respect to the positively oriented canonical basis $\frac{\partial}{\partial u^0}, \frac{\partial}{\partial u^1}, \frac{\partial}{\partial u^2}, \frac{\partial}{\partial u^3}$, the metric g is given by the matrix

$$(g_{i,k}) = \begin{pmatrix} 1 & 0 & 0 & 0 \\ 0 & -a(u^0)^{\frac{4}{3}} & 0 & 0 \\ 0 & 0 & -a(u^0)^{\frac{4}{3}} & 0 \\ 0 & 0 & 0 & -a(u^0)^{\frac{4}{3}} \end{pmatrix}$$

for a positive number a. The vector $\frac{\partial}{\partial u^0}$ is defined to be future pointing.

Every tangent space of a spacetime is a Lorentzian space. The following two theorems, which are formulated here in the context of Lorentzian spaces, are essential for the spacetime model of special relativity. An element x of a Lorentzian space is called **timelike** if $g(x, x) > 0$. In a time-oriented Lorentzian space, the vectors belonging to the chosen cone are said to be **future pointing**. The dimension does not matter here.

Theorem 4.5 *For every timelike vector x of a Lorentzian space, the subspace*

$$x^{\perp} = \{y \in E : g(x, y) = 0\},$$

equipped with the metric $-g$, is Euclidean.

Proof For $y \in x^{\perp}$ with $y \neq 0$, we must show that $-g(y, y) > 0$. We choose a Lorentzian basis and express x and y in terms of their components as the n-tuples $(x^0, x^1, \ldots, x^{n-1})$, respectively $(y^0, y^1, \ldots, y^{n-1})$. The fact that $g(x, y) = 0$ means that

$$x^0 y^0 = x^1 y^1 + \cdots + x^{n-1} y^{n-1}$$

and this implies

$$|x^0||y^0| \leq \sqrt{(x^1)^2 + \cdots + (x^{n-1})^2}\sqrt{(y^1)^2 + \cdots + (y^{n-1})^2}.$$

As $g(x, x) > 0$, it follows that

$$(x^1)^2 + \cdots + (x^{n-1})^2 < (x^0)^2.$$

The components $y^1, \ldots, y^{n-1}$ are not all zero, since otherwise $g(x, y) = 0$ gives then $y^0 = 0$ and so $y = 0$. These inequalities then give

$$|x^0||y^0| < |x^0|\sqrt{(y^1)^2 + \cdots + (y^{n-1})^2},$$

and so

$$(y^0)^2 < (y^1)^2 + \cdots + (y^{n-1})^2$$

i.e., $g(y, y) < 0$.

Theorem 4.6 *If x and z are two future pointing vectors in a Lorentzian space with $g(x, x) = g(z, z) = 1$, then $g(x, z) \geq 1$. All timelike vectors y satisifying $g(x, y) > 0$ are also future pointing.*

Proof We choose an orthonormal basis $e_1, \ldots, e_{n-1}$ in the Euclidean space $x^\perp$ and represent z using the Lorentzian basis $x, e_1, \ldots, e_{n-1}$. Then it can be seen that

$$z = g(z, x)x - g(z, e_1)e_1 - \cdots - g(z, e_{n-1})e_{n-1}.$$

Use unknown coefficients, and solve for them by applying g. It follows that

$$g(z, z) = \big(g(z, x)\big)^2 - \big(g(z, e_1)\big)^2 - \cdots - \big(g(z, e_{n-1})\big)^2.$$

As $g(z, z) = 1$, we have $(g(z, x))^2 - 1$, being the sum of squares, is nonnegative, and so

$$\big(g(x, z)\big)^2 - 1 \geq 0.$$

Hence $|g(x, z)| \geq 1$. If $g(x, y) > 0$, then $g(x, -y) < 0$, and so $-y$ cannot be future pointing, and so y muste be future pointing.

The notion of orthogonality in a Lorentzian space in the sense of $g(x, y) = 0$ depends of course essentially on the metric g, and in the case of $\mathbb{R}^2$ for example,

Fig. 4.3 Orthogonal lines

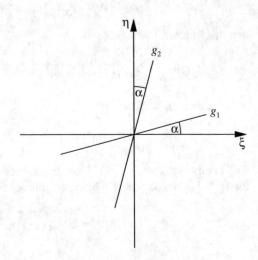

must not be confused with the geometric notion of orthogonality in a plane. In Fig. 4.3, the plane $\mathbb{R}^2$ is equipped with the metric

$$g\big((\xi_1, \eta_1), (\xi_2, \eta_2)\big) = \xi_1 \xi_2 - \eta_1 \eta_2 \, .$$

Two straight lines passing through $(0, 0)$ are orthogonal if and only if $\xi = \eta$ is their angle bisector.

Theory of Special Relativity

5.1   Kinematics

In Mechanics, one first investigates the motion of particles. In classical Newtonian mechanics, a path is a map from $\mathbb{R}$ to $\mathbb{R}^3$ which gives the position $\bar{r}(t)$ of the particle at time t. A path is thus a parameter representation of a curve in $\mathbb{R}^3$. This is depicted in Fig. 5.1, but one dimension is suppressed. In order to be able to compare this Newtonian viewpoint with the relativistic one, we rephrase the above. We understand the world as a Cartesian product, where the first component of its elements is time, and the second is position, and instead of the path $\bar{r}(t)$, we use the map $\hat{r} : \mathbb{R} \to \mathbb{R} \times \mathbb{R}^3$ given by $\hat{r}(t) = (t, \bar{r}(t))$. This creates a curve in $\mathbb{R} \times \mathbb{R}^3$ (Fig. 5.2). The velocity vector then has the form $\hat{r}'(t) = (1, \bar{r}'(t))$. This again lies in $\mathbb{R} \times \mathbb{R}^3$, and so every tangent space can also be identified with $\mathbb{R} \times \mathbb{R}^3$. We introduce the metric

$$g((s_1, v_1), (s_2, v_2)) = s_1 s_2 - v_1 v_2 .$$

Then the vector $\hat{r}'(t)$ has the magnitude $\sqrt{1 - (\bar{r}'(t))^2}$, which is almost one for small $\bar{r}'(t)$ (speed of light c $= 1$). At time t, the set $\{(t, v) : v \in \mathbb{R}^3\}$ plays the role of the three-dimensional visual space. The vector $\hat{r}'(t)$ is almost orthogonal to this hyperspace.

This is the Newtonian viewpoint on the motion of a particle (or observer) in a somewhat unusual formulation. The relativistic point of view differs from this for small speeds only very slightly. The role of the basic set $\mathbb{R} \times \mathbb{R}^3$ is now played by a spacetime M, i.e., a four-dimensional oriented and time-oriented Lorentzian manifold.

Definition 5.1 An **observer** is a continuously differentiable map γ from an interval to M (i.e., a curve), where tangent vector $\gamma'(t) \in M_{\gamma(t)}$ is future-pointing and is such that $g(\gamma'(t), \gamma'(t)) = 1$.   ◆

© The Author(s), under exclusive license to Springer Nature Switzerland AG 2023
R. Oloff, *The Geometry of Spacetime*, Graduate Texts in Physics,
https://doi.org/10.1007/978-3-031-16139-1_5

Fig. 5.1 Curve in $\mathbb{R}^3$

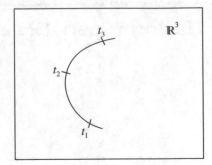

Fig. 5.2 Curve in $\mathbb{R} \times \mathbb{R}^3$

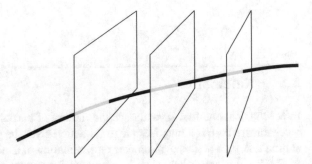

Many considerations only relate to an instant. Thus, we also consider the motion of an observer at an instant.

Definition 5.2 An **instantaneous observer** at a point $P \in M$ is a future pointing timelike tangent vector $x \in M_P$ such that $g(x, x) = 1$.   ♦

The same definitions are also given for a particle. However, a particle is also subject to laws of motion, where its mass and electric charge play a role.

Definition 5.3 A **particle** at a point $P \in M$ is a future-pointing timelike tangent vector $z \in M_P$, such that $g(z, z) = 1$, additionally equipped with a nonnegative number m **(rest mass)** and a real number e **(charge)**.   ♦

In Newtonian physis, an instantaneous observer moving at a speed x perceives a particle moving at speed z, to be moving at a relative speed v, defined via $x + v = z$. This viewpoint is modified in the theory of relativity. For the instantaneous observer x, the instantaneous action takes place in the Euclidean space $x^\perp$ with the metric $-g$.

Definition 5.4 An instantaneous observer x measures a particle z to have the speed to be the spacelike tangent vector $v \in x^\perp$, uniquely determined by $z = \lambda(x + v)$ with $\lambda > 0$. Then v is called the **relative speed of z with respect to x**.   ♦

Fig. 5.3 The relative speed v
of z for the observer x

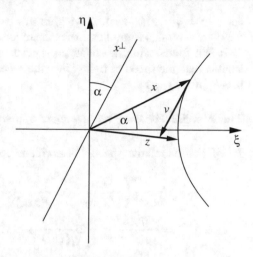

If we replace the four-dimensional tangent space by a two-dimensional Lorentzian
space $\mathbb{R}^2$, with the metric

$$g\big((\xi_1, \eta_1), (\xi_2, \eta_2)\big) = \xi_1 \xi_2 - \eta_1 \eta_2 \,,$$

then Fig. 5.3 gives a geometric illustration. The tips of all future-pointing timelike
vectors form the right branch of the hyperbola $\xi^2 - \eta^2 = 1$. The subspace $x^\perp$ belongs
to the observer x. In order to construct the velocity vector v that x perceives the
particle z has, the straight line $x^\perp$ has to be shifted parallel to itself so that it runs
through the tip of x. This shifted line is tangential to the branch of the hyperbola.
The velocity vector v then starts at the tip of x and ends at the intersection of the
shifted line with the straight line generated by z. In the three-dimensional case, the
branch of the hyperbola is replaced by one sheet of a double-sheeted hyperboloid,
and the shifted version of $x^\perp$ is tangential to the hyperboloid sheet at the tip of x. The
velocity vector v again starts at the tip of x and ends at the point of intersection of the
straight line generated by z and the shifted plane. Unfortunately we can't visualise
the four-dimensional case.

In applications, the following relative speed formula is more manageable than the
wording given in Definition 5.4. It also clarifies the existence and uniqueness, which
was not immedtialy apparent earlier.

Theorem 5.1 *For an observer x, a particle z has the relative speed*

$$v = \frac{z}{g(x, z)} - x \,.$$

Proof From $z = \lambda(x + v)$, it follows that

$$g(z, x) = \lambda(g(x, x) + g(v, x)) = \lambda$$

and so $z = g(x, z)(x + v)$. Solving for v, we obtain the above formula. Conversely, the v given by this formula satisfies the requirements placed in Definition 5.4.

It is a fundamental postulate in special relativity that all occuring speeds are smaller than the speed of light. Since the speed of light is normalised to 1, we must have $-g(v, v) < 1$.

Theorem 5.2 *For every relative speed v, there holds $0 \geq g(v, v) > -1$.*

Proof For the relative speed v of z with respect to x, we have

$$g(v, v) = g\left(\frac{z}{g(x, z)} - x, \frac{z}{g(x, z)} - x\right)$$
$$= \frac{g(z, z)}{(g(x, z))^2} - 2\frac{g(x, z)}{g(x, z)} + g(x, x) = \frac{1}{(g(x, z))^2} - 1 > -1.$$

The last inequality follows from Theorem 4.6. The inequality $0 \geq g(v, v)$ is a consequence of Theorem 4.5.

Theorem 5.3 *An observer x perceives the relative speed v as a particle*

$$z = \frac{x + v}{\sqrt{1 + g(v, v)}} .$$

Proof According to Definition 5.4, the sum $x + v$ should be normalised to unit length, and this corresponds exactly to the given formula.

The focus of the special theory of relativity is the conversion of physical quantities between different 'reference systems'. A reference system here means an observer x with an orthonormal basis for the Euclidean space $x^\perp$. A different observer cannot use this basis because he is observing a different Euclidean space. How the bases can be modified is given in the following theorem and sketched in Fig. 5.4 when $\beta = 2/3$.

Fig. 5.4 Observers x and x', normalised spacelike vectors $e \in x^\perp$ and $e' \in (x')^\perp$

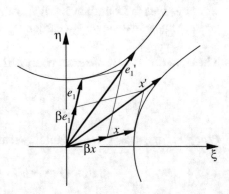

Theorem 5.4 *An observer x perceives another observer x' with the relative speed βe_1. Here $e_1 \in x^{\perp}$, $g(e_1, e_1) = -1$ and $0 < \beta < 1$. Then for two vectors e_2 and e_3 such that x, e_1, e_2, e_3 form a Lorentzian basis, we have that*

$$x' = \frac{x + \beta e_1}{\sqrt{1 - \beta^2}}, \qquad e_1' = \frac{\beta x + e_1}{\sqrt{1 - \beta^2}}, \qquad e_2' = e_2, \qquad e_3' = e_3$$

also forms a Lorentzian basis.

Proof The formula for x' follows from Theorem 5.3. Also, it can be easily seen that $e_1', e_2', e_3' \in (x')^{\perp}$ and that they are orthogonal to each other.

The Lorentzian basis x, e_1, e_2, e_3 in the tangent space M_P induces a coordinate system $\xi^0, \xi^1, \xi^2, \xi^3$. For transition to the other Lorentzian system, the new coordinates $\eta^0, \eta^1, \eta^2, \eta^3$ are characterised by

$$\eta^0 \frac{x + \beta e_1}{\sqrt{1 - \beta^2}} + \eta^1 \frac{\beta x + e_1}{\sqrt{1 - \beta^2}} + \eta^2 e_2 + \eta^3 e_3 = \xi^0 x + \xi^1 e_1 + \xi^2 e_2 + \xi^3 e_3.$$

Equating coefficients, we obtain

$$\frac{1}{\sqrt{1 - \beta^2}} \eta^0 + \frac{\beta}{\sqrt{1 - \beta^2}} \eta^1 = \xi^0$$

and

$$\frac{\beta}{\sqrt{1 - \beta^2}} \eta^0 + \frac{1}{\sqrt{1 - \beta^2}} \eta^1 = \xi^1.$$

These give

$$\eta^0 = \frac{\xi^0 - \beta \xi^1}{\sqrt{1 - \beta^2}}$$

and

$$\eta^1 = \frac{\xi^1 - \beta \xi^0}{\sqrt{1 - \beta^2}}.$$

Together with $\eta^2 = \xi^2$ and $\eta^3 = \xi^3$, this is the Lorentz transformation (see the Introduction), since β is the e_1-component of the relative speed of x' with respect to x.

For $f \in \mathcal{F}(P)$, an observer x at P perceives the number xf as the partial derivative of f with respect to time. For the observer x' as seen by x having relative speed βe_1, Theorem 5.4 gives

$$x'f = \frac{1}{\sqrt{1 - \beta^2}} xf + \frac{\beta}{\sqrt{1 - \beta^2}} e_1 f.$$

The prefactor $1/\sqrt{1 - \beta^2}$, as with the Lorentz transformation, can be interpreted as time dilation (see also Sect. 10.1).

Theorem 5.5 *An observer x sees a particle with the relative speed having compo-nents v^1, v^2, v^3 with respect to the orthonormal basis e_1, e_2, e_3 for $x^\perp$ (orthonor-mal with respect to $-g$) and another observer x' with the relative speed βe_1. Then x' sees the particle with the relative speed, with respect to the basis $(\beta x + e_1)/\sqrt{1 - \beta^2}, e_2, e_3$ in $(x')^\perp$, as having the components*

$$(v')^1 = \frac{v^1 - \beta}{1 - \beta v^1}, \qquad (v')^2 = \frac{v^2\sqrt{1 - \beta^2}}{1 - \beta v^1}, \qquad (v')^3 = \frac{v^3\sqrt{1 - \beta^2}}{1 - \beta v^1}.$$

Proof From Theorem 5.3, x perceives the other observer

$$x' = \frac{x + \beta e_1}{\sqrt{1 - \beta^2}}$$

and the particle

$$z = \frac{x + v^i e_i}{\sqrt{1 - (v^1)^2 - (v^2)^2 - (v^3)^2}}.$$

By Theorem 5.1, the relative speed of z with respect to x' is

$$\frac{z}{g(x', z)} - x' = \frac{x + v^i e_i}{g\left(\dfrac{x + \beta e_1}{\sqrt{1 - \beta^2}}, x + v^i e_i\right)} - \frac{x + \beta e_1}{\sqrt{1 - \beta^2}} = \frac{\sqrt{1 - \beta^2}}{1 - \beta v^1}(x + v^i e_i) - \frac{x + \beta e_1}{\sqrt{1 - \beta^2}}.$$

The coefficients $(v')^2$ and $(v')^3$ can now be read off and what remains is

$$\frac{\sqrt{1 - \beta^2}}{1 - \beta v^1}(x + v^1 e_1) - \frac{x + \beta e_1}{\sqrt{1 - \beta^2}} = \frac{v^1 - \beta}{1 - \beta v^1} \cdot \frac{\beta x + e_1}{\sqrt{1 - \beta^2}}.$$

Thus

$$(1 - \beta^2)(x + v^1 e_1) - (1 - \beta v^1)(x + \beta e_1) = (v^1 - \beta)(\beta x + e_1).$$

5.2   Dynamics

In Newtonian physics, the momentum of a particle is the product of *mass times its speed*. This point of view is now modified as follows: *The momentum of a particle with mass m and velocity v is the linear form correspond to the vector mv*, i.e., the linear form $g(mv, \cdot) = mg(v, \cdot)$. In the relativistic point of view, this linear form is created on the Euclidean space by restricting a linear form given on the four-dimensional tangential space.

Definition 5.5 The **energy-momentum form** p of a particle $z \in M_P$ with the rest mass m is the linear form on M_P given by

$$p = mg(z, \cdot).$$

◆

Definition 5.6 The **momentum** measured by the observer x for the particle z is the restriction of $-p$ on $x^{\perp}$, where p is the energy-momentum form. For the observer x, the particle z has the **relative mass**

$$E = p(x) = mg(z, x) = mg\left(\frac{x + v}{\sqrt{1 + g(v, v)}}, x\right) = \frac{m}{\sqrt{1 + g(v, v)}},$$

where v is the relative speed of z with respect to x.

◆

The notation E indicates the interpretation of the relative mass as energy. In fact, in Newtonian mechanics, the kinetic energy approximates the difference between the relative mass and the rest mass at low speeds. This is based on the Taylor expansion of the function $f(w) = 1/\sqrt{1 - w^2}$ up to the quadratic term, and since $f'(0) = 0$ and $f''(0) = 1$, we obtain

$$\frac{1}{\sqrt{1 - w^2}} \approx 1 + \frac{1}{2}w^2,$$

and for the relative mass, we thus have

$$E = \frac{m}{\sqrt{1 - \left(\sqrt{-g(v, v)}\right)^2}} \approx m + \frac{m}{2}\left(\sqrt{-g(v, v)}\right)^2.$$

With respect to an orthonormal basis, vector components correspond to the components of the corresponding linear form. There is a very simple relationship between the components of the relative speed and the momentum.

Theorem 5.6 *Let e_1, e_2, e_3 be an orthonormal basis for $x^{\perp}$ where x is an observer. A particle with rest mass m and relative velocity $v = v^i e_i$ with respect to x has a momentum with the components*

$$-p_i = \frac{mv^i}{\sqrt{1 + g(v, v)}}.$$

Proof The particle is

$$z = \frac{x + v^i e_i}{\sqrt{1 + g(v, v)}},$$

and by inserting the basis vectors e_i into the energy-momentum form p, we obtain

$$p_i = p(e_i) = mg\left(\frac{x + v^k e_k}{\sqrt{1 + g(v, v)}}, e_i\right) = -\frac{mv^i}{\sqrt{1 + g(v, v)}}.$$

Thus the relative mass has the same meaning in the relationship between the relative speed and the measured momentum as the mass in Newtonian mechanics.

In Newtonian physics, the concept of energy is based on the calculation of the work integral, defined as the line integral of the force acting on the particle in question. According to Newton, the force is equal to the time rate of change of momentum it causes. When moving from point P_1 to point P_2 along a path $r(t)$, the work is given by

$$\int_{P_1}^{P_2} (mr'(t))' dr = m\int_{t_1}^{t_2} r''(t) r'(t)\, dt = \frac{m}{2}\int_{t_1}^{t_2} ((r'(t))^2)'\, dt = \frac{m}{2}(r'(t_2))^2 - \frac{m}{2}(r'(t_1))^2.$$

In the framework of the theory of relativity, the mass m is to be replaced by the relative mass $m/\sqrt{1 - (r'(t))^2}$. That the relative mass is also called energy in the theory of relativity can be motivated by the following consideration: In the work integral, the relative mass takes place of the mass. As

$$\left(\frac{r'(t)}{\sqrt{1 - (r'(t))^2}}\right)' = \frac{r''(t)}{\sqrt{1 - (r'(t))^2}^{\,3}}$$

and

$$\left(\frac{1}{\sqrt{1 - (r'(t))^2}}\right)' = \frac{r'(t) r''(t)}{\sqrt{1 - (r'(t))^2}^{\,3}}$$

it follows that

$$\int_{P_1}^{P_2} \left(\frac{mr'(t)}{\sqrt{1 - (r'(t))^2}}\right)' dr = m\int_{t_1}^{t_2} \left(\frac{r'(t)}{\sqrt{1 - (r'(t))^2}}\right)' r'(t)\, dt$$

$$= m\int_{t_1}^{t_2} \left(\frac{1}{\sqrt{1 - (r'(t))^2}}\right)' dt = \frac{m}{\sqrt{1 - (r'(t_2))^2}} - \frac{m}{\sqrt{1 - (r'(t_1))^2}}.$$

Thus in the theory of relativity, the work done is the difference of the relative masses after and before motion.

Theorem 5.7 *Let e_1, e_2, e_3 be an orthonormal basis for $x^\perp$, where x is an observer. The observer x sees another observer x' with a relative velocity βe_1, and a particle with a relative mass E and momentum components $-p_1$, $-p_2$, $-p_3$. Then the other observer measures the relative mass*

$$E' = \frac{E - \beta(-p_1)}{\sqrt{1 - \beta^2}}$$

and the momentum components

$$-p_1' = \frac{(-p_1) - \beta E}{\sqrt{1 - \beta^2}}, \quad -p_2' = -p_2, \quad -p_3' = -p_3$$

with respect to the basis $(\beta x + e_1)/\sqrt{1 - \beta^2}$, e_2, e_3 in $(x')^\perp$.

Proof By inserting the vectors

$$x' = \frac{x + \beta e_1}{\sqrt{1 - \beta^2}} \quad \text{and} \quad e_1' = \frac{\beta x + e_1}{\sqrt{1 - \beta^2}}$$

in the energy-momentum form gives

$$E' = p(x') = p\left(\frac{x + \beta e_1}{\sqrt{1 - \beta^2}}\right) = \frac{p(x) + \beta p(e_1)}{\sqrt{1 - \beta^2}} = \frac{E - \beta(-p_1)}{\sqrt{1 - \beta^2}}$$

and

$$-p_1' = -p\left(\frac{\beta x + e_1}{\sqrt{1 - \beta^2}}\right) = \frac{-\beta p(x) - p(e_1)}{\sqrt{1 - \beta^2}} = \frac{(-p_1) - \beta E}{\sqrt{1 - \beta^2}}.$$

Moreover, $p_2' = p(e_2) = p_2$ and $p_3' = p(e_3) = p_3$.

5.3 Electrodynamics

The orientation of a spacetime also induces an orientation on the space $x^\perp$ for each observer x.

Definition 5.7 Let x be an observer at a point P of a spacetime M. A basis e_1, e_2, e_3 for $x^\perp$ is said to be **positively oriented** if the basis x, e_1, e_2, e_3 lies in the orientation of the tangent space M_P. ♦

An electromagnetic field is described by a skew-symmetric $(0, 2)$-tensor field F on spacetime, called the **field strength tensor** or **Faraday tensor**. Other properties of F, which also describe the relationship with the charge distribution and correspond to the classical Maxwell equations, will only be formulated later. Right now we just clarify how the field strengths can be obtained from the field strength tensor.

Definition 5.8 An observer x at a point P measures from the field strength tensor F the **electric field strength** $E \in x^{\perp}$ defined by

$$F(y, x) = g(y, E) \quad \text{for} \quad y \in x^{\perp},$$

and the **magnetic field strength** $B \in x^{\perp}$ defined by

$$F(y, z) = g(y \times z, B) \quad \text{for} \quad y, z \in x^{\perp}.$$

$\blacklozenge$

Thus the electric field strength is the vector $E \in M_P$ which represents the restriction of the linear form $F(., x)$ to the subspace $x^{\perp}$. To understand the definition of the magnetic field strength B, it should be pointed out that any skew-symmetric bilinear form f on the oriented three-dimensional Euclidean space $x^{\perp}$ can be generated by a uniquely determined vector w by

$$f(y, z) = g(y \times z, w).$$

It is known from classical electrodynamics that the field strengths exert the force $e(E + v \times B)$ on a particle moving with velocity v and having the charge e. It creates a change in its momentum. With this momentum change in mind, it is therefore better to understand the force as a linear form. In this terminology this means that the linear form $e(-g)(E + v \times B, .)$ describes the change in the momentum. What a temporal change means for a linear form is only clarified later on in Chap. 11. As preparation and for the interpretation of F, E and B, the following result is already of interest.

Theorem 5.8 *An observer x observes a particle z with relative velocity v in the field F. Then the linear forms $F(z, .)$ and $-g(., E + v \times B)/\sqrt{1 + g(v, v)}$ agree on $x^{\perp}$.*

Proof For $y \in x^{\perp}$, we have

$$F(z, y) = F\left(\frac{x + v}{\sqrt{1 + g(v, v)}}, y\right) = \frac{F(x, y) + F(v, y)}{\sqrt{1 + g(v, v)}} = \frac{-g(y, E) + g(v \times y, B)}{\sqrt{1 + g(v, v)}}.$$

The triple product $(y, v, B) \longmapsto g(y \times v, B)$ is invariant under a cyclic permutation of the variables, and so we obtain

$$F(z, y) = \frac{-g(y, E + v \times B)}{\sqrt{1 + g(v, v)}}.$$

Theorem 5.9 *The field strength tensor F has, with respect to a positively oriented basis $e_0 = x$, e_1, e_2, e_3, the components*

$$(F_{ik}) = \begin{pmatrix} 0 & E^1 & E^2 & E^3 \\ -E^1 & 0 & -B^3 & B^2 \\ -E^2 & B^3 & 0 & -B^1 \\ -E^3 & -B^2 & B^1 & 0 \end{pmatrix},$$

where the triples (E^1, E^2, E^3) and (B^1, B^2, B^3) are the components of E and B, respectively, with respect to the orthonormal (in the sense of $-g$) basis e_1, e_2, e_3.

Proof For $k = 1, 2, 3$, we have

$$F_{0k} = F(e_0, e_k) = -F(e_k, x) = -g(e_k, E) = E^k.$$

Furthermore,

$$F_{12} = F(e_1, e_2) = g(e_1 \times e_2, B) = g(e_3, B) = -B^3$$

and analogously $F_{13} = B^2$ and $F_{23} = -B^1$. The remaining matrix elements can be obtained using the skew-symmetry of F.

Theorem 5.10 *An observer measures the electric and magnetic field strength with the components E^1, E^2, E^3 and B^1, B^2, B^3 with respect to a positively oriented orthonormal basis e_1, e_2, e_3, and sees another observer $\hat{x}$ with relative velocity βe_1. Then the other observer, with respect to the basis $(\beta x + e_1)/\sqrt{1 - \beta^2}$, e_2, e_3, measures the field strength component as*

$$\hat{E}^1 = E^1 \qquad\qquad \hat{B}^1 = B^1$$

$$\hat{E}^2 = \frac{E^2 - \beta B^3}{\sqrt{1 - \beta^2}} \qquad\qquad \hat{B}^2 = \frac{B^2 + \beta E^3}{\sqrt{1 - \beta^2}}$$

$$\hat{E}^3 = \frac{E^3 + \beta B^2}{\sqrt{1 - \beta^2}} \qquad\qquad \hat{B}^3 = \frac{B^3 - \beta E^2}{\sqrt{1 - \beta^2}}.$$

Proof We must insert the vectors

$$\hat{e}_0 = \hat{x} = \frac{x + \beta e_1}{\sqrt{1 - \beta^2}}, \quad \hat{e}_1 = \frac{\beta x + e_1}{\sqrt{1 - \beta^2}}, \quad \hat{e}_2 = e_2 \quad \text{and} \quad \hat{e}_3 = e_3$$

in the field strength tensor. Then we obtain

$$\hat{E}^1 = F(\hat{x}, \hat{e}_1) = \frac{F(x + \beta e_1, \beta x + e_1)}{1 - \beta^2} = \frac{F(x, e_1) + \beta^2 F(e_1, x)}{1 - \beta^2} = F(x, e_1) = E^1$$

$$\hat{E}^2 = F(\hat{x}, e_2) = \frac{F(x + \beta e_1, e_2)}{\sqrt{1 - \beta^2}} = \frac{E^2 - \beta B^3}{\sqrt{1 - \beta^2}}$$

$$\hat{E}^3 = F(\hat{x}, e_3) = \frac{F(x + \beta e_1, e_3)}{\sqrt{1 - \beta^2}} = \frac{E^3 + \beta B^2}{\sqrt{1 - \beta^2}}$$

$$\hat{B}^1 = F(e_3, e_2) = B^1$$

$$\hat{B}^2 = F(\hat{e}_1, e_3) = \frac{F(\beta x + e_1, e_3)}{\sqrt{1 - \beta^2}} = \frac{B^2 + \beta E^3}{\sqrt{1 - \beta^2}}$$

$$\hat{B}^3 = F(e_2, \hat{e}_1) = \frac{F(e_2, \beta x + e_1)}{\sqrt{1 - \beta^2}} = \frac{B^3 - \beta E^2}{\sqrt{1 - \beta^2}} \; .$$

Differential Forms

6

6.1 p-forms

The discussion in Sects. 6.1 to 6.3 of this chapter refer to a finite-dimensional real vector space E, which will later on be replaced by the tangent spaces of a manifold.

Definition 6.1 A p-**form** on E is a skew-symmetric $(0, p)$-tensor on E.   ◆

The skew-symmetry for a $(0, p)$-tensor f naturally means

$$f(..., x_{i-1}, x_i, x_{i+1}, ..., x_{k-1}, x_k, x_{k+1}, ...) = -f(..., x_{i-1}, x_k, x_{i+1}, ..., x_{k-1}, x_i, x_{k+1}, ...).$$

This property implies that

$$f(x_{\mathcal{P}(1)}, \ldots, x_{\mathcal{P}(p)}) = \chi(\mathcal{P}) f(x_1, \ldots, x_p)$$

for every permutation $\mathcal{P}$. Instead of "skew-symmetry" one also uses the term "**alternating**". Every linear form is a 1-form, and the 0-forms are the numbers.

Example. For linear forms $a^1, \ldots, a^p \in E^*$, the tensor defined by

$$a^1 \wedge \cdots \wedge a^p = \sum_{\mathcal{P}} \chi(\mathcal{P}) a^{\mathcal{P}(1)} \otimes \cdots \otimes a^{\mathcal{P}(p)}$$

(summed over all permutations $\mathcal{P}$) defines a p-form. The proof of the skew-symmetry mostly follows along the same lines of the proof that the determinant defined by Leibniz's fomula changes sings when two rows are swapped. With

© The Author(s), under exclusive license to Springer Nature Switzerland AG 2023
R. Oloff, *The Geometry of Spacetime*, Graduate Texts in Physics,
https://doi.org/10.1007/978-3-031-16139-1_6

$$\mathcal{P}_{i,k} = \begin{pmatrix} 1 \cdots i \cdots k \cdots n \\ 1 \cdots k \cdots i \cdots n \end{pmatrix}$$

and $\mathcal{Q} = \mathcal{P} \circ \mathcal{P}_{i,k}$, we have

$$
\begin{aligned}
a^1 \wedge \cdots &\wedge a^p(x_1, \ldots, x_k, \ldots, x_i, \ldots, x_p) \\
&= \sum_{\mathcal{P}} \chi(\mathcal{P}) \langle x_1, a^{\mathcal{P}(1)} \rangle \cdots \langle x_k, a^{\mathcal{P}(i)} \rangle \cdots \langle x_i, a^{\mathcal{P}(k)} \rangle \cdots \langle x_p, a^{\mathcal{P}(p)} \rangle \\
&= \sum_{\mathcal{P}} \chi(\mathcal{P}) \left(\cdots \langle x_k, a^{\mathcal{P} \circ \mathcal{P}_{i,k}(k)} \rangle \cdots \langle x_i, a^{\mathcal{P} \circ \mathcal{P}_{i,k}(i)} \rangle \cdots \right) \\
&= \sum_{\mathcal{Q}} (-\chi(\mathcal{Q})) \langle x_1, a^{\mathcal{Q}(1)} \rangle \cdots \langle x_p, a^{\mathcal{Q}(p)} \rangle \\
&= -a^1 \wedge \cdots \wedge a^p(x_1, \ldots, x_p) .
\end{aligned}
$$

The set of all p-forms on E is obviously a subspace of $E_p^0 = E^* \otimes \cdots \otimes E^*$, and is denoted by $E^* \wedge \cdots \wedge E^*$ or by $\wedge^p E^*$. As is well-known, the map that assigns to n elements of $\mathbb{R}^n$ the determinant formed by them, is an n-form, and every other n-form differs from it by only a factor. Thus the space $\wedge^n(\mathbb{R}^n)$ is one-dimensional. The following result generalises this.

Theorem 6.1 *Let $x_1, \ldots, x_n$ be a basis of E and $a^1, \ldots, a^n$ be its dual basis for E^*. The $\binom{n}{p}$ p-forms $a^{i_1} \wedge \cdots \wedge a^{i_p}$ with $1 \le i_1 < \ldots < i_p \le n$ form a basis for $\wedge^p E^*$. The coefficient of the p-form $a^{i_1} \wedge \cdots \wedge a^{i_p}$ in the representation of $f \in \wedge^p E^*$ is*

$$f_{i_1 \ldots i_p} = f(x_{i_1}, \ldots, x_{i_p}) .$$

Each $f \in \wedge^p E^*$ can also be represented by

$$f = \frac{1}{p!} f_{i_1 \ldots i_p} a^{i_1} \wedge \ldots \wedge a^{i_p}$$

where all the indices run from 1 to n independently of each other, in the sense of the summation convention.

Proof The linear independence of these $\binom{n}{p}$ p-forms follows from the observation that application on basis vectors $x_{k_1}, \ldots, x_{k_p}$ with $1 \le k_1 < \cdots < k_p \le n$ is non-zero only for the p-form $a^{k_1} \wedge \cdots \wedge a^{k_p}$ among these p-forms. We now express any p-form f as a linear combination of the $\binom{n}{p}$ p-forms $a^{i_1} \wedge \cdots \wedge a^{i_p}$ with $1 \le i_1 < \cdots < i_p \le n$. We have

$$f = f_{j_1 \ldots j_p} a^{j_1} \otimes \cdots \otimes a^{j_p}$$

$$= \sum_{1 \leq i_1 < \cdots < i_p \leq n} \sum_{\mathcal{P}} f_{i_{\mathcal{P}(1)} \ldots i_{\mathcal{P}(p)}} a^{i_{\mathcal{P}(1)}} \otimes \cdots \otimes a^{i_{\mathcal{P}(p)}}$$

$$= \sum_{1 \leq i_1 < \cdots < i_p \leq n} \sum_{\mathcal{P}} \chi(\mathcal{P}) f_{i_1 \ldots i_p} a^{i_{\mathcal{P}(1)}} \otimes \cdots \otimes a^{i_{\mathcal{P}(p)}}$$

$$= \sum_{1 \leq i_1 < \cdots < i_p \leq n} f_{i_1 \ldots i_p} a^{i_1} \wedge \cdots \wedge a^{i_p} .$$

Since for every permutation $\mathcal{P}$

$$f_{i_{\mathcal{P}(1)} \ldots i_{\mathcal{P}(p)}} a^{\mathcal{P}(1)} \wedge \cdots \wedge a^{\mathcal{P}(p)} = f_{i_1 \ldots i_p} a^{i_1} \wedge \cdots \wedge a^{i_p}$$

(no sum meant here), and because this applies to each permutation, we have

$$f_{i_1 \ldots i_p} a^{i_1} \wedge \cdots \wedge a^{i_p} = \frac{1}{p!} \sum_{\mathcal{P}} f_{i_{\mathcal{P}(1)} \ldots i_{\mathcal{P}(p)}} a^{i_{\mathcal{P}(1)}} \wedge \cdots \wedge a^{i_{\mathcal{P}(p)}} .$$

Setting $j_k = i_{\mathcal{P}(k)}$, we get the representation

$$f = \frac{1}{p!} \sum_{\substack{j_1, \ldots, j_p \\ j_k \neq j_l}} f_{j_1 \ldots j_p} a^{j_1} \wedge \cdots \wedge a^{j_p} ,$$

but the requirement that the indices $j_1, \ldots, j_p$ are pairwise distinct can also be dropped, because the coefficients of the summands with nondistinct j_k will be 0. This also proves the second representation in the theorem.

6.2   The Wedge Product

The wedge product (also called the **outer product**) gives the theoretical background for the expression "$\wedge$" already used by us in the previous section. The name comes from the shape of this symbol. We will define the wedge product in two steps.

Definition 6.2 The **alternatisation** A assigns to each $(0, p)$-tensor f the p-form Af given by

$$Af(x_1, \ldots, x_p) = \frac{1}{p!} \sum_{\mathcal{P}} \chi(\mathcal{P}) f(x_{\mathcal{P}(1)}, \ldots, x_{\mathcal{P}(p)}) .$$

Of course, Af is multilinear and skew-symmetric (or alternating), which can be shown as in the example considered in the previous section. Also, the map A is clearly linear on E_p^0. As A leaves the p-forms invariant, we have $A \circ A = A$.

Example. For linear forms $a^1, \ldots, a^p$, we have

$$A(a^1 \otimes \cdots \otimes a^p) = \frac{1}{p!} a^1 \wedge \cdots \wedge a^p .$$

Definition 6.3 The **wedge product** $f \wedge g$ of a p-form f and a q-form g is the $(p+q)$-form

$$f \wedge g = \frac{(p+q)!}{p!q!} A(f \otimes g) .$$

$\blacklozenge$

Example. The p-forms $a^1 \wedge \cdots \wedge a^p$ defined in the previous section can be regarded as iterated wedge products. We will show

$$(a^1 \wedge \cdots \wedge a^p) \wedge (a^{p+1} \wedge \cdots \wedge a^{p+q}) = a^1 \wedge \cdots \wedge a^p \wedge a^{p+1} \wedge \cdots \wedge a^{p+q} ,$$

where the wedge between the two brackets on the left-hand side is meant in the sense of Definition 6.3, while the other wedges are meant in the sense of the Example after Definition 6.2. We have

$$(a^1 \wedge \cdots \wedge a^p) \wedge (a^{p+1} \wedge \cdots \wedge a^{p+q})$$

$$= \left(\sum_{\mathcal{P}} \chi(\mathcal{P}) a^{\mathcal{P}(1)} \otimes \cdots \otimes a^{\mathcal{P}(p)} \right) \wedge \left(\sum_{\mathcal{Q}} \chi(\mathcal{Q}) a^{\mathcal{Q}(p+1)} \otimes \cdots \otimes a^{\mathcal{Q}(p+q)} \right)$$

$$= \frac{(p+q)!}{p!q!} \sum_{\mathcal{P},\mathcal{Q}} \chi(\mathcal{P})\chi(\mathcal{Q}) A \left(a^{\mathcal{P}(1)} \otimes \cdots \otimes a^{\mathcal{P}(p)} \otimes a^{\mathcal{Q}(p+1)} \otimes \cdots \otimes a^{\mathcal{Q}(p+q)} \right) ,$$

where the summation is over all permutations $\mathcal{P}$ of $\{1, \ldots, p\}$ and $\mathcal{Q}$ of $\{p+1, \ldots, p+q\}$. The definition of A now gives

$$(a^1 \wedge \cdots \wedge a^p) \wedge (a^{p+1} \wedge \cdots \wedge a^{p+q})$$

$$= \frac{1}{p!q!} \sum_{\mathcal{P},\mathcal{Q},\mathcal{R}} \chi(\mathcal{P})\chi(\mathcal{Q})\chi(\mathcal{R}) a^{\mathcal{R}(\mathcal{P}(1))} \otimes \cdots \otimes a^{\mathcal{R}(\mathcal{Q}(p+q))} .$$

Here $\mathcal{R}$ runs over all permutations of $\{1, \ldots, p+q\}$. For the modified permutations

$$\tilde{\mathcal{P}} = \begin{pmatrix} 1 & \cdots & p & p+1 & \cdots & p+q \\ \mathcal{P}(1) & \cdots & \mathcal{P}(p) & p+1 & \cdots & p+q \end{pmatrix}$$

and

$$\tilde{\mathcal{Q}} = \begin{pmatrix} 1 \cdots p & p+1 & \cdots & p+q \\ 1 \cdots p & \mathcal{Q}(p+1) & \cdots & \mathcal{Q}(p+q) \end{pmatrix}$$

we obviously have $\chi(\tilde{\mathcal{P}}) = \chi(\mathcal{P})$ and $\chi(\tilde{\mathcal{Q}}) = \chi(\mathcal{Q})$. So with $\mathcal{S} = \mathcal{R} \circ \tilde{\mathcal{Q}} \circ \tilde{\mathcal{P}}$, we obtain

$$(a^1 \wedge \cdots \wedge a^p) \wedge (a^{p+1} \wedge \cdots \wedge a^{p+q}) = \frac{1}{p!q!} \sum_{\mathcal{P},\mathcal{Q},\mathcal{R}} \chi(\mathcal{S}) a^{\mathcal{S}(1)} \otimes \cdots \otimes a^{\mathcal{S}(p+q)} .$$

By running through all the $p!q!$ possibilities for $\mathcal{P}$ and $\mathcal{Q}$, $\mathcal{S}$ runs through all the permutations of $\{1, \ldots, p+q\}$ when $\mathcal{R}$ runs through all the permutations, finally proving the desired equality.

Theorem 6.2

(1) *The wedge product is a bilinear map from $(\wedge^p E^*) \times (\wedge^q E^*)$ to $\wedge^{p+q} E^*$.*
(2) *The wedge product is associative.*
(3) *For $f \in \wedge^p E^*$ and $g \in \wedge^q E^*$, $f \wedge g = (-1)^{pq} g \wedge f$.*

Proof The bilinearity of the wedge product results from the tensor product and the linearity of the alternatisation map. Thanks to the bilinearity, the associativity needs to be checked only for the basis elements. However, in these special cases, it is fulfilled, in light of the previous example. (3) only needs to be shown for the basis elements. Let f and g be wedge products of linear forms. Then the swapping of f and g can be done by swapping pq adjacent linear forms, with the sign changing each time, accounting for the overall factor of $(-1)^{pq}$.

6.3   The Hodge-Star Operator

In this section, E will be an oriented n-dimensional real vector space, which is equipped with an unspecified symmetric bilinear form g. By the Sylvester law of inertia (Theorem 4.2) there exists an orthogonal basis $x_1, \ldots, x_n$ with the property that $g(x_i, x_i) = \varepsilon_i = \pm 1$. We will call such a basis **semi-orthonormal**.

Theorem 6.3 *For two positive oriented semi-orthonormal bases $x_1, \ldots, x_n$ and $y_1, \ldots, y_n$ related by $y_i = \lambda_i^j x_j$, there holds $\det(\lambda_i^j) = 1$.*

Proof With $\varepsilon_i = g(x_i, x_i)$ and $\varepsilon_i' = g(y_i, y_i)$, we have

$$\varepsilon_i' = g(\lambda_i^j x_j, \lambda_i^k x_k) = \sum_{j=1}^n (\lambda_i^j)^2 \varepsilon_j$$

and for $i \neq k$

$$0 = g(y_i, y_k) = g(\lambda_i^j x_j, \lambda_k^l x_l) = \sum_{j=1}^{n} \lambda_i^j \lambda_k^j \varepsilon_j .$$

In matrix notation this means

$$\begin{pmatrix} \varepsilon_1' & 0 & \cdots & 0 \\ 0 & & & \vdots \\ \vdots & & & 0 \\ 0 & \cdots & 0 & \varepsilon_n' \end{pmatrix} = \begin{pmatrix} \lambda_1^1 & \cdots & \lambda_1^n \\ \vdots & \ddots & \vdots \\ \lambda_n^1 & \cdots & \lambda_n^n \end{pmatrix} \begin{pmatrix} \varepsilon_1 & 0 & \cdots & 0 \\ 0 & & & \vdots \\ \vdots & & & 0 \\ 0 & \cdots & 0 & \varepsilon_n \end{pmatrix} \begin{pmatrix} \lambda_1^1 & \cdots & \lambda_n^1 \\ \vdots & \ddots & \vdots \\ \lambda_1^n & \cdots & \lambda_n^n \end{pmatrix}$$

which implies $(\det (\lambda_i^j))^2 = 1$. Since the bases are co-oriented, it follows that $\det (\lambda_i^j) = 1$.

Theorem 6.4 *An n-form has the same value on all positively oriented semi-orthonormal bases.*

Proof Let $x_1, \ldots, x_n$ and $y_1, \ldots, y_n$ be such bases. Inserting $y_i = \lambda_i^j x_j$ into the n-form f, we obtain

$$f(y_1, \ldots, y_n) = f(\lambda_1^{j_1} x_{j_1}, \cdots, \lambda_n^{j_n} x_{j_n}) = \lambda_1^{j_1} \cdots \lambda_n^{j_n} f(x_{j_1}, \ldots, x_{j_n})$$

$$= \sum_{\mathcal{P}} \lambda_1^{\mathcal{P}(1)} \cdots \lambda_n^{\mathcal{P}(n)} f(x_{\mathcal{P}(1)}, \ldots, x_{\mathcal{P}(n)})$$

$$= \sum_{\mathcal{P}} \lambda_1^{\mathcal{P}(1)} \cdots \lambda_n^{\mathcal{P}(n)} \chi(\mathcal{P}) f(x_1, \ldots, x_n)$$

$$= f(x_1, \ldots, x_n) \det (\lambda_i^j) = f(x_1, \ldots, x_n) .$$

The vector space of all n-forms on the n-dimensional space E is one-dimensional according to Theorem 6.1. Thus every n-form on E is a multiple of the n-form which assumes the value 1 on every positive oriented semi-orthonormal basis.

Definition 6.4 The **volume form** V is the n-form which has the value 1 on every positive oriented semi-orthonormal basis.　　　　　　　　　　　　　　　◆

For $p = 0, 1, \ldots, n$, the spaces $\wedge^p E^*$ and $\wedge^{n-p} E^*$ have the equal dimensions $\binom{n}{p}$ (Theorem 6.1), and consequently, they are isomorphic to each other. The Hodge-star operator is an isomorphism that does not depend on the choice of bases in the two spaces, although a positive oriented semi-orthonormal basis $x_1, \ldots, x_n$ is used for the definition. We recall that $\varepsilon_1 J x_1, \ldots, \varepsilon_n J x_n$ is a dual basis, where $\varepsilon_i = g(x_i, x_i)$ and $J : E \longrightarrow E^*$ is the isomorphism characterised by $\langle y, Jx \rangle = g(x, y)$ (Theorems 4.3 and 4.4).

Definition 6.5 Let $x_1, \ldots, x_n$ be a positive oriented semi-orthonormal basis for E. The **Hodge-star operator** $*$ maps the p-form $Jx_{i_1} \wedge \cdots \wedge Jx_{i_p}$ ($i_1, \ldots, i_p$ pairwise distinct) to the $(n - p)$-form

$$*(Jx_{i_1} \wedge \cdots \wedge Jx_{i_p}) = \varepsilon_{i_{p+1}} \cdots \varepsilon_{i_n} Jx_{i_{p+1}} \wedge \cdots \wedge Jx_{i_n}$$

where

$$\begin{pmatrix} 1 & \cdots & p & p+1 & \cdots & n \\ i_1 & \cdots & i_p & i_{p+1} & \cdots & i_n \end{pmatrix}$$

is an even permutation.   ◆

The order of the numbers $i_{p+1}, \ldots, i_n$ is of course not uniquely determined by the choice of $i_1, \ldots, i_p$, but this ambiguity has no effect on the expression $Jx_{i_{p+1}} \wedge \cdots \wedge Jx_{i_n}$. For the extreme cases $p = 0$ and $p = n$, we have

$$*1 = \varepsilon_1 \cdots \varepsilon_n Jx_1 \wedge \cdots \wedge Jx_n$$

and

$$*(Jx_1 \wedge \cdots \wedge Jx_n) = 1 .$$

As $\varepsilon_1 Jx_1 \wedge \cdots \wedge \varepsilon_n Jx_n$ is the volume form V, we have $*1 = V$ and $*V = \varepsilon_1 \cdots \varepsilon_n$. In general, the image of $*$ can also be characterised by the property

$$*(Jx_{\mathcal{P}(1)} \wedge \cdots \wedge Jx_{\mathcal{P}(p)}) = \varepsilon_{\mathcal{P}(p+1)} \cdots \varepsilon_{\mathcal{P}(n)} \chi(\mathcal{P}) Jx_{\mathcal{P}(p+1)} \wedge \cdots \wedge Jx_{\mathcal{P}(n)}$$

for every permutation $\mathcal{P}$. From this it follows that, in particular,

$$*(Jx_{\mathcal{P}(1)} \wedge \cdots \wedge Jx_{\mathcal{P}(p)})(x_{\mathcal{P}(p+1)}, \ldots, x_{\mathcal{P}(n)}) = \chi(\mathcal{P}) .$$

We now show that the construction of the Hodge-star operator does not depend on the choice of the basis $x_1, \ldots, x_n$. Let this operator be defined on the basis elements $Jx_{i_1} \wedge \cdots \wedge Jx_{i_p}$ according to Definition 6.5 and extended linearly to the space $\wedge^p E^*$. We show that the equation

$$*(Jy_{\mathcal{P}(1)} \wedge \cdots \wedge Jy_{\mathcal{P}(p)}) = \varepsilon'_{\mathcal{P}(p+1)} \cdots \varepsilon'_{\mathcal{P}(n)} \chi(\mathcal{P}) Jy_{\mathcal{P}(p+1)} \wedge \cdots \wedge Jy_{\mathcal{P}(n)}$$

holds also for every other positive oriented semi-orthonormal basis $y_1, \ldots, y_n$ with $\varepsilon'_i = g(y_i, y_i)$ and for every permutation $\mathcal{P}$. We check that

$$*(Jy_{\mathcal{P}(1)} \wedge \cdots \wedge Jy_{\mathcal{P}(p)})(y_{\mathcal{P}(p+1)}, \ldots, y_{\mathcal{P}(n)}) = \chi(\mathcal{P})$$

and for indices $i_{p+1}, \ldots, i_n$ with $\{\mathcal{P}(1), \ldots, \mathcal{P}(p)\} \cap \{i_{p+1}, \ldots, i_n\} \neq \emptyset$

$$*(Jy_{\mathcal{P}(1)} \wedge \cdots \wedge Jy_{\mathcal{P}(p)})(y_{i_{p+1}}, \ldots, y_{i_n}) = 0 .$$

By setting $y_i = \lambda_i^j x_j$ we obtain

$$* (J y_{\mathcal{P}(1)} \wedge \cdots \wedge J y_{\mathcal{P}(p)})(y_{\mathcal{P}(p+1)}, \ldots, y_{\mathcal{P}(n)})$$

$$= \lambda_{\mathcal{P}(1)}^{j_1} \cdots \lambda_{\mathcal{P}(p)}^{j_p} \lambda_{\mathcal{P}(p+1)}^{j_{p+1}} \cdots \lambda_{\mathcal{P}(n)}^{j_n} * (J x_{j_1} \wedge \cdots \wedge J x_{j_p})(x_{j_{p+1}}, \ldots, x_{j_n})$$

$$= \sum_{\mathcal{Q}} \lambda_{\mathcal{P}(1)}^{\mathcal{Q}(1)} \cdots \lambda_{\mathcal{P}(n)}^{\mathcal{Q}(n)} \chi(\mathcal{Q}) = \chi(\mathcal{P}) \det(\lambda_i^j) = \chi(\mathcal{P})$$

by Theorem 6.3. For indices $i_{p+1}, \ldots, i_n$, one has a similar calculation

$$* (J y_{\mathcal{P}(1)} \wedge \cdots \wedge J y_{\mathcal{P}(p)})(y_{i_{p+1}}, \ldots, y_{i_n})$$

$$= \lambda_{\mathcal{P}(1)}^{j_1} \cdots \lambda_{\mathcal{P}(p)}^{j_p} \lambda_{i_{p+1}}^{j_{p+1}} \cdots \lambda_{i_n}^{j_n} * (J x_{j_1} \wedge \cdots \wedge J x_{j_p})(x_{j_{p+1}}, \ldots, x_{j_n})$$

$$= \sum_{\mathcal{Q}} \lambda_{\mathcal{P}(1)}^{\mathcal{Q}(1)} \cdots \lambda_{\mathcal{P}(p)}^{\mathcal{Q}(p)} \lambda_{i_{p+1}}^{\mathcal{Q}(p+1)} \cdots \lambda_{i_n}^{\mathcal{Q}(n)} \chi(\mathcal{Q}) .$$

This is the determinant of a matrix with at least two identical columns, and so this expression is zero.

Thus the Hodge-star operation is independent of the choice of basis. In Theorem 6.6, we will formulate a basis-independent characterisation. Linear forms on $\wedge^p E^*$ will be used for this. Such a linear form in particular also assigns a number to each p-form $J x_1 \wedge \cdots \wedge J x_p$ in a linear manner and can thus itself be thought of as a p-form.

Theorem 6.5 *The linear map* $K : (\wedge^p E^*)^* \longrightarrow \wedge^p E^*$, *which maps every linear form* c *on* $\wedge^p E^*$ *to the* p-*form*

$$K c(x_1, \ldots, x_p) = \langle J x_1 \wedge \cdots \wedge J x_p, c \rangle$$

is an isomorphism.

Proof Since the dimensions of the two spaces match, it is enough to show that K is injective. Let $K c = 0$. Then for all $x_1, \ldots, x_p \in E$, we have

$$0 = K c(x_1, \ldots, x_p) = \langle J x_1 \wedge \cdots \wedge J x_p, c \rangle .$$

Since a basis for $\wedge^p E^*$ can be formed by such p-forms, we conclude that $\langle f, c \rangle = 0$ for every p-form f, and so $c = 0$.

Example. Let $x_1, \ldots, x_n$ be an orthonormal basis in E with $\varepsilon_i = g(x_i, x_i)$. Let linear forms $c_{j_1 \cdots j_p}$ $(1 \le j_1 < \cdots < j_p \le n)$ in $\wedge^p E^*$, satisfy

$$\langle \varepsilon_{i_1} J x_{i_1} \wedge \cdots \wedge \varepsilon_{i_p} J x_{i_p}, c_{j_1 \cdots j_n} \rangle = \begin{cases} 1 & \text{for } i_k = j_k \\ 0 & \text{for } \{i_1, \ldots, i_p\} \ne \{j_1, \ldots, j_p\} \end{cases} .$$

Then we have

$$Kc_{i_1\cdots i_p} = Jx_{i_1} \wedge \cdots \wedge Jx_{i_p}$$

because

$$Kc_{i_1\cdots i_p}(x_{i_1}, \ldots, x_{i_p}) = \langle Jx_{i_1} \wedge \cdots \wedge Jx_{i_p}, c_{i_1\cdots i_p}\rangle$$
$$= \varepsilon_{i_1} \cdots \varepsilon_{i_p} = Jx_{i_1} \wedge \cdots \wedge Jx_{i_p}(x_{i_1}, \ldots, x_{i_p})$$

and

$$Kc_{i_1\cdots i_p}(x_{j_1}, \ldots, x_{j_p}) = \langle Jx_{j_1} \wedge \cdots \wedge Jx_{j_p}, c_{i_1\cdots i_p}\rangle$$
$$= 0 = Jx_{i_1} \wedge \cdots \wedge Jx_{i_p}(x_{j_1}, \ldots, x_{j_p})$$

for other index systems $j_1, \ldots, j_p$.

Theorem 6.6 *For $f \in \wedge^p E^*$ and $c \in (\wedge^p E^*)^*$, and with the isomorphism K as in Theorem 6.5 and the volume form V*

$$Kc \wedge (*f) = \langle f, c\rangle V .$$

Proof Let $x_1, \ldots, x_n$ be a positive oriented semi-orthonormal basis in E. Since both sides of the equation to be proven depend bilinearly on f and c, it is sufficient to show this equation for p-forms $\varepsilon_{i_1} Jx_{i_1} \wedge \cdots \wedge \varepsilon_{i_p} Jx_{i_p}$ and the dual linear forms $c_{j_1\cdots j_p}$. We have

$$Kc_{i_1\cdots i_p} \wedge (*(\varepsilon_{i_1} Jx_{i_1} \wedge \cdots \wedge \varepsilon_{i_p} Jx_{i_p}))$$
$$= (Jx_{i_1} \wedge \cdots \wedge Jx_{i_p}) \wedge \varepsilon_{i_1} \cdots \varepsilon_{i_p}\varepsilon_{i_{p+1}} \cdots \varepsilon_{i_n}(Jx_{i_{p+1}} \wedge \cdots \wedge Jx_{i_n}) = V$$

and for other indices $j_1, \ldots, j_p$

$$Kc_{j_1\cdots j_p} \wedge (*(\varepsilon_{i_1} Jx_{i_1} \wedge \cdots \wedge \varepsilon_{i_p} Jx_{i_p}))$$
$$= \varepsilon_{i_1} \cdots \varepsilon_{i_n} Jx_{j_1} \wedge \cdots \wedge Jx_{j_p} \wedge Jx_{i_{p+1}} \wedge \cdots \wedge Jx_{i_n} = 0 .$$

6.4   Outer Derivative

Definition 6.6 A differential form F of order p on an n-dimensional C^∞-manifold M is a $C^\infty - (0, p)$-tensor field on M, such that for each point P, the p-fold covariant tensor field $F(P)$ is skew-symmetric.   ♦

Such a differential form F thus assigns a p-form $F(P) \in \wedge^p M_P^*$ at each point P. Differentials forms can be added pointwise and mulitplied by a number. Thus differential forms of order p form a vector space, which we denote by Λ_p. The Hodge-star operation can also be done pointwise giving a map $\Lambda_p \longrightarrow \Lambda_{n-p}$ defined by $*F(P) = *(F(P))$. The wedge product $(F \wedge G)(P) = F(P) \wedge G(P)$ gives a map $\Lambda_p \times \Lambda_q \longrightarrow \Lambda_{p+q}$. The main objective of this section, however, is to introduce the linear map $d : \Lambda_p \longrightarrow \Lambda_{p+1}$ defined below, which initially refers to a chart, but is shown to be independent of the choice of the chart subsequently. The covector fields du^i are as defined in Definition 4.3.

Definition 6.7 The **outer derivative** of the differential form

$$F = \frac{1}{p!} F_{i_1 \dots i_p} du^{i_1} \wedge \cdots \wedge du^{i_p}$$

is the differential form

$$dF = \frac{1}{p!} \frac{\partial}{\partial u^{i_0}} F_{i_1 \dots i_p} du^{i_0} \wedge du^{i_1} \wedge \cdots \wedge du^{i_p}.$$

$\blacklozenge$

This definition actually does not depend on the choice of the chart with the coordinate vector fields $\frac{\partial}{\partial u^i}$. Let us consider another chart with the coordinate vector fields $\frac{\partial}{\partial v^i}$ with corresponding covector fields $dv^1, \dots, dv^n$. Then we have

$$dF = \frac{1}{p!} \frac{\partial}{\partial u^{i_0}} F\left(\frac{\partial}{\partial u^{i_1}}, \dots, \frac{\partial}{\partial u^{i_p}} \right) du^{i_0} \wedge du^{i_1} \wedge \cdots \wedge du^{i_p}$$

$$= \frac{1}{p!} \frac{\partial v^{k_0}}{\partial u^{i_0}} \frac{\partial}{\partial v^{k_0}} F\left(\frac{\partial v^{k_1}}{\partial u^{i_1}} \frac{\partial}{\partial v^{k_1}}, \dots, \frac{\partial v^{k_p}}{\partial u^{i_p}} \frac{\partial}{\partial v^{k_p}} \right) \left(\frac{\partial u^{i_0}}{\partial v^{l_0}} dv^{l_0} \right) \wedge \cdots \wedge \left(\frac{\partial u^{i_p}}{\partial v^{l_p}} dv^{l_p} \right).$$

Thanks to the multilinearity of the wedge product and the differential form, and using the product rule for partial derivatives, we obtain

$$dF = \frac{1}{p!} \frac{\partial}{\partial v^{k_0}} F\left(\frac{\partial}{\partial v^{k_1}}, \dots, \frac{\partial}{\partial v^{k_p}} \right) dv^{k_0} \wedge \cdots \wedge dv^{k_p}$$

$$+ \frac{1}{p!} \sum_{r=1}^{p} \frac{\partial u^{i_r}}{\partial v^{l_r}} \frac{\partial}{\partial v^{k_0}} \frac{\partial v^{k_r}}{\partial u^{i_r}} F\left(\frac{\partial}{\partial v^{k_1}}, \dots, \frac{\partial}{\partial v^{k_p}} \right) dv^{k_0} \wedge \cdots \wedge dv^{l_r} \wedge \cdots \wedge dv^{k_p}.$$

The last line is zero, since, first of all because of the product rule, we have

$$\frac{\partial u^{i_r}}{\partial v^{l_r}} \frac{\partial}{\partial v^{k_0}} \frac{\partial v^{k_r}}{\partial u^{i_r}} = \frac{\partial}{\partial v^{k_0}} \left(\frac{\partial u^{i_r}}{\partial v^{l_r}} \frac{\partial v^{k_r}}{\partial u^{i_r}} \right) - \frac{\partial v^{k_r}}{\partial u^{i_r}} \frac{\partial^2 u^{i_r}}{\partial v^{k_0} \partial v^{l_r}}$$

and secondly the expression in brackets being either $=$ or 1, is in any case a constant, making that term 0, and thirdly

$$\frac{\partial^2 u^{i_r}}{\partial v^{k_0} \partial v^{l_r}} dv^{k_0} \wedge \cdots \wedge dv^{k_{r-1}} \wedge dv^{l_r} \wedge dv^{k_{r+1}} \wedge \cdots \wedge dv^{k_p} = 0$$

since the summands cancel in pairs. Thus

$$dF = \frac{1}{p!} \frac{\partial}{\partial v^{k_0}} F\left(\frac{\partial}{\partial v^{k_1}}, \ldots, \frac{\partial}{\partial v^{k_p}}\right) dv^{k_0} \wedge dv^{k_1} \wedge \cdots \wedge dv^{k_p}.$$

A differential form F of order 0 is a real valued function. In this case $(p = 0)$, Definition 6.7 gives

$$dF = \frac{\partial F}{\partial u^i} du^i.$$

This differential form of first order is called the **differential** of F. Another example $(p = 2, \ n = 4)$ is explained in detail in the next section.

The map $F \longrightarrow dF$ is clearly linear, and in particular for any two differential forms F, G of the same order, we have $d(F + G) = dF + dG$.

Theorem 6.7 *For any two differential forms F and G, we have*

$$d(F \wedge G) = dF \wedge G + (-1)^p F \wedge dG,$$

where p is the order of F. In particular for a real function φ, we have the product rule

$$d(\varphi F) = d\varphi \wedge F + \varphi dF.$$

Proof The wedge product of

$$F = \frac{1}{p!} F_{i_1 \ldots i_p} du^{i_1} \wedge \cdots \wedge du^{i_p}$$

and

$$G = \frac{1}{q!} G_{i_{p+1} \ldots i_{p+q}} du^{i_{p+1}} \wedge \cdots \wedge du^{i_{p+q}}$$

is

$$F \wedge G = \frac{1}{p!q!} F_{i_1 \ldots i_p} G_{i_{p+1} \ldots i_{p+q}} du^{i_1} \wedge \cdots \wedge du^{i_{p+q}},$$

and its outer derivative is

$$d(F \wedge G) = \frac{1}{p!q!} \frac{\partial}{\partial u^{i_0}} (F_{i_1 \ldots i_p} G_{i_{p+1} \ldots i_{p+q}}) du^{i_0} \wedge du^{i_1} \wedge \cdots \wedge du^{i_{p+q}}.$$

Using the usual product rule for partial differentiation, and upon exchanging du^{i_0} with $du^{i_1} \wedge \cdots \wedge du^{i_p}$ in the second expression, the sign changes p times, giving

$$d(F \wedge G) =$$

$$= \left(\frac{1}{p!} \frac{\partial}{\partial u^{i_0}} F_{i_1 \dots i_p} du^{i_0} \wedge du^{i_1} \wedge \cdots \wedge du^{i_p} \right) \wedge \left(\frac{1}{q!} G_{i_{p+1} \dots i_{p+q}} du^{i_{p+1}} \wedge \cdots \wedge du^{i_{p+q}} \right)$$

$$+ (-1)^p \left(\frac{1}{p!} F_{i_1 \dots i_p} du^{i_1} \wedge \cdots \wedge du^{i_p} \right) \wedge \left(\frac{1}{q!} \frac{\partial}{\partial u^{i_0}} G_{i_{p+1} \dots i_{p+q}} du^{i_0} \wedge du^{i_{p+1}} \wedge \cdots \wedge du^{i_{p+q}} \right)$$

$$= dF \wedge G + (-1)^p F \wedge dG.$$

The following result is called the **first Poincare-lemma**.

Theorem 6.8 *For every differential form F, we have $d(dF) = 0$.*

Proof Applying the definition twice gives

$$ddF = \frac{1}{p!} \frac{\partial}{\partial u^{i-1}} \left(\frac{\partial}{\partial u^{i_0}} F_{i_1 \dots i_p} \right) du^{i-1} \wedge du^{i_0} \wedge du^{i_1} \wedge \cdots \wedge du^{i_p}.$$

Since

$$\frac{\partial^2}{\partial u^{i_0} \partial u^{i-1}} = \frac{\partial^2}{\partial u^{i-1} \partial u^{i_0}}$$

and

$$du^{i_0} \wedge du^{i-1} = -du^{i-1} \wedge du^{i_0}$$

the summands cancel in pairs.

Theorem 6.9 *The components of the differential form dF are calculated from the components of F via the formula*

$$(dF)_{k_0 \dots k_p} = \frac{\partial}{\partial u^{k_0}} F_{k_1 \dots k_p} - \frac{\partial}{\partial u^{k_1}} F_{k_0 k_2 \dots k_p}$$

$$+ \frac{\partial}{\partial u^{k_2}} F_{k_0 k_1 k_3 \dots k_p} - + \cdots + (-1)^p \frac{\partial}{\partial u^{k_p}} F_{k_0 \dots k_{p-1}}.$$

Proof We have

$$
\begin{aligned}
(dF)_{k_0 \cdots k_p} &= dF\left(\frac{\partial}{\partial u^{k_0}}, \ldots, \frac{\partial}{\partial u^{k_p}}\right) \\
&= \frac{1}{p!} \sum_{i_0, \ldots, i_p} \frac{\partial}{\partial u^{i_0}} F_{i_1 \cdots i_p} \sum_{\mathcal{Q}^{-1}} \chi(\mathcal{Q}^{-1}) \left\langle \frac{\partial}{\partial u^{k_0}}, du^{i_{\mathcal{Q}^{-1}(0)}} \right\rangle \cdots \left\langle \frac{\partial}{\partial u^{k_p}}, du^{i_{\mathcal{Q}^{-1}(p)}} \right\rangle \\
&= \frac{1}{p!} \sum_{\mathcal{Q}} \chi(\mathcal{Q}) \sum_{i_0, \ldots, i_p} \frac{\partial}{\partial u^{i_0}} F_{i_1 \cdots i_p} \left\langle \frac{\partial}{\partial u^{k_{\mathcal{Q}(0)}}}, du^{i_0} \right\rangle \cdots \left\langle \frac{\partial}{\partial u^{k_{\mathcal{Q}(p)}}}, du^{i_p} \right\rangle \\
&= \frac{1}{p!} \sum_{\mathcal{Q}} \chi(\mathcal{Q}) \frac{\partial}{\partial u^{k_{\mathcal{Q}(0)}}} F_{k_{\mathcal{Q}(1)} \cdots k_{\mathcal{Q}(p)}} .
\end{aligned}
$$

The sum is over all permutations of the numbers 0 to p. We classify these permutations according to the number $\mathcal{Q}(0)$. For $r \in \{0, 1, \ldots, p\}$ and $\mathcal{Q}$ with $\mathcal{Q}(0) = r$, there exists exactly one permutation $\mathcal{P}$ of the numbers 1 to p such that

$$
\begin{aligned}
\mathcal{Q} \circ \begin{pmatrix} 0 & 1 & \cdots & p \\ 0 & \mathcal{P}(1) & \cdots & \mathcal{P}(p) \end{pmatrix} \\
= \begin{pmatrix} 0 & 1 & \cdots & p \\ r & \mathcal{Q}(\mathcal{P}(1)) & \cdots & \mathcal{Q}(\mathcal{P}(p)) \end{pmatrix} = \begin{pmatrix} 0 & 1 & \cdots & r & r+1 & \cdots & p \\ r & 0 & \cdots & r-1 & r+1 & \cdots & p \end{pmatrix} .
\end{aligned}
$$

Clearly, we have $\chi(\mathcal{Q})\chi(\mathcal{P}) = (-1)^r$. By the skew-symmetry of F, we have

$$
F_{k_{\mathcal{Q}(1)} \cdots k_{\mathcal{Q}(p)}} = \chi(\mathcal{P}) F_{k_{\mathcal{Q}(\mathcal{P}(1))} \cdots k_{\mathcal{Q}(\mathcal{P}(p))}} = \chi(\mathcal{P}) F_{k_0 \cdots k_{r-1} k_{r+1} \cdots k_p} ,
$$

and we can continue the chain of equations above as

$$
\begin{aligned}
(dF)_{k_0 \cdots k_p} &= \frac{1}{p!} \sum_{r=0}^{p} \sum_{\mathcal{Q}(0)=r} \chi(\mathcal{Q})\chi(\mathcal{P}) \frac{\partial}{\partial u^{k_r}} F_{k_0 \cdots k_{r-1} k_{r+1} \cdots k_p} \\
&= \sum_{r=0}^{p} (-1)^r \frac{\partial}{\partial u^{k_r}} F_{k_0 \cdots k_{r-1} k_{r+1} \cdots k_p} .
\end{aligned}
$$

The $(p+1)$-flag covaraint tensor field dF maps $p+1$ vector fields to a scalar field. By the previous theorem, the coordinate vector fields $\partial_{k_0}, \ldots, \partial_{k_p}$ are mapped to the scalar field

$$
dF(\partial_{k_0}, \ldots, \partial_{k_p}) = \sum_{r=0}^{p} (-1)^r \partial_{k_r} F(\partial_{k_0}, \ldots, \partial_{k_{r-1}}, \partial_{k_{r+1}}, \ldots, \partial_{k_p}) .
$$

For other vector fields, the result is a little more complicated.

Theorem 6.10 *For a differential form of order p and C^∞-vector fields $X_0, \ldots, X_p$,*
we have

$$
dF(X_0, X_1, \ldots, X_p) = \sum_{r=0}^{p} (-1)^r X_r F(X_0, \ldots, X_{r-1}, X_{r+1}, \ldots, X_p)
$$

$$
+ \sum_{0 \leq r < s \leq p} (-1)^{r+s} F([X_r, X_s], X_0, \ldots, X_{r-1}, X_{r+1}, \ldots, X_{s-1}, X_{s+1}, \ldots, X_p) \, .
$$

Proof We will verify the claim in the neighborhood of an arbitrary chosen point
P. Both sides of the equality are clearly multilinear maps on $\mathcal{X}(P) \times \cdots \times \mathcal{X}(P)$
to $\mathcal{F}(P)$. The left side is $\mathcal{F}$-homogeneous because it is generated by a tensor field
(see Definition 4.6). Both sides coincide for coordinate vector fields. If the right side
is also shown to be $\mathcal{F}$-homogeneous, then it follows that the equality holds for all
vector fields. Let

$$
G : \mathcal{X}(P) \times \cdots \times \mathcal{X}(P) \longrightarrow \mathcal{F}(P)
$$

be the multilinear mapping on the right hand side. We wish to show that

$$
G(X_0, \ldots, X_{l-1}, \varphi X_l, X_{l+1}, \ldots, X_p) = \varphi G(\ldots, X_l, \ldots) \, .
$$

Definition 2.1 (P) and Theorem 2.9 imply

$$
G(X_0, \ldots, \varphi X_l, \ldots, X_p) = \varphi G(X_0, \ldots, X_l, \ldots, X_p)
$$

$$
+ \sum_{r=0}^{l-1} (-1)^r (X_r \varphi) F(\cdots, X_{r-1}, X_{r+1}, \cdots) + \sum_{r=l+1}^{p} (-1)^r (X_r \varphi) F(\cdots, X_{r-1}, X_{r+1}, \cdots)
$$

$$
+ \sum_{r=0}^{l-1} (-1)^{r+l} F((X_r \varphi) X_l, X_0, \ldots, X_{r-1}, X_{r+1}, \ldots, X_{l-1}, X_{l+1}, \ldots, X_p)
$$

$$
+ \sum_{r=l+1}^{p} (-1)^{l+r} F(-(X_r \varphi) X_l, X_0, \ldots, X_{l-1}, X_{l+1}, \ldots, X_{r-1}, X_{r+1}, \ldots, X_p) \, .
$$

Finally, the skew-symmetry of F causes the sums to cancel each other out.

Now that Theorem 6.10 has been proven in general, let us note in particular the
special cases $p = 0, 1, 2$. They are

$$
dF(X) = X F \, ,
$$

$$
dF(X, Y) = X F(Y) - Y F(X) - F([X, Y])
$$

and

$$dF(X, Y, Z)$$

$$= X F(Y, Z) - Y F(X, Z) + Z F(X, Y) - F([X, Y], Z) + F([X, Z], Y) - F([Y, Z], X).$$

The first Poincare-lemma (Theorem 6.8) states the outer derivative of a differential form, if it is derived again becomes zero. The question arises whether conversly, every differential form, whose outer derivative is zero, is the outer derivative of some differential form. We first try to find an answer in the special case when $M = \mathbb{R}^3$ and $p = 1$. A first-order differential form $F = F_i du^i$ on $\mathbb{R}^3$ is the outer derivative of a zero-order differential form G, if and only if the components F_i are the partial derivatives of G. The outer derivative of $F = F_i du^i$ is

$$dF = \frac{\partial F_k}{\partial u^j} du^j \wedge du^k$$

$$= \left(\frac{\partial F_3}{\partial u^2} - \frac{\partial F_2}{\partial u^3} \right) du^2 \wedge du^3 + \left(\frac{\partial F_1}{\partial u^3} - \frac{\partial F_3}{\partial u^1} \right) du^3 \wedge du^1 + \left(\frac{\partial F_2}{\partial u^1} - \frac{\partial F_1}{\partial u^2} \right) du^1 \wedge du^2 .$$

Thus, in the language of vector calculus, the requirement that $dF = 0$ corresponds to the vector field with components F_1, F_2, F_3 being curl-free. As is well-known, a curl-free vector field on $\mathbb{R}^3$ possesses a potential. However, if the vector field is only present in a partial area M of $\mathbb{R}^3$, and if no further demands are placed on M (such as being star-shaped), then a potential may only exist locally. Hence, in general, for a differential form F with $dF = 0$, the question of the existence of a differential form G such that $dG = F$ may therefore only be answered positive, if at all, in the local sense. The concept of homotopy operator underlies the constructive proof.

Definition 6.8 Let M be an open ball in $\mathbb{R}^n$ with center 0. The **homotopy operator of pth order** H_p is the map which sends a pth order differential form F on M to the $(p - 1)$st order differential form $H_p F$ given by

$$H_p F(x)(x_1, \ldots, x_{p-1}) = \int_0^1 t^{p-1} F(tx)(x, x_1, \ldots, x_{p-1}) \, dt .$$

$\blacklozenge$

Example. For a basis $x_1, \ldots, x_n$ in $\mathbb{R}^n$, let us calculate the image $H_p F$ of the p-form

$$F(x) = f(x) \, dx^{i_1} \wedge \cdots \wedge dx^{i_p} = f(x) \sum_{\cdot_\mu} \chi(\mathcal{P}) \, dx^{i_{\mathcal{P}(1)}} \otimes \cdots \otimes dx^{i_{\mathcal{P}(p)}} .$$

With $x^1, \ldots, x^n$ as the components of x with respect to the given basis, we have

$$H_p F(x) = \sum_{k=1}^{p} \sum_{\mathcal{P}(1)=k} \chi(\mathcal{P}) \int_0^1 t^{p-1} f(tx)\, dt\, x^{i_k} dx^{i_{\mathcal{P}(2)}} \otimes \cdots \otimes dx^{i_{\mathcal{P}(p)}}$$

$$= \int_0^1 t^{p-1} f(tx)\, dt \sum_{k=1}^{p} (-1)^{k-1} x^{i_k}\, dx^{i_1} \wedge \cdots \wedge dx^{i_{k-1}} \wedge dx^{i_{k+1}} \wedge \cdots \wedge dx^{i_p} ,$$

since the numbers $\mathcal{P}(2), \ldots, \mathcal{P}(p)$ in this arrangement contain $k-1$ fewer transpositions than the permutation $\mathcal{P}$.

Theorem 6.11 *For every differential form F on an open ball of $\mathbb{R}^n$ with center 0, we have*

$$d H_p F + H_{p+1} d F = F .$$

Proof The map $H_p : \Lambda_p \longrightarrow \Lambda_{p-1}$ is clearly linear. So it is enough to show the claimed equality for p-forms of the type

$$F(x) = f(x)\, dx^{i_1} \wedge \cdots \wedge dx^{i_p} .$$

From the expression just obtained for $H_p F$, we have

$$d H_p F(x) =$$

$$= \frac{\partial}{\partial x^{i_0}} \int_0^1 t^{p-1} f(tx)\, dt \sum_{k=1}^{p} (-1)^{k-1} x^{i_k}\, dx^{i_0} \wedge dx^{i_1} \wedge \cdots \wedge dx^{i_{k-1}} \wedge dx^{i_{k+1}} \wedge \cdots \wedge dx^{i_p}$$

$$+ \int_0^1 t^{p-1} f(tx)\, dt \sum_{k=1}^{p} (-1)^{k-1} dx^{i_k} \wedge dx^{i_1} \wedge \cdots \wedge dx^{i_{k-1}} \wedge dx^{i_{k+1}} \wedge \cdots \wedge dx^{i_p}$$

$$= \int_0^1 t^{p} \frac{\partial f}{\partial x^{i_0}}(tx)\, dt \sum_{k=1}^{p} (-1)^{k-1} x^{i_k} dx^{i_0} \wedge \cdots \wedge dx^{i_{k-1}} \wedge dx^{i_{k+1}} \wedge \cdots \wedge dx^{i_p}$$

$$+ \int_0^1 t^{p-1} f(tx)\, dt\, p\, dx^{i_1} \wedge \cdots \wedge dx^{i_p} .$$

The corresponding expression for H_{p+1} acting on

$$d F(x) = \frac{\partial f}{\partial x^{i_0}}(x)\, dx^{i_0} \wedge dx^{i_1} \wedge \cdots \wedge dx^{i_p}$$

gives

$$H_{p+1} d F(x) = \int_0^1 t^{p} \frac{\partial f}{\partial x^{i_0}}(tx)\, dt \sum_{k=0}^{p} (-1)^{k} x^{i_k} dx^{i_0} \wedge \cdots \wedge dx^{i_{k-1}} \wedge dx^{i_{k+1}} \wedge \cdots \wedge dx^{i_p} .$$

Adding these, we obtain

$$dH_p F(x) + H_{p+1} dF(x)$$

$$= \int_0^1 \left[t^p \frac{\partial f}{\partial x^{i_0}} (tx) x^{i_0} + p t^{p-1} f(tx) \right] dt\, dx^{i_1} \wedge \cdots \wedge dx^{i_p} = f(x)\, dx^{i_1} \wedge \cdots \wedge dx^{i_p} = F(x).$$

Now we are ready to prove the aforementioned converse of Theorem 6.8, the so-called **second Poincare-lemma**.

Theorem 6.12 *For a differential form F with $dF = 0$, there exists a differential form G in a neighborhoud of any point P with $dG = F$.*

Proof We choose a chart γ with $\gamma(P) = 0$ and apply Theorem 6.11 to the restriction F' of $F \circ \gamma^{-1}$ to a ball around 0. Since the bijection γ also induces isomorphisms between the corresponding tangent spaces of M and $\mathbb{R}^n$, F' is indeed again a differential form, and we have $dF' = 0$. With the homotopy operators H_p and H_{p+1}, we have

$$F' = dH_p F' + H_{p+1} dF',$$

and so the differential form G on M corresponding to the differential form $H_p F'$ on $\mathbb{R}^n$ is the one which is sought, satisfying $dG = F$.

6.5   The Maxwell Equations in Vacuum

The electromagnetic field is described by the field strength tensor F (see Sect. 5.3). If the Hodge-star operator is applied pointwise to this second order differential form, then we get another second order differential form $*F$, which is called the **Maxwell-tensor**. Since the Hodge-star operator is an isomorphism, the Maxwell-tensor also contains full information about the electromagnetic field.

Theorem 6.13 *The Maxwell-tensor $*F$ has the following components with respect to a positive oriented Lorentz basis*

$$(*F_{ik}) = \begin{pmatrix} 0 & -B^1 & -B^2 & -B^3 \\ B^1 & 0 & -E^3 & E^2 \\ B^2 & E^3 & 0 & -E^1 \\ B^3 & -E^2 & E^1 & 0 \end{pmatrix}.$$

Proof We choose a chart, whose coordinate vector fields $\frac{\partial}{\partial u^i}$, $i = 0, 1, 2, 3$ pointwise form a positive oriented Lorentz basis in the respective tangent spaces. From Theorem 5.9, we have the representation

$$F = E^1 du^0 \wedge du^1 + E^2 du^0 \wedge du^2 + E^3 du^0 \wedge du^3 - B^1 du^2 \wedge du^3 + B^2 du^1 \wedge du^3 - B^3 du^1 \wedge du^2$$

which can be read from the component matrix of the field strength tensor. To use Definition 6.5, we note that here $\varepsilon_0 = 1$, $J\partial_0 = du^0$ and $\varepsilon_i = -1$, $J\partial_i = -du^i$ for $i = 1, 2, 3$. It follows that

$$*(du^0 \wedge du^1) = (-1) * (J\partial_0 \wedge J\partial_1) = (-1)^3 J\partial_2 \wedge J\partial_3 = -du^2 \wedge du^3$$

and analogously

$$*(du^0 \wedge du^2) = du^1 \wedge du^3$$
$$*(du^0 \wedge du^3) = -du^1 \wedge du^2$$
$$*(du^1 \wedge du^2) = du^0 \wedge du^3$$
$$*(du^1 \wedge du^3) = -du^0 \wedge du^2$$
$$*(du^2 \wedge du^3) = du^0 \wedge du^1,$$

so that

$$*F = -E^1 du^2 \wedge du^3 + E^2 du^1 \wedge du^3 - E^3 du^1 \wedge du^2$$
$$- B^1 du^0 \wedge du^1 - B^2 du^0 \wedge du^2 - B^3 du^0 \wedge du^3,$$

from which we can read off the matric of components.

A comparison of the component matrices of F and $*F$ shows that the vectors $-B$ and E play the same role in the Maxwell-tensor, as the vectors E and B in the Faraday-tensor. If Definition 5.8 is modified accordingly, then we arrive at the following result.

Theorem 6.14 *For an observer* x *measuring the field strengths as* E *and* B, *the following hold:*

$$*F(y, z) = g(y \times z, E) \quad for \quad y, z \in x^{\perp}$$

and

$$*F(x, y) = g(y, B) \quad for \quad y \in x^{\perp}.$$

In charge-free space, non-relativistic electrodynamics is based on the four Maxwell equations

$$\operatorname{div} E = 0, \quad \operatorname{rot} E = -\frac{\partial B}{\partial t}, \quad \operatorname{div} B = 0, \quad \operatorname{rot} B = \frac{\partial E}{\partial t}.$$

These can now be formulated much more elegantly.

Theorem 6.15 *The equations* $\operatorname{div} B = 0$ *and* $\operatorname{rot} E = -\frac{\partial B}{\partial t}$ *are equivalent to* $dF = 0$, *and the equations* $\operatorname{div} E = 0$ *and* $\operatorname{rot} B = \frac{\partial E}{\partial t}$ *are equivalent to* $d*F = 0$.

Proof From the representation of F given in the proof of Theorem 6.13, and the results given after Definition 6.7, we obtain

$$dF = \left(-\frac{\partial B^1}{\partial u^1} - \frac{\partial B^2}{\partial u^2} - \frac{\partial B^3}{\partial u^3} \right) du^1 \wedge du^2 \wedge du^3$$

$$+ \left(\frac{\partial E^2}{\partial u^3} - \frac{\partial E^3}{\partial u^2} - \frac{\partial B^1}{\partial u^0} \right) du^0 \wedge du^2 \wedge du^3$$

$$+ \left(\frac{\partial E^1}{\partial u^3} - \frac{\partial E^3}{\partial u^1} + \frac{\partial B^2}{\partial u^0} \right) du^0 \wedge du^1 \wedge du^3$$

$$+ \left(\frac{\partial E^1}{\partial u^2} - \frac{\partial E^2}{\partial u^1} - \frac{\partial B^3}{\partial u^0} \right) du^0 \wedge du^1 \wedge du^2 ,$$

from which we get the first claim of the theorem. The second claim is obtained analogously.

The tangent vector $\frac{\partial}{\partial u^0}$ is the observer x, and so the application of x to the scalar field B^i, is for the observer, the partial differentiation of B^i with respect to time, giving $\frac{\partial B^i}{\partial u^0} = \frac{\partial B^i}{\partial t}$. The spacelike vectors $\frac{\partial}{\partial u^i}$, $i = 1, 2, 3$ form a positive oriented orthonormal basis for $x^\perp$. The application of $\frac{\partial}{\partial u^i}$ is partial differentiation with respect to the ith spatial coordinate. If all four coefficients in the above representation are zero, then this means, $dF = 0$, that is, $\operatorname{div} B = 0$ and $\operatorname{rot} E = -\frac{\partial B}{\partial t}$. Analogously, we have $d*F = 0$, $\operatorname{div} E = 0$ and $\operatorname{rot}(-B) = -\frac{\partial E}{\partial t}$.

The condition $dF = 0$ ensures, at least locally, the existence of a differential form A of order one, satisfying $dA = F$, which is called the **four-potential**. However, it is uniquely determined only up to the addition of a term $d\varphi$. The following calculation shows that the four-potential contains the non-relativistic concepts of electrostatic potential and the vector potential. The outer derivative of $A = A_i du^i$ is

$$dA = (\partial_0 A_1 - \partial_1 A_0) du^0 \wedge du^1 + (\partial_0 A_2 - \partial_2 A_0) du^0 \wedge du^2$$

$$+ (\partial_0 A_3 - \partial_3 A_0) du^0 \wedge du^3 + (\partial_1 A_2 - \partial_2 A_1) du^1 \wedge du^2$$

$$+ (\partial_1 A_3 - \partial_3 A_1) du^1 \wedge du^3 + (\partial_2 A_3 - \partial_3 A_2) du^2 \wedge du^3 .$$

By a comparison with the representation of F, we get $E_i = \partial_0 A_i - \partial_i A_0$ for $i = 1, 2, 3$ and

$$B_1 = \partial_3 A_2 - \partial_2 A_3 , \qquad B_2 = \partial_1 A_3 - \partial_3 A_1 , \qquad B_3 = \partial_2 A_1 - \partial_1 A_2 ,$$

so that

$$E = -\frac{\partial}{\partial t}(-A_1, -A_2, -A_3) - \operatorname{grad} A_0$$

and

$$B = \operatorname{rot}(-A_1, -A_2, -A_3) .$$

The Covariant Differentiation of Vector Fields

7

7.1   The Directional Derivative in $\mathbb{R}^n$

The subject of this chapter is the description of the change in a vector field Y under a small displacement of the point P. At the point P, we must get a vector in M_P from the vector field and a vector $x \in M_P$ describing the direction of differentiation. In the specialcase of $M = \mathbb{R}^3$, for a point P, avector x and a vector field Y, we define the directional derivative

$$D_x Y = \lim_{h \longrightarrow 0} \frac{Y(\gamma(h)) - Y(\gamma(0))}{h}$$

where γ is a curve with $\gamma(0) = P$ and $\gamma'(0) = x$. The curve γ given by $\gamma(t) = P + tx$ can be used (Fig. 7.1). If, on the other hand, M is a curved surface in $\mathbb{R}^3$, then only tangential vector fields are permissible as the vector field Y. A curve γ in M is chosen for the tangent vector. The vectors $Y(\gamma(h))$ can be transported to the point $P = \gamma(0)$ in general, but cannot be guaranteed to be tangent vectors to M after transportation. So the difference quotient would only be formable in the sense of $\mathbb{R}^3$, and the difference quotient would then be a vector in general, but is not guaranteed to be tangential to the surface.

This method is not transferable to the general case of a manifold because it uses concepts that are not available in the general case. Instead, we will take properties of the directional derivative in $\mathbb{R}^n$ and axiomatically place them as defining properties of the covariant derivative introduced in the next section. It will be shown later that in the case of a curved surface in $\mathbb{R}^3$, the covariant derivative is the orthogonal projection of the directional derivative in the sense of $\mathbb{R}^3$ onto the tangential plane.

For $P \in \mathbb{R}^n$, $x \in \mathbb{R}^n$ and $Y \in \mathcal{X}(\mathbb{R}^n)$, let $D_x Y$ be defined as in the special case $n = 3$ mentioned above. For $X \in \mathcal{X}(\mathbb{R}^n)$ instead of $x \in \mathbb{R}^n$, we define

© The Author(s), under exclusive license to Springer Nature Switzerland AG 2023
R. Oloff, *The Geometry of Spacetime*, Graduate Texts in Physics,
https://doi.org/10.1007/978-3-031-16139-1_7

Fig. 7.1 Directional
derivative $D_x Y$ of Y
in the direction x

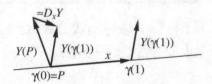

$D_X Y \in \mathcal{X}(\mathbb{R}^n)$ by $D_X Y(P) = D_{X(P)} Y$. Since the directional derivative is calculated componentwise, we have

$$D_X Y = X^i \partial_i Y^k \partial_k \,.$$

Using this representation, the following result can be shown.

Theorem 7.1 *The operation* $(X, Y) \longrightarrow D_X Y$ *in* $\mathcal{X}(\mathbb{R}^n)$ *has the following properties:*

(D1) $D_{fX+gY} Z = f D_X Z + g D_Y Z$
(D2) $D_X(\lambda Y + \mu Z) = \lambda D_X Y + \mu D_X Z$
(D3) $D_X(fY) = (Xf)Y + f D_X Y$
(D4) $Z(X \cdot Y) = D_Z X \cdot Y + X \cdot D_Z Y$
(D5) $D_X Y - D_Y X = [X, Y]$

for $X, Y, Z \in \mathcal{X}(\mathbb{R}^n)$, $f, g \in \mathcal{F}(\mathbb{R}^n)$ *and real* λ *and* μ.

7.2   The Levi-Civita Connection

We will now generalise the concept of the directional derivative from $\mathbb{R}^n$ to a manifold M.

Definition 7.1 An operation $(X, Y) \longrightarrow \nabla_X Y$ with the properties

(D1) $\nabla_{fX+gY} Z = f \nabla_X Z + g \nabla_Y Z$
(D2) $\nabla_X(\lambda Y + \mu Z) = \lambda \nabla_X Y + \mu \nabla_X Z$
(D3) $\nabla_X(fY) = (Xf)Y + f \nabla_X Y$

for $f, g \in \mathcal{F}(M)$ and numbers λ and μ, is called a **connection**.   ◆

Such an operation induces an isomorphism between two tangential spaces M_P and M_Q in the form of parallel displacement. So the individual tangent spaces are no longer coexisting side by side in isolation, but are connected in this sense, hence providing the rationale behind the name.

The properties (D4) and (D5) of the operation $(X, Y) \longrightarrow D_X Y$ in $\mathbb{R}^n$ hold with the dot product in $\mathbb{R}^n$. So we now need a fundamental tensor on the manifold M as an additional structure. The following so-called **fundamental theorem of Riemannian geometry** is the basis for the definition of the covariant derivative.

Theorem 7.2 *On a semi-Riemannian manifold* $[M, g]$ *there exists a unique connection* ∇ *with the additional properties*

(D4) $Zg(X, Y) = g(\nabla_Z X, Y) + g(X, \nabla_Z Y)$
(D5) $\nabla_X Y - \nabla_Y X = [X, Y]$.

Proof For a connection ∇ possessing the properties (D4) and (D5), an explicit expression for $\nabla_X Y$ can be obtained for vector fields X and Y as follows: In the equation

$$Xg(Y, Z) = g(\nabla_X Y, Z) + g(Y, \nabla_X Z)$$

we use the equation

$$\nabla_X Z = \nabla_Z X + [X, Z].$$

This yields

$$Xg(Y, Z) = g(\nabla_X Y, Z) + g(Y, \nabla_Z X) + g(Y, [X, Z])$$

to which we add an analogous expression for $Yg(Z, X)$ and subtract an analogous expression for $Zg(X, Y)$. This results in the so-called **Koszul formula** (K)

$$2g(\nabla_X Y, Z) = Xg(Y, Z) + Yg(X, Z) - Zg(X, Y) - g(X, [Y, Z])$$
$$- g(Y, [X, Z]) + g(Z, [X, Y]).$$

Given the vector fields X and Y, the right hand side depends linearly on the vector field Z, and Z is in particular $\mathcal{F}$-homogenous (see Definition 4.6) since the disruptive additional summands cancel each other out. The value of the function on the right hand side at the point P is determined by $Z(P)$ uniquely, and so the vector $\nabla_X Y(P)$ is uniquely determined by (K).

We now suppose that the vector field $\nabla_X Y$ has been defined by (K), and show that (D1)–(D5) hold for the operation ∇. Since the right hand side of (K) is additive with respect to X, we have

$$\nabla_{X+Z} Y = \nabla_X Y + \nabla_Z Y.$$

Furthermore, for $Z \in \mathcal{X}(M)$, we have, by (K), that

$$2g(\nabla_{fX} Y, Z) = 2fg(\nabla_X Y, Z),$$

since the disruptive additional summands cancel each other out, and from this it follows that

$$\nabla_{fX} Y = f \nabla_X Y,$$

hence showing (D1). (D2) holds since the right hand side of (K) depends linearly on Y. In order to prove the product rule (D3), we verify

$$2g(\nabla_X(fY), Z) = 2(Xf)g(Y, Z) + 2fg(\nabla_X Y, Z),$$

but this can be done using Theorem 2.9 and (K). The so-called **Ricci identity** (D4) follows from (K) in the form

$$2g(\nabla_Z X, Y) + 2g(\nabla_Z Y, X) = 2Zg(X, Y),$$

because all other summands cancel each other out in pairs. Analogously, we have

$$2g(\nabla_X Y, Z) - 2g(\nabla_Y X, Z) = 2g([X, Y], Z)$$

and so (D5) holds.

Definition 7.2 The operation ∇ on $\mathcal{X}(M)$ for a semi-Riemannian manifold $[M, g]$ with the properties (D1)–(D5) is called the **Levi-Civita connection**, and in the special case of a Riemannian manifold, the **Riemannian connection**. For vector fields X and Y, the vector field $\nabla_X Y$ is called the **covariant derivative of** Y **with respect to** X. The **covariant derivative of** Y is the $(1, 1)$-tensor field ∇Y, which assigns to a point $P \in M$ and a tangent vector $x \in M_P$, the tangent vector $\nabla_X Y(P)$, where X satisfies $X(P) = x$.   ◆

In the case of covariant derivation, the $(1, 0)$-tensor field Y becomes the $(1, 1)$-tensor field ∇Y, and so the degree of covariance increases by 1, justifying the name. In Chap. 11, we will also be able to covariantly differentiate general tensor fields, so that a (p, q)-tensor field will then become a $(p, q+1)$-tensor field.

7.3   Christoffel Symbols

Definition 7.3 The representation

$$\nabla_{\partial_i} \partial_k = \Gamma_{ik}^j \partial_j$$

defines the real valued functions Γ_{ik}^j, called the **Christoffel symbols**.   ◆

Theorem 7.3 *The symmetry* $\Gamma_{ik}^j = \Gamma_{ki}^j$ *holds for Christoffel symbols, and for vector fields* $X = X^i \partial_i$ *and* $Y = Y^k \partial_k$, *the vector field* $\nabla_X Y$ *has the components*

$$(\nabla_X Y)^j = X^i \partial_i Y^j + X^i Y^k \Gamma_{ik}^j.$$

Proof The symmetry follows from

$$(\Gamma^j_{ik} - \Gamma^j_{ki})\partial_j = \nabla_{\partial_i}\partial_k - \nabla_{\partial_k}\partial_i = [\partial_i, \partial_k] = 0.$$

From the calculational rules for the covariant derivative, we have

$$\nabla_X Y = \nabla_{X^i\partial_i} Y^k\partial_k = X^i(\partial_i Y^k\partial_k + Y^k\nabla_{\partial_i}\partial_k) = X^i(\partial_i Y^j + Y^k\Gamma^j_{ik})\partial_j.$$

Theorem 7.4 *The Christoffel symbols are calculated from the components of the metric and the contravariant metric tensor using the formula*

$$\Gamma^r_{ij} = \frac{1}{2}g^{kr}(\partial_i g_{jk} + \partial_j g_{ik} - \partial_k g_{ij}).$$

Proof By (K) we have

$$2\Gamma^l_{ij}g_{lk} = 2g(\nabla_{\partial_i}\partial_j, \partial_k) = \partial_i g_{jk} + \partial_j g_{ik} - \partial_k g_{ij},$$

and consequently,

$$(\partial_i g_{jk} + \partial_j g_{ik} - \partial_k g_{ij})g^{kr} = 2\Gamma^l_{ij}g_{lk}g^{kr} = 2\Gamma^r_{ij}.$$

The following transformation formula shows that the Christoffel symbols cannot be interpreted as the components of a tensor field.

Theorem 7.5 *Let two charts have coordinate vector fields ∂_j and $\bar{\partial}_i$ related by $\bar{\partial}_i = \alpha^j_i\partial_j$ and $\partial_j = \beta^i_j\bar{\partial}_i$. The Christoffel symbols calculated for these charts are related by the formula $\partial_j = \beta^i_j\bar{\partial}_i$ durch die Formel*

$$\bar{\Gamma}^r_{ij} = \beta^r_t(\alpha^l_i\alpha^v_j\Gamma^t_{lv} + \bar{\partial}_i\alpha^t_j).$$

Proof The formula can be read from

$$\nabla_{\bar{\partial}_i}\bar{\partial}_j = \nabla_{\alpha^l_i\partial_l}\alpha^v_j\partial_v = \alpha^l_i(\alpha^v_j\Gamma^t_{lv}\partial_t + \partial_l\alpha^v_j\partial_t)$$

$$= \alpha^l_i(\alpha^v_j\Gamma^t_{lv} + \partial_l\alpha^t_j)\beta^r_t\bar{\partial}_r = \beta^r_t(\alpha^l_i\alpha^v_j\Gamma^t_{lv} + \bar{\partial}_i\alpha^t_j)\bar{\partial}_r.$$

As an example, we now calculate the Christoffel symbols of a curved surface. Because of the symmetry mentioned in Theorem 7.3, there are six numbers. The fundamental tensor is the usual scalar product, and has the components $g_{ik} = \partial_i \cdot \partial_k$ (see also Fig. 2.1). By Theorem 7.4,

$$\Gamma^1_{11} = \frac{g^{11}}{2}\partial_1 g_{11} + \frac{g^{12}}{2}(2\partial_1 g_{12} - \partial_2 g_{11})$$

or in the classical Gaussian notation (see the example following Definition 4.7)

$$\Gamma_{11}^1 = \frac{GE_u - 2FF_u + FE_v}{2(EG - F^2)}$$

and analogously

$$\Gamma_{12}^1 = \frac{GE_v - FG_u}{2(EG - F^2)}, \qquad \Gamma_{22}^1 = \frac{2GF_v - GG_u - FG_v}{2(EG - F^2)},$$

$$\Gamma_{11}^2 = \frac{2EF_u - EE_v - FE_u}{2(EG - F^2)}, \qquad \Gamma_{12}^2 = \frac{EG_u - FE_v}{2(EG - F^2)}, \qquad \Gamma_{22}^2 = \frac{EG_v - 2FF_v + FG_u}{2(EG - F^2)}.$$

The two parameters are denoted by u and v here. The index terms are the corresponding partial derivatives.

Example. The parameters u and v for the surface of a sphere in $\mathbb{R}^3$ with radius r are taken as the spherical coordinates ϑ and φ. In this case, we obtain $E = r^2$, $G = r^2 \sin^2 \vartheta$ and $F = 0$. The above formulae give the Christoffel symbols

$$\Gamma_{22}^1 = -\sin \vartheta \cos \vartheta \qquad \text{and} \qquad \Gamma_{12}^2 = \Gamma_{21}^2 = \cot \vartheta$$

and the others are zero. Thus for the coordinate vector fields ∂_ϑ and ∂_φ, we have

$$\nabla_{\partial_\vartheta} \partial_\vartheta = 0, \qquad \nabla_{\partial_\varphi} \partial_\varphi = -\sin \vartheta \cos \vartheta \, \partial_\vartheta, \qquad \nabla_{\partial_\varphi} \partial_\vartheta = \nabla_{\partial_\vartheta} \partial_\varphi = \cot \vartheta \, \partial_\varphi.$$

These formulae also appear in the corresponding statements about the covariant derivatives in the Schwarzschild spacetime in the section following the next one.

7.4   The Covariant Derivative on Hypersurfaces

The aim of this section is to prove the assertion made earlier in Sect. 7.1, that the covariant derivative of a vector field on a curved surface is the orthogonal projection of the directional derivative in the sense of $\mathbb{R}^3$ onto the tangent plane (Fig. 7.2). We examine the situation more generally for a hypersurface in $\mathbb{R}^n$, since the considerations there are also not complicated.

Let M be an $(n-1)$-dimensional submanifold of $\mathbb{R}^n$. The choice of a chart induces $n-1$ coordinate vector fields $\partial n - 1$. In addition to the covariant derivative $\nabla_X Y$ along the tangent vector field X, also the derivative $D_X Y$ is available.

Theorem 7.6 *For tangent vector fields X and Y on an $(n-1)$-dimensional submanifold M of $\mathbb{R}^n$, the tangent vector field $\nabla_X Y$ is the orthogonal projection of the vector field $D_X Y$ on the respective tangent space.*

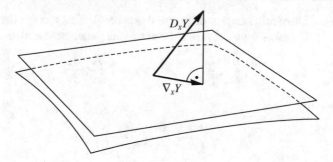

Fig. 7.2 Covariant derivative as a projection

Proof It is sufficient to prove the statement for coordinate vector fields $X = \partial_i$ and $Y = \partial_j$. A normal vector field satisfying $\vec{n} \cdot \vec{n} = 1$ can be introduced on M, at least locally. This normal vector field completes the coordinate vector fields pointwise to form a basis for $\mathbb{R}^n$. The vector field $D_{\partial_i}\partial_j$ can thus be expressed as

$$D_{\partial_i}\partial_j = \lambda_{ij}^l \partial_l + \lambda_{ij}\vec{n} \, .$$

The connection D on the Riemannian manifold $\mathbb{R}^n$ also satisfies the axioms (D1)–(D5), and in particular, also the Koszul formula (K). As

$$2\lambda_{ij}^l \partial_l \cdot \partial_k = 2D_{\partial_i}\partial_j \cdot \partial_k \, ,$$

the triply-indexed coefficients λ_{ij}^l thus satisfy the characterising equation for the Christoffel symbols,

$$2\lambda_{ij}^l \partial_l \cdot \partial_k = \partial_i(\partial_j \cdot \partial_k) + \partial_j(\partial_i \cdot \partial_k) - \partial_k(\partial_i \cdot \partial_j) \, .$$

The above representation is thus

$$D_{\partial_i}\partial_j = \Gamma_{ij}^l \partial_l + \lambda_{ij}\vec{n} = \nabla_{\partial_i}\partial_j + \lambda_{ij}\vec{n} \, ,$$

and so the difference between the derivatives in question is a multiple of the normal vector, which is actually orthogonal to the tangential hyperplane. In the special case of $n = 3$, that is, for a curved surface, the three equations

$$D_{\partial_1}\partial_1 = \Gamma_{11}^1 \partial_1 + \Gamma_{11}^2 \partial_2 + L\vec{n}$$

$$D_{\partial_1}\partial_2 = \Gamma_{12}^1 \partial_1 + \Gamma_{12}^2 \partial_2 + M\vec{n}$$

$$D_{\partial_2}\partial_2 = \Gamma_{22}^1 \partial_1 + \Gamma_{22}^2 \partial_2 + N\vec{n}$$

are called the **Gauss equations**. The notations L, M and N for the coefficients of the normal vector field $\vec{n}$, just like $E = \partial_1 \cdot \partial_1$, $F = \partial_1 \cdot \partial_2$ and $G = \partial_2 \cdot \partial_2$, are standard in the classical Gaussian differential geometry. If x, y, z are the Cartesian coordinates in $\mathbb{R}^3$ and u, v are the variables used in the parametrisation, then we have

$$\partial_1 = \left(\frac{\partial x}{\partial u}, \frac{\partial y}{\partial u}, \frac{\partial z}{\partial u}\right), \qquad \partial_2 = \left(\frac{\partial x}{\partial v}, \frac{\partial y}{\partial v}, \frac{\partial z}{\partial v}\right)$$

and consequently

$$D_{\partial_1}\partial_1 = \left(\frac{\partial^2 x}{\partial u^2}, \frac{\partial^2 y}{\partial u^2}, \frac{\partial^2 z}{\partial u^2}\right)$$

$$D_{\partial_1}\partial_2 = \left(\frac{\partial^2 x}{\partial u \partial v}, \frac{\partial^2 y}{\partial u \partial v}, \frac{\partial^2 z}{\partial u \partial v}\right)$$

$$D_{\partial_2}\partial_2 = \left(\frac{\partial^2 x}{\partial v^2}, \frac{\partial^2 y}{\partial v^2}, \frac{\partial^2 z}{\partial v^2}\right).$$

The normal vector $\vec{n}$ is the vector product

$$\vec{n} = \frac{\partial_1 \times \partial_2}{\sqrt{EG - F^2}},$$

and from this formula, we obtain the triple product

$$L = D_{\partial_1}\partial_1 \cdot \vec{n} = \frac{1}{\sqrt{EG - F^2}} \begin{vmatrix} \dfrac{\partial x}{\partial u} & \dfrac{\partial y}{\partial u} & \dfrac{\partial z}{\partial u} \\[2mm] \dfrac{\partial x}{\partial v} & \dfrac{\partial y}{\partial v} & \dfrac{\partial z}{\partial v} \\[2mm] \dfrac{\partial^2 x}{\partial u^2} & \dfrac{\partial^2 y}{\partial u^2} & \dfrac{\partial^2 z}{\partial u^2} \end{vmatrix}$$

$$M = \frac{1}{\sqrt{EG - F^2}} \begin{vmatrix} \dfrac{\partial x}{\partial u} & \dfrac{\partial y}{\partial u} & \dfrac{\partial z}{\partial u} \\[2mm] \dfrac{\partial x}{\partial v} & \dfrac{\partial y}{\partial v} & \dfrac{\partial z}{\partial v} \\[2mm] \dfrac{\partial^2 x}{\partial u \, \partial v} & \dfrac{\partial^2 y}{\partial u \, \partial v} & \dfrac{\partial^2 z}{\partial u \, \partial v} \end{vmatrix}$$

$$N = \frac{1}{\sqrt{EG - F^2}} \begin{vmatrix} \dfrac{\partial x}{\partial u} & \dfrac{\partial y}{\partial u} & \dfrac{\partial z}{\partial u} \\[2mm] \dfrac{\partial x}{\partial v} & \dfrac{\partial y}{\partial v} & \dfrac{\partial z}{\partial v} \\[2mm] \dfrac{\partial^2 x}{\partial v^2} & \dfrac{\partial^2 y}{\partial v^2} & \dfrac{\partial^2 z}{\partial v^2} \end{vmatrix}.$$

Fig. 7.3 $D_{\partial_\varphi}\partial_\varphi$ and $\nabla_{\partial_\varphi}\partial_\varphi$ for the sphere

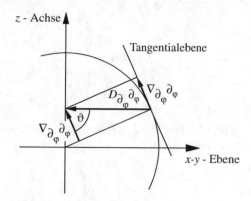

Example. On a spherical surface in $\mathbb{R}^3$ with the radius r, we have the coordinate vector fields ∂_ϑ and ∂_φ with the representations

$$\partial_\vartheta = r\cos\vartheta\cos\varphi\,\partial_x + r\cos\vartheta\sin\varphi\,\partial_y - r\sin\vartheta\,\partial_z$$
$$\partial_\varphi = -r\sin\vartheta\sin\varphi\,\partial_x + r\sin\vartheta\cos\varphi\,\partial_y$$

and upon partial differentiation,

$$D_{\partial_\vartheta}\partial_\vartheta = -r\sin\vartheta\cos\varphi\,\partial_x - r\sin\vartheta\sin\varphi\,\partial_y - r\cos\vartheta\,\partial_z$$
$$D_{\partial_\varphi}\partial_\varphi = -r\sin\vartheta\cos\varphi\,\partial_x - r\sin\vartheta\sin\varphi\,\partial_y$$
$$D_{\partial_\vartheta}\partial_\varphi = -r\cos\vartheta\cos\varphi\,\partial_x + r\cos\vartheta\cos\varphi\,\partial_y = D_{\partial_\varphi}\partial_\vartheta\,.$$

The vector $D_{\partial_\vartheta}\partial_\vartheta$ joins the point to the center of the sphere, and is thus perpendicular to the tangent plane. Thus, its orthogonal projection to the tangent plane, which is the covariant derivative $\nabla_{\partial_\vartheta}\partial_\vartheta$, is zero. The vector $D_{\partial_\vartheta}\partial_\varphi = D_{\partial_\varphi}\partial_\vartheta = \cot\vartheta\,\partial_\varphi$ lies in the tangent plane and so it coincides with its orthogonal projection onto the tangent plane, giving $\nabla_{\partial_\vartheta}\partial_\varphi = \nabla_{\partial_\varphi}\partial_\vartheta = \cot\vartheta\,\partial_\varphi$. The vector $D_{\partial_\varphi}\partial_\varphi$ lies in a plane parallel to the x-y-plane (Fig. 7.3). Its tangential part $\nabla_{\partial_\varphi}\partial_\varphi$ has the length $|r\sin\vartheta\cos\vartheta|$, is parallel to ∂_ϑ, but points in the opposite direction. Therefore $\nabla_{\partial_\varphi}\partial_\varphi = -\sin\vartheta\cos\vartheta\,\partial_\vartheta$. We had already obtained these results in the previous section by calculating the Christoffel symbols.

7.5   The Covariant Derivative in the Schwarzschild Spacetime

The Schwarzschild spacetime is the set $\mathbb{R} \times (2\mu, \infty) \times S_2$ (see Example 1 in Sect. 4.5). If the first two coordinates are t and r and the usual angular coordinates ϑ and φ are used on the surface of the unit sphere S_2, then the component matrix of the metric g has the diagonal form

$$(g_{ik}) = \begin{pmatrix} h(r) & 0 & 0 & 0 \\ 0 & -1/h(r) & 0 & 0 \\ 0 & 0 & -r^2 & 0 \\ 0 & 0 & 0 & -r^2 \sin^2 \vartheta \end{pmatrix}$$

where $h(r) = 1 - 2\,\mu/r$, and its inverse give the component-matrix for the contravariant metric:

$$(g^{ik}) = \begin{pmatrix} \dfrac{1}{h(r)} & 0 & 0 & 0 \\ 0 & -h(r) & 0 & 0 \\ 0 & 0 & -\dfrac{1}{r^2} & 0 \\ 0 & 0 & 0 & -\dfrac{1}{r^2 \sin^2 \vartheta} \end{pmatrix}$$

The formulae for the Christoffel symbols given in Theorem 7.4 acquire the following simplified form for orthogonal coordinates:

$$\Gamma^i_{ii} = \frac{1}{2} g^{ii} \partial_i g_{ii}$$

$$\Gamma^k_{ii} = -\frac{1}{2} g^{kk} \partial_k g_{ii} \quad \text{for} \ \ i \neq k$$

$$\Gamma^j_{ij} = \frac{1}{2} g^{jj} \partial_i g_{jj} \quad \text{for} \ \ i \neq j$$

$$\Gamma^k_{ij} = 0 \quad \text{for pairwise distinct} \ \ i, j, k \,.$$

For the covariant derivatives of the coordinate fields, one then obtains:

$$\nabla_{\partial_i} \partial_i = \frac{1}{2} \left(g^{ii} \partial_i g_{ii} \partial_i - \sum_{k \neq i} g^{kk} \partial_k g_{ii} \partial_k \right)$$

and for $i \neq j$

$$\nabla_{\partial_i} \partial_j = \frac{1}{2} (g^{ii} \partial_j g_{ii} \partial_i + g^{jj} \partial_i g_{jj} \partial_j) \,.$$

For the Schwarzschild spacetime, the covariant derivatives of the coordinate vector fields $\partial_t, \partial_r, \partial_\theta, \partial_\varphi$ are given by

$$\nabla_{\partial_t}\partial_t = \frac{h(r)\mu}{r^2}\,\partial_r\,, \qquad \nabla_{\partial_r}\partial_r = -\frac{\mu}{r^2 h(r)}\,\partial_r\,, \qquad \nabla_{\partial_\vartheta}\partial_\vartheta = -rh(r)\,\partial_r\,,$$

$$\nabla_{\partial_\varphi}\partial_\varphi = -rh(r)\sin^2\vartheta\,\partial_r - \sin\vartheta\cos\vartheta\,\partial_\vartheta\,, \qquad \nabla_{\partial_t}\partial_r = \frac{\mu}{r^2 h(r)}\,\partial_t\,,$$

$$\nabla_{\partial_t}\partial_\vartheta = 0 = \nabla_{\partial_t}\partial_\varphi\,, \qquad \nabla_{\partial_r}\partial_\vartheta = \frac{1}{r}\,\partial_\vartheta\,, \qquad \nabla_{\partial_r}\partial_\varphi = \frac{1}{r}\,\partial_\varphi\,, \qquad \nabla_{\partial_\vartheta}\partial_\varphi = \cot\vartheta\,\partial_\varphi\,.$$

Curvature

8

8.1 The Curvature Tensor

Here we adopt an abstract approach, where initially nothing of what may be imagined of a curved surface in $\mathbb{R}^3$ is evident. We will use the concept of the covariant derivative, and so we work with a semi-Riemannian manifold $[M, g]$.

Definition 8.1 For $Y, Z \in \mathcal{X}(M)$, the **curvature operator** $R(Y, Z) : \mathcal{X}(M) \longrightarrow \mathcal{X}(M)$ maps the vector field X to the vector field

$$R(Y, Z)X = \nabla_Y \nabla_Z X - \nabla_Z \nabla_Y X - \nabla_{[Y,Z]} X.$$

$\blacklozenge$

Altogether, we take three vector fields to produce a fourthone in a trilinear manner. One could equivalently say that one takes three vector fields X, Y, Z and one covector field A to obtain the scalar field $\langle R(Y, Z)X, A \rangle$. This map is multilinear and clearly $\mathcal{F}$-homogeneous with respect to A. We now show that it is $\mathcal{F}$-homogeneous also with respect to the three vector fields X, Y, Z. According to Theorem 4.1, it is then a $(1, 3)$-tensor field.

Theorem 8.1 *The trilinear map*

$$R(Y, Z)X = \nabla_Y \nabla_Z X - \nabla_Z \nabla_Y X - \nabla_{[Y,Z]} X$$

from $\mathcal{X}(M) \times \mathcal{X}(M) \times \mathcal{X}(M)$ *to* $\mathcal{X}(M)$ *is* $\mathcal{F}$-*homogeneous.*

© The Author(s), under exclusive license to Springer Nature Switzerland AG 2023
R. Oloff, *The Geometry of Spacetime*, Graduate Texts in Physics,
https://doi.org/10.1007/978-3-031-16139-1_8

Proof For the first vector variable

$$\nabla_Y \nabla_Z(fX) = \nabla_Y(f\nabla_Z X) + \nabla_Y((Zf)X)$$
$$= f\nabla_Y \nabla_Z X + (Yf)\nabla_Z X + (Zf)\nabla_Y X + (Y(Zf))X$$
$$\nabla_Z \nabla_Y(fX) = f\nabla_Z \nabla_Y X + (Zf)\nabla_Y X + (Yf)\nabla_Z X + (Z(Yf))X$$
$$\nabla_{[Y,Z]}(fX) = f\nabla_{[Y,Z]}X + ([Y,Z]f)X$$

and so we obtain

$$R(Y,Z)(fX) = fR(Y,Z)X .$$

For the second variable, we have

$$R(fY,Z)X = \nabla_{fY}\nabla_Z X - \nabla_Z \nabla_{fY}X - \nabla_{[fY,Z]}X$$
$$= f\nabla_Y \nabla_Z X - f\nabla_Z \nabla_Y X - (Zf)\nabla_Y X - f\nabla_{[Y,Z]}X + \nabla_{(Zf)Y}X = fR(Y,Z)X$$

and for the third, analogously we have

$$R(Y,fZ)X = fR(Y,Z)X .$$

Definition 8.2 The **Riemannian curvature tensor** R is the $(1,3)$-tensor field that assigns to the point $P \in M$ and the tangent vectors $x, y, z \in M_P$, the tangent vector $R(Y,Z)X(P)$, where $X(P) = x,\ \ Y(P) = y,\ \ Z(P) = z$.   ♦

Example. Let $M = \mathbb{R}^n$. Then the covariant derivative is the usual directional derivative, the Cartesian coordinate vector fields ∂_i are constant in this sense and thus also in the sense of the covariant derivative. Application of ∇_{∂_i} means partial differentiation of the components along the ith variable. Since the order of partial differentiation can be changed, ∇_{∂_i} and ∇_{∂_j} commute. Consequently, the curvature operator $R(\partial_i, \partial_j)$ is zero, and more generally for vector fields Y and Z, $R(Y,Z) = 0$. So the curvature tensor is 0.

Theorem 8.2 *The Riemann curvature tensor R has the components*

$$R^s_{ijk} = \partial_j \Gamma^s_{ki} - \partial_k \Gamma^s_{ji} + \Gamma^s_{jr}\Gamma^r_{ki} - \Gamma^s_{kr}\Gamma^r_{ji} .$$

Proof We need to compute the sth component of the vector field

$$R(\partial_j, \partial_k)\partial_i = \nabla_{\partial_j}\nabla_{\partial_k}\partial_i - \nabla_{\partial_k}\nabla_{\partial_j}\partial_i = \nabla_{\partial_j}(\Gamma^s_{ki}\partial_s) - \nabla_{\partial_k}(\Gamma^s_{ji}\partial_s)$$

$$= (\partial_j \Gamma^s_{ki})\partial_s + \Gamma^r_{ki}\nabla_{\partial_j}\partial_r - (\partial_k \Gamma^s_{ji})\partial_s - \Gamma^r_{ji}\nabla_{\partial_k}\partial_r .$$

The claim can now be read off from here.

For a four-dimensional semi-Riemannian manifold, the curvature tensor has $4^4 = 256$ components. Fortunately, there is redundancy owing to symmetry, and there is no need to do as many calculations.

Theorem 8.3 *For vector fields* X, Y, Z*, we have*

(1) $R(Y, Z)X = -R(Z, Y)X$
(2) $R(Y, Z)X + R(Z, X)Y + R(X, Y)Z = 0$ (**first Bianchi identity**).

For the components of the Riemann curvature tensor R, there holds

$$R^s_{ijk} = -R^s_{ikj}$$

and

$$R^s_{ijk} + R^s_{jki} + R^s_{kij} = 0.$$

Proof As $[Y, Z] = -[Z, Y]$, (1) follows. With axiom (D5) for the covariant derivative, (2) can be obtained using the Jacobi-identity:

$$R(Y, Z)X + R(Z, X)Y + R(X, Y)Z$$
$$= \nabla_X(\nabla_Y Z - \nabla_Z Y) + \nabla_Y(\nabla_Z X - \nabla_X Z) + \nabla_Z(\nabla_X Y - \nabla_Y X)$$
$$- \nabla_{[Y,Z]}X - \nabla_{[Z,X]}Y - \nabla_{[X,Y]}Z$$
$$= \nabla_X[Y, Z] - \nabla_{[Y,Z]}X + \nabla_Y[Z, X] - \nabla_{[Z,X]}Y + \nabla_Z[X, Y] - \nabla_{[X,Y]}Z$$
$$= \Big[X, [Y, Z]\Big] + \Big[Y, [Z, X]\Big] + \Big[Z, [X, Y]\Big] = 0.$$

Further symmetries become apparent when the Riemann curvature tensor is modified using index pulling (see Sect. 3.5). Besides the curvature operator and the Riemann curvature tensor, we also denote the modified curvature tensor by R. It is always clear from context which tensor is actually meant.

Definition 8.3 The **covariant curvature tensor** is the $(0, 4)$-tensor field R, characterised by

$$R(V, X, Y, Z) = g(V, R(Y, Z)X)$$

for vector fields V, X, Y, Z.   ◆

Theorem 8.4 *The components of the covariant curvature tensor are calculated from those of the Riemann curvature tensor by*

$$R_{rijk} = g_{rs} R^s_{ijk}.$$

Proof There holds

$$R_{rijk} = R(\partial_r, \partial_i, \partial_j, \partial_k) = g(\partial_r, R(\partial_j, \partial_k)\partial_i) = g(\partial_r, R^s_{ijk}\partial_s) = g(\partial_r, \partial_s)R^s_{ijk}.$$

Theorem 8.5 *The covariant curvature tensor R has the symmetries*

(1) $R(V, X, Y, Z) = -R(V, X, Z, Y)$
(2) $R(V, X, Y, Z) + R(V, Y, Z, X) + R(V, Z, X, Y) = 0$
(3) $R(V, X, Y, Z) = -R(X, V, Y, Z)$
(4) $R(V, X, Y, Z) = R(Y, Z, V, X)$

and in terms of components

(1) $R_{rijk} = -R_{rikj}$
(2) $R_{rijk} + R_{rjki} + R_{rkij} = 0$
(3) $R_{rijk} = -R_{irjk}$
(4) $R_{rijk} = R_{jkri}.$

Proof (1), (2) are simply new formulations of (1), (2) of Theorem 8.3. (3) says that there is skew-symmetry with respect to the first two vector arguments, and in light of

$$R(V + X, V + X, Y, Z) = R(V, V, Y, Z) + R(X, X, Y, Z)$$
$$+ R(V, X, Y, Z) + R(X, V, Y, Z)$$

it is enough to prove the special case that $R(X, X, Y, Z) = 0$. From the definition,

$$R(X, X, Y, Z) = g(X, R(Y, Z)X)$$
$$= g(X, \nabla_Y \nabla_Z X) - g(X, \nabla_Z \nabla_Y X) - g(X, \nabla_{[Y,Z]}X).$$

Using the axiom (D4) for the covariant derivative, we obtain

$$g(X, \nabla_Y \nabla_Z X) = Y g(X, \nabla_Z X) - g(\nabla_Y X, \nabla_Z X)$$

and with an analogous conversion for $g(X, \nabla_Z X)$, we get

$$g(X, \nabla_Z X) = \frac{1}{2} Z g(X, X).$$

After further use of (D4), we finally arrive at

$$R(X, X, Y, Z) = Y \frac{1}{2} Z g(X, X) - Z \frac{1}{2} Y g(X, X) - \frac{1}{2}[Y, Z] g(X, X) = 0.$$

The proof of (4) is a combination of (1), (2), (3). Using (1) and (2),

$$R(V, X, Y, Z) = -R(V, X, Z, Y) = R(V, Z, Y, X) + R(V, Y, X, Z),$$

and by (3) and (2)

$$R(V, X, Y, Z) = -R(X, V, Y, Z) = R(X, Y, Z, V) + R(X, Z, V, Y).$$

Adding these gives

$$2R(V, X, Y, Z) = R(V, Z, Y, X) + R(V, Y, X, Z)$$
$$+ R(X, Y, Z, V) + R(X, Z, V, Y)$$

and by swapping the vector fields also

$$2R(Y, Z, V, X) = R(Y, X, V, Z) + R(Y, V, Z, X)$$
$$+ R(Z, V, X, Y) + R(Z, X, Y, V).$$

By (1) and (3), the simultaneous swapping of the first two and the last two vectors does not change the value of the covariant curvature tensor, and so it can be seen that the four summands on the right hand side of the equation match pairwise with those on the right hand side of the second to last equation.

8.2   The Weingarten Map

In this section, we will learn about the Weingarten map to describe the curvature of a surface. However, the fundamental considerations also apply more generally to the case of a hypersurface, i.e., an $(n-1)$-dimensional submanifold of $\mathbb{R}^n$.

For a flat surface, the normal vector changes only in the sense of parallel displacement if the base point on the surface varies. In the case of a curved surface, a change in the direction of the normal vector can also be expected when the base point changes (Fig. 8.1). This effect can easily be quantified with the directional derivative.

Theorem 8.6 *Let M be an oriented $(n-1)$-dimensional submanifold of $\mathbb{R}^n$ with a normal vector field $\vec{n}$. Then for every tangent vector x, $D_x\vec{n}$ is also a tangent vector.*

Proof Crucially, we will use $\vec{n} \cdot \vec{n} = 1$. Differentiation using the tangent vector x gives

$$D_x\vec{n} \cdot \vec{n} + \vec{n} \cdot D_x\vec{n} = 0,$$

and so $D_x\vec{n} \cdot \vec{n} = 0$.

Definition 8.4 For M as above and $P \in M$, the **Weingarten map** W on M_P sends the tangent vector x to the tangent vector $D_x\vec{n}$, that is, $Wx = D_x\vec{n}$.   ◆

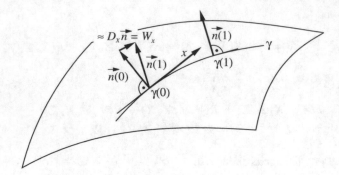

Fig. 8.1 Weingarten map W

The Weingarten map is obviously linear and therefore a $(1, 1)$-tensor at every point, giving altogether a $(1, 1)$-tensor field. With $WX = D_X\vec{n}$, the vector field X is mapped to the vector field WX.

We had seen that the covariant derivative is the orthogonal projection of the Euclidean directional derivative onto the tangential hyperplane (Theorem 7.6). The difference can now be formulated using the Weingarten map.

Theorem 8.7 *For vector fields X and Y on an oriented $(n-1)$-dimensional submanifold of $\mathbb{R}^n$,*

$$\nabla_X Y = D_X Y + (WX \cdot Y)\vec{n}.$$

Proof As the vector field $\nabla_X Y$ is tangential, we have that the coefficient λ in the decomposition $D_X Y = \nabla_X Y + \lambda\vec{n}$, is given by

$$\lambda = D_X Y \cdot \vec{n} = D_X(Y \cdot \vec{n}) - Y \cdot D_X\vec{n} = -WX \cdot Y.$$

Theorem 8.8 *The curvature operator is given in terms of the Weingarten map by*

$$R(Y, Z)X = (WZ \cdot X)WY - (WY \cdot X)WZ.$$

Proof We use the definition

$$R(Y, Z)X = \nabla_Y \nabla_Z X - \nabla_Z \nabla_Y X - \nabla_{[Y,Z]}X$$

and replace the covariant derivatives using Theorem 8.7. For the first summand we obtain

$$\begin{aligned}
\nabla_Y \nabla_Z X &= \nabla_Y(D_Z X + (WZ \cdot X)\vec{n}) = \\
&= D_Y(D_Z X + (WZ \cdot X)\vec{n}) + (WY \cdot (D_Z X + (WZ \cdot X)\vec{n}))\vec{n} \\
&= D_Y D_Z X + Y(WZ \cdot X)\vec{n} + (WZ \cdot X)WY + (WY \cdot D_Z X)\vec{n},
\end{aligned}$$

and analogously for the second summand

$$\nabla_Z \nabla_Y X = D_Z D_Y X + Z(WY \cdot X)\vec{n} + (WY \cdot X)WZ + (WZ \cdot D_Y X)\vec{n},$$

for the third summand, we have

$$\nabla_{[Y,Z]} X = D_{[Y,Z]} X + (W[Y, Z] \cdot X)\vec{n}.$$

As

$$D_Y D_Z X - D_Z D_Y X - D_{[Y,Z]} X = 0$$

(see the example following Definition 8.2), it remains to show that

$$Y(WZ \cdot X) + WY \cdot D_Z X - Z(WY \cdot X) - WZ \cdot D_Y X - W[Y, Z] \cdot X = 0.$$

But this will follow from

$$D_Y(WZ) - D_Z(WY) = W[Y, Z],$$

which is

$$D_Y D_Z \vec{n} - D_Z D_Y \vec{n} = D_{[Y,Z]} \vec{n}$$

and this is indeed true.

In classical differential geometry, the curvature of a surface is reduced to the curvature of curves that are created by intersecting the surface with planes orthogonal to the tangent plane. We now describe the curvature of such **normal sections** using the Weingarten map. A curve γ is said to be in its **natural parameter representation** if the parameter s is the length of the arc from a fixed point on the curve to the point $\gamma(s)$, and then the tangent vector $\gamma'(s)$ has length 1. The **unit tangent vector** $\gamma'(s)$ together with the second derivative $\gamma''(s)$ span the **osculating plane**. The **osculating circle** with **radius of curvature** R has as its tangent the unit tangent vector, and they lie in this plane (Fig. 8.2). The number $|\gamma''(s)|$ is called the **curvature** and is the reciprocal of the radius of curvature R. The vector $\gamma''(s)$ is orthogonal to the unit tangent vector $\gamma'(s)$ and points to the center of the osculating circle.

Theorem 8.9 *Let P be a point on the oriented surface M with normal vector field $\vec{n}$, $x \in M_P$ be such that $|x| = 1$ and γ be a normal section curve with $\gamma(0) = P$ and $\gamma'(0) = x$. The in the natural parameter representation*

$$\vec{n} \cdot \gamma''(0) = -Wx \cdot x.$$

Proof $\vec{n}(s)$ is the normal vector at the point $\gamma(s)$. As $\vec{n}(s) \cdot \gamma'(s) = 0$ it follows upon differentiation that

$$\vec{n}(0)\gamma''(0) = -\vec{n}'(0) \cdot \gamma'(0) = -D_x \vec{n} \cdot x = -Wx \cdot x.$$

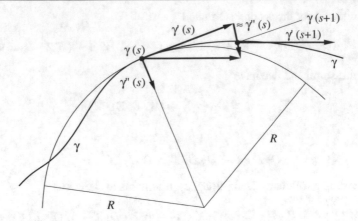

Fig. 8.2 The radius of curvature R and curvature $|\gamma''|$

Given the definition of the curvature of a curve, Theorem 8.9 has the following interpretation:

Theorem 8.10 *The curvature of a curve, resulting from the intersection of a surface with a plane orthogonal to the tangent plane, is given by $Wx \cdot x$, where x is one of the two tangent vectors of length 1, which lie in the section plane. If $Wx \cdot x < 0$, then the number $-1/Wx \cdot x$ is the radius of curvature, and the normal vector $\vec{n}$ of the surface points to the center of the osculating circle. If $Wx \cdot x > 0$, then $1/Wx \cdot x$ is the radius of curvature and $-\vec{n}$ points to the center.*

To calculate the expression $Wx \cdot x$ for the vector $x^1\partial_1 + x^2\partial_2$, we extend x to a vector field X with constant coefficients with respect to the coordinate vector fields ∂_1 and ∂_2. In the notation from Sect. 7.4, we then have

$$Wx \cdot x = D_x\vec{n} \cdot x = -\vec{n} \cdot D_x X = -\vec{n} \cdot \left[(x^1)^2 D_{\partial_1}\partial_1 + 2x^1x^2 D_{\partial_1}\partial_2 + (x^2)^2 D_{\partial_2}\partial_2 \right]$$

$$= -\left[L(x^1)^2 + 2Mx^1x^2 + N(x^2)^2 \right].$$

If we divide $Wx \cdot x$ by

$$x \cdot x = E(x^1)^2 + 2Fx^1x^2 + G(x^2)^2,$$

then one can also drop the additional condition $|x| = 1$. This results in the following theorem.

Theorem 8.11 *The radius of curvature of the normal section, which contains the vector $x = x^1\partial_1 + x^2\partial_2$, is given by*

$$\frac{1}{R} = \frac{L(x^1)^2 + 2Mx^1x^2 + N(x^2)^2}{E(x^1)^2 + 2Fx^1x^2 + G(x^2)^2},$$

where a negative radius R means that $-\vec{n}$ *points to the center of the osculating circle.*

In order to investigate the effect of the vector x on the curvature $1/R = -Wx \cdot x/(x \cdot x)$ of the corresponding normal section, we consider the eigenvalue problem for the Weingarten map. According to the following theorem, which is also valid in higher dimensional cases, one can rely on the major axis transformation.

Theorem 8.12 *The Weingarten map* W *is symmetric in the sense that* $Wx \cdot y = x \cdot Wy$ *for tangent vectors* x *and* y.

Proof For tangent vector fields X and Y with $X(P) = x$ and $Y(P) = y$, we have

$$Wx \cdot y = D_x\vec{n} \cdot y = -\vec{n} \cdot D_x Y$$

and

$$x \cdot Wy = x \cdot D_y\vec{n} = -D_y X \cdot \vec{n},$$

and for their difference we obtain

$$Wx \cdot y - x \cdot Wy = \vec{n} \cdot \left(D_Y X - D_X Y\right)(P) = -\vec{n} \cdot [X, Y](P) = 0,$$

since the Lie-bracket is again tangential.

For a point P on a surface M, the Weingarten map has real eigenvalues $\lambda_1 \leq \lambda_2$, which can also be the same. If $\lambda_1 < \lambda_2$, then we can use orthogonal vectors as the basis, and one can see that the expression $(Wx \cdot x)/(x \cdot x)$ is constrained by

$$\lambda_1 \leq \frac{Wx \cdot x}{x \cdot x} \leq \lambda_2$$

and assumes all values of the closed interval $[\lambda_1, \lambda_2]$. For the curvature of normal sections, this means that when the point is fixed, the curvatures lie between the two extreme curvature $1/R_1$ and $1/R_2$ (signed), called the **principal curvatures**. Their arithmetic mean $H = \frac{1}{2}(1/R_1 + 1/R_2)$ is called the **mean curvature**, and their product $K = (1/R_1) \cdot (1/R_2)$ is called the **Gaussian curvature**. These last two numbers can easily be calculated from the Weingarten map.

Theorem 8.13 *The Weingarten map has the matrix*

$$W = \frac{-1}{EG - F^2} \begin{pmatrix} G & -F \\ -F & E \end{pmatrix} \begin{pmatrix} L & M \\ M & N \end{pmatrix}.$$

Proof In the representation

$$W = \begin{pmatrix} w_1^1 & w_2^1 \\ w_1^2 & w_2^2 \end{pmatrix}$$

the elements in the first column have the meaning

$$W\partial_1 = w_1^1 \partial_1 + w_1^2 \partial_2 \,.$$

Scalar multiplication with ∂_1 gives

$$w_1^1 E + w_1^2 F = D_{\partial_1}\vec{n} \cdot \partial_1 = -\vec{n} \cdot D_{\partial_1}\partial_1 = -L$$

and with ∂_2 yields

$$w_1^1 F + w_1^2 G = -M \,.$$

Analogously considering

$$W\partial_2 = w_2^1 \partial_1 + w_2^2 \partial_2$$

imply two more equations, which can be summarised with the previous ones in the matrix equation

$$\begin{pmatrix} E & F \\ F & G \end{pmatrix} \begin{pmatrix} w_1^1 & w_2^1 \\ w_1^2 & w_2^2 \end{pmatrix} = \begin{pmatrix} -L & -M \\ -M & -N \end{pmatrix}$$

The assertion follows by premultiplying with the inverse of

$$\begin{pmatrix} E & F \\ F & G \end{pmatrix}$$

Theorem 8.14 *The mean curvature H and the Gaussian curvature K are given by*

$$H = \frac{EN - 2FM + GL}{2(EG - F^2)}$$

and

$$K = \frac{LN - M^2}{EG - F^2} \,.$$

Proof The number $-2H$ is the sum of the two eigenvalues of the matrix given in Theorem 8.13, i.e., its trace. K is the product of the two eigenvalues of $-W$, i.e., the determinant

$$\det(-W) = \frac{1}{(EG - F^2)^2} \begin{vmatrix} G & -F \\ -F & E \end{vmatrix} \begin{vmatrix} L & M \\ M & N \end{vmatrix} = \frac{LN - M^2}{EG - F^2} \,.$$

8.3   The Ricci Tensor

Definition 8.5 The **Ricci tensor** Ric is the contraction of the Riemann curvature tensor with respect to the middle vector variable, i.e., it has the components

$$\text{Ric}_{ik} = R^j_{ijk}.$$
◆

The vector $R(Z, Y)X(P)$ depends trilinearly on $X(P), Z(P), Y(P)$, and in particular, linearly on $Z(P)$. By the example following Definition 3.3, the number $\text{Ric}(X, Y)(P)$ is the trace of this linear map, and this trace depends linearly on $X(P)$ and $Y(P)$.

Theorem 8.15 *The Ricci tensor is symmetric.*

Proof The Ricci tensor can be linked to the covariant curvature tensor, and its symmetry (4) from Theorem 8.5 can be used, to obtain

$$\text{Ric}_{ik} = R^j_{ijk} = g^{jl} R_{lijk} = g^{jl} R_{jkli} = R^l_{kli} = \text{Ric}_{ki}.$$

Definition 8.6 The **mixed Ricci tensor**, again denoted by Ric, is derived from the Ricci tensor by pulling an index up, that is, it has the components

$$\text{Ric}^j_i = g^{jk}\text{Ric}_{ik}.$$
◆

The mixed Ricci tensor can be interpreted pointwise as a linear map: *The mixed Ricci tensor assigns to the vector field X the unique vector field Z representing the covector field* $\text{Ric}(X, \cdot)$, *in the sense that for all Y* $g(Z, Y) = \text{Ric}(X, Y)$. Because

$$g_{ik}Z^i Y^k = \text{Ric}_{jk}X^j Y^k$$

gives

$$g_{ik}Z^i = \text{Ric}_{kj}X^j,$$

and the components of Z can be isolated as follows:

$$Z^l = g^{lk}g_{ki}Z^i = g^{lk}\text{Ric}_{kj}X^j = \text{Ric}^l_j X^j.$$

Definition 8.7 The **scalar curvature** S is the trace of the mixed Ricci tensor, i.e.,

$$S = \text{Ric}^i_i.$$
◆

Theorem 8.16 *The scalar curvature of a surface is twice the Gaussian curvature, that is, $S = 2K$.*

Proof We use a chart for which the two coordinate vector fields ∂_1 and ∂_2 are orthogonal at the point under consideration. The sum

$$S = g^{ik} \mathrm{Ric}_{ik} = g^{ik} R^j_{ijk}$$

then gets reduced, since $g^{12} = g^{21} = 0$ and $g^{11} = g^{22} = 1$, to the four summands

$$S = R^1_{111} + R^2_{121} + R^1_{212} + R^2_{222}.$$

Owing to the skew-symmetry with respect to the second and third vector variables (Theorem 8.3), we have $R^1_{111} = R^2_{222} = 0$, and so $S = R^2_{121} + R^1_{212}$. The first summand is the second component of

$$R(\partial_2, \partial_1)\partial_1 = (W\partial_1 \cdot \partial_1)W\partial_2 - (W\partial_2 \cdot \partial_1)W\partial_1$$

(Theorem 8.8), and so

$$R^2_{121} = (W\partial_1 \cdot \partial_1)(W\partial_2 \cdot \partial_2) - (W\partial_2 \cdot \partial_1)(W\partial_1 \cdot \partial_2) = \det W.$$

Similarly

$$R^1_{212} = (W\partial_2 \cdot \partial_2)(W\partial_1 \cdot \partial_1) - (W\partial_1 \cdot \partial_2)(W\partial_2 \cdot \partial_1) = \det W,$$

so that altogether $S = 2 \det W$. The Gaussian curvature K is the product of the two eigenvalues of $-W$, and thus

$$K = \det(-W) = \det W.$$

The previous theorem has far-reaching consequences. While the Weingarten map describes the shape of the surface as a subset of space, other quantities, such as the numbers $E.F$ and G, and thus the length surrounding ambient space. Bending the surface does not change the length of curves on the surface. All concepts that are not influenced by bending belong to the inner **intrinsic geometry** of the surface. This also includes the covariant derivative and thus also the curvature tensor, Ricci tensor and scalar curvature. In contrast, the Weingarten map and also the principal curvatures belong to the outer **extrinsic geometry** of the surface. In his famous **theorema egregium**, Gauss formulated that the Gaussian curvature, which is the product of the pricipal curvatures, is bending-invariant, by expressing K in terms of E, F, G and their partial derivatives. The relationship $K = \frac{1}{2}S$ in the theorem above shows that K is bending-invariant since the scalar curvature S is intrinsic and bending-invariant.

Theorem 8.17 *The Gaussian curvature is bending-invariant.*

8.4   The Curvature of the Schwarzschild Spacetime

In Sect. 7.5, we had determined the covariant derivative for the coordinate vector fields $\partial_0 = \partial_t$, $\partial_1 = \partial_r$, $\partial_2 = \partial_\vartheta$, $\partial_3 = \partial_\varphi$. Now, using these results, we will compute the matrices describing the curvature operators. Since the curvature operator depends bilinearly and skew-symmetrically on the two vector fields, only the six matrices (R^k_{i01}), (R^k_{i02}), (R^k_{i03}), (R^k_{i12}), (R^k_{i13}) and (R^k_{i23}) need to be determined. For ∂_j and ∂_l the curvature operator $R(\partial_j, \partial_l)$ is, according to Definition 8.1, characterised by

$$R(\partial_j, \partial_l)\partial_i = \nabla_{\partial_j}\nabla_{\partial_l}\partial_i - \nabla_{\partial_l}\nabla_{\partial_j}\partial_i \,.$$

The four components of this vector build the ith column of the matrix $(R^k_{ijl})_{ki}$. The 0th column of (R^k_{i01}) consists of the components of the vector

$$\nabla_{\partial_0}\nabla_{\partial_1}\partial_0 - \nabla_{\partial_1}\nabla_{\partial_0}\partial_0 = (\nabla_{\partial_t}\nabla_{\partial_r} - \nabla_{\partial_r}\nabla_{\partial_t})\partial_t$$

$$= \nabla_{\partial_t}\left(\frac{\mu}{r^2 h(r)}\partial_t\right) - \nabla_{\partial_r}\left(\frac{\mu h(r)}{r^2}\partial_r\right)$$

$$= \frac{\mu}{r^2 h(r)} \cdot \frac{\mu h(r)}{r^2}\partial_r - \left(\frac{6\mu^2}{r^4} - \frac{2\mu}{r^3}\right)\partial_r - \frac{\mu h(r)}{r^2}\left(-\frac{\mu}{r^2 h(r)}\right)\partial_r = \frac{2\mu h(r)}{r^3}\partial_r \,,$$

which are the numbers 0, $2\mu h(r)/r^3$, 0, 0. From

$$(\nabla_{\partial_t}\nabla_{\partial_r} - \nabla_{\partial_r}\nabla_{\partial_t})\partial_r = \frac{2\mu}{r^3 h(r)}\partial_t$$

$$(\nabla_{\partial_t}\nabla_{\partial_r} - \nabla_{\partial_r}\nabla_{\partial_t})\partial_\vartheta = 0$$

$$(\nabla_{\partial_t}\nabla_{\partial_r} - \nabla_{\partial_r}\nabla_{\partial_t})\partial_\varphi = 0$$

the other three columns are obtained. The matrix is

$$(R^k_{i01}) = \begin{pmatrix} 0 & \dfrac{2\mu}{r^3 h(r)} & 0 & 0 \\ \dfrac{2\mu h(r)}{r^3} & 0 & 0 & 0 \\ 0 & 0 & 0 & 0 \\ 0 & 0 & 0 & 0 \end{pmatrix} = -(R^k_{i10}) \,.$$

The other five matrices are calculated in an analogous manner

$$(R^k_{i02}) = \begin{pmatrix} 0 & 0 & -\dfrac{\mu}{r} & 0 \\ 0 & 0 & 0 & 0 \\ -\dfrac{\mu h(r)}{r^3} & 0 & 0 & 0 \\ 0 & 0 & 0 & 0 \end{pmatrix}$$

$$(R^k_{i03}) = \begin{pmatrix} 0 & 0 & 0 & -\dfrac{\mu \sin^2 \vartheta}{r} \\[2ex] 0 & 0 & 0 & 0 \\[2ex] 0 & 0 & 0 & 0 \\[2ex] -\dfrac{\mu h(r)}{r^3} & 0 & 0 & 0 \end{pmatrix}$$

$$(R^k_{i12}) = \begin{pmatrix} 0 & 0 & 0 & 0 \\[2ex] 0 & 0 & -\dfrac{\mu}{r} & 0 \\[2ex] 0 & \dfrac{\mu}{r^3 h(r)} & 0 & 0 \\[2ex] 0 & 0 & 0 & 0 \end{pmatrix}$$

$$(R^k_{i13}) = \begin{pmatrix} 0 & 0 & 0 & 0 \\[2ex] 0 & 0 & 0 & -\dfrac{\mu \sin^2 \vartheta}{r} \\[2ex] 0 & 0 & 0 & 0 \\[2ex] 0 & \dfrac{\mu}{r^3 h(r)} & 0 & 0 \end{pmatrix}$$

$$(R^k_{i23}) = \begin{pmatrix} 0 & 0 & 0 & 0 \\[2ex] 0 & 0 & 0 & 0 \\[2ex] 0 & 0 & 0 & \dfrac{2\mu \sin^2 \vartheta}{r} \\[2ex] 0 & 0 & -\dfrac{2\mu}{r} & 0 \end{pmatrix}$$

The calculation of the components of the Ricci tensor, according to Definition 8.5, results in the zero matrix.

8.5   Connection Forms and Curvature Forms

The calculation of the curvature tensor of an n-dimensional semi-Riemannian manifold with its n^4 components is formidable organisational problem. The so-called **Cartan-calculus** of forms offers an alternative for effective and clear computation.

According to the definition of the Christoffel symbols (Definition 7.3), we have

$$\nabla_{X^i \partial_i} \partial_j = X^i \Gamma^k_{ij} \partial_k .$$

The components of the vector field $\nabla_X \partial_j$ at the point P depend linearly on X and so they are 1-forms.

Definition 8.8 The n^2 1-forms

$$\omega_j^k(X) = X^i \Gamma_{ij}^k$$

are called **connection forms**.   ◆

Theorem 8.18 *The curvature operator is determined by the connection forms via*

$$R(Y, Z)X = X^k(d\omega_k^j(Y, Z) + \omega_l^j \wedge \omega_k^l(Y, Z))\partial_j \,.$$

Proof We have

$$R(Y, Z)X = \nabla_Y \nabla_Z X - \nabla_Z \nabla_Y X - \nabla_{[Y,Z]} X$$

$$= \nabla_Y((ZX^j + X^k \omega_k^j(Z))\partial_j) - \nabla_Z((YX^j + X^k \omega_k^j(Y))\partial_j)$$
$$\quad - [Y, Z]X^j \partial_j - X^k \omega_k^j([Y, Z])\partial_j$$

$$= Y(ZX^j + X^k \omega_k^j(Z))\partial_j + (ZX^l + X^k \omega_k^l(Z))\omega_l^j(Y)\partial_j$$
$$\quad - Z(YX^j + X^k \omega_k^j(Y))\partial_j - (YX^l + X^k \omega_k^l(Y))\omega_l^j(Z)\partial_j$$
$$\quad - [Y, Z]X^j \partial_j - X^k \omega_k^j([Y, Z])\partial_j$$

$$= \{(YZ - ZY - [Y, Z])X^j + X^k(Y\omega_k^j(Z) - Z\omega_k^j(Y) - \omega_k^j([Y, Z]))$$
$$\quad + (YX^k)\omega_k^j(Z) - (YX^l)\omega_l^j(Z) - (ZX^k)\omega_k^j(Y) + (ZX^l)\omega_l^j(Y)$$
$$\quad + X^k(\omega_k^l(Z)\omega_l^j(Y) - \omega_k^l(Y)\omega_l^j(Z))\}\partial_j$$

$$= \{X^k d\omega_k^j(Y, Z) + X^k(\omega_l^j \wedge \omega_k^l(Y, Z))\}\partial_j \,.$$

Definition 8.9 The n^2 2-forms

$$\Omega_k^j = d\omega_k^j + \omega_l^j \wedge \omega_k^l$$

are called **curvature forms**.   ◆

Theorem 8.19 *The curvature tensor R can be obtained from the curvature forms Ω_j^i via*

$$R = \partial_i \otimes \omega^j \otimes \Omega_j^i \,,$$

where the forms $\omega^1, \ldots, \omega^n$ are dual to the coordinate vector fields $\partial_1, \ldots, \partial_n$.

Proof For a covector field A and vector fields X, Y, Z, we have

$$\partial_i \otimes \omega^j \otimes \Omega^i_j(A, X, Y, Z) = A_i X^j \Omega^i_j(Y, Z),$$

and by Theorem 8.18, we obtain for the curvature tensor that

$$R(A, X, Y, Z) = \langle R(Y, Z)X, A \rangle = \langle X^j \Omega^i_j(Y, Z)\partial_i, A_l\omega^l \rangle = X^j \Omega^i_j(Y, Z)A_i.$$

We now illustrate this method doing calculations for the Einstein-de Sitter space-time introduced in Example 4 of Sect. 4.5. On the set

$$M = (0, \infty) \times \mathbb{R}^3 = \{(u^0, u^1, u^2, u^3) \in \mathbb{R}^4 : u^0 > 0\},$$

we have the metric g with the matrix of components given by

$$(g_{ik}) = \begin{pmatrix} 1 & 0 & 0 & 0 \\ 0 & -c(u^0)^{\frac{4}{3}} & 0 & 0 \\ 0 & 0 & -c(u^0)^{\frac{4}{3}} & 0 \\ 0 & 0 & 0 & -c(u^0)^{\frac{4}{3}} \end{pmatrix}$$

with respect to the basis $\dfrac{\partial}{\partial u^0}, \dfrac{\partial}{\partial u^1}, \dfrac{\partial}{\partial u^2}, \dfrac{\partial}{\partial u^3}$. The contravariant metric is given by the matrix

$$(g^{ik}) = \begin{pmatrix} 1 & 0 & 0 & 0 \\ 0 & -c^{-1}(u^0)^{-\frac{4}{3}} & 0 & 0 \\ 0 & 0 & -c^{-1}(u^0)^{-\frac{4}{3}} & 0 \\ 0 & 0 & 0 & -c^{-1}(u^0)^{-\frac{4}{3}} \end{pmatrix},$$

with respect to the dual basis $\omega^0, \omega^1, \omega^2, \omega^3$. Because of the diagonal structure of these matrices, the formula for computing the Christoffel symbols is simplified, as already mentioned in Sect. 7.5, given by

$$\Gamma^j_{ij} = \frac{1}{2} g^{jj} \partial_i g_{jj} \quad \text{and} \quad \Gamma^k_{ii} = -\frac{1}{2} g^{kk} \partial_k g_{ii} \quad \text{for } i \neq k.$$

In particular, we obtain here, for $j = 1, 2, 3$, that

$$\Gamma^j_{0j} = \Gamma^j_{j0} = \frac{2}{3} \cdot \frac{1}{u^0}$$

and furthermore

$$\Gamma_{11}^0 = \Gamma_{22}^0 = \Gamma_{33}^0 = \frac{2}{3}c(u^0)^{\frac{1}{3}}.$$

All other Christoffel symbols are zero. For the connection forms this implies, according to their definition, that for $j = 1, 2, 3$

$$\omega_0^j(\partial_j) = \omega_j^j(\partial_0) = \frac{2}{3u^0} \quad \text{and} \quad \omega_j^0(\partial_j) = \frac{2}{3}c(u^0)^{\frac{1}{3}},$$

and the connection forms are zero on the other basis elements. It can be seen that the seven connection forms ω_0^0 and ω_i^k with $i, k \neq 0$ and $i \neq k$ are zero and the remaining nine are multiples of the dual basis elements $\omega^0, \omega^1, \omega^2, \omega^3$. We have

$$\omega_j^j = \frac{2}{3u^0}\omega^0 \quad \text{for} \quad j = 1, 2, 3$$

$$\omega_0^k = \frac{2}{3u^0}\omega^k \quad \text{for} \quad k = 1, 2, 3$$

$$\omega_i^0 = \frac{2}{3}c(u^0)^{\frac{1}{3}}\omega^i \quad \text{for} \quad i = 1, 2, 3.$$

The outer derivatives are given by

$$d\omega_j^j = 0$$

$$d\omega_0^k = -\frac{2}{3(u^0)^2}\omega^0 \wedge \omega^k$$

$$d\omega_i^0 = \frac{2c}{9(u^0)^{\frac{2}{3}}}\omega^0 \wedge \omega^i.$$

In accordace with Definition 8.9, the curvature forms Ω_k^j can now be determined. We have for $j, k = 1, 2, 3$

$$\Omega_0^0 = \Omega_k^k = 0$$

$$\Omega_0^j = d\omega_0^j + \omega_j^j \wedge \omega_0^j = \left(-\frac{2}{3(u^0)^2} + \left(\frac{2}{3u^0}\right)^2\right)\omega^0 \wedge \omega^j = -\frac{2}{9}\frac{1}{(u^0)^2}\omega^0 \wedge \omega^j$$

$$\Omega_k^0 = d\omega_k^0 + \omega_k^0 \wedge \omega_k^k = \left(\frac{2c}{9(u^0)^{\frac{2}{3}}} - \frac{2}{3}c(u^0)^{\frac{1}{3}} \cdot \frac{2}{3u^0}\right)\omega^0 \wedge \omega^k = -\frac{2}{9}\frac{c}{(u^0)^{\frac{2}{3}}}\omega^0 \wedge \omega^k$$

and for $j \neq k$,

$$\Omega_k^j = \omega_0^j \wedge \omega_k^0 = \frac{4}{9}\frac{c}{(u^0)^{\frac{2}{3}}}\omega^j \wedge \omega^k = -\Omega_j^k.$$

The curvature tensor $R = \partial_i \otimes \omega^j \otimes \Omega^i_j$ can now be calculated. We have

$$
\begin{aligned}
R =\ & \partial_0 \otimes \omega^1 \otimes \Omega^0_1 + \partial_0 \otimes \omega^2 \otimes \Omega^0_2 + \partial_0 \otimes \omega^3 \otimes \Omega^0_3 \\
& + \partial_1 \otimes \omega^0 \otimes \Omega^1_0 + \partial_2 \otimes \omega^0 \otimes \Omega^2_0 + \partial_3 \otimes \omega^0 \otimes \Omega^3_0 \\
& + (\partial_1 \otimes \omega^2 - \partial_2 \otimes \omega^1) \otimes \Omega^1_2 + (\partial_1 \otimes \omega^3 - \partial_3 \otimes \omega^1) \otimes \Omega^1_3 \\
& + (\partial_2 \otimes \omega^3 - \partial_3 \otimes \omega^2) \otimes \Omega^2_3 \\
=\ & -\frac{2}{9} \frac{c}{(u^0)^{\frac{2}{3}}} A_1 - \frac{2}{9} \frac{1}{(u^0)^2} A_2 + \frac{4}{9} \frac{c}{(u^0)^{\frac{2}{3}}} A_3
\end{aligned}
$$

where

$$
\begin{aligned}
A_1 =\ & \partial_0 \otimes \omega^1 \otimes (\omega^0 \otimes \omega^1 - \omega^1 \otimes \omega^0) \\
& + \partial_0 \otimes \omega^2 \otimes (\omega^0 \otimes \omega^2 - \omega^2 \otimes \omega^0) \\
& + \partial_0 \otimes \omega^3 \otimes (\omega^0 \otimes \omega^3 - \omega^3 \otimes \omega^0) \\
A_2 =\ & \partial_1 \otimes \omega^0 \otimes (\omega^0 \otimes \omega^1 - \omega^1 \otimes \omega^0) \\
& + \partial_2 \otimes \omega^0 \otimes (\omega^0 \otimes \omega^2 - \omega^2 \otimes \omega^0) \\
& + \partial_3 \otimes \omega^0 \otimes (\omega^0 \otimes \omega^3 - \omega^3 \otimes \omega^0) \\
A_3 =\ & (\partial_1 \otimes \omega^2 - \partial_2 \otimes \omega^1) \otimes (\omega^1 \otimes \omega^2 - \omega^2 \otimes \omega^1) \\
& + (\partial_1 \otimes \omega^3 - \partial_3 \otimes \omega^1) \otimes (\omega^1 \otimes \omega^3 - \omega^3 \otimes \omega^1) \\
& + (\partial_2 \otimes \omega^3 - \partial_3 \otimes \omega^2) \otimes (\omega^2 \otimes \omega^3 - \omega^3 \otimes \omega^2) \ .
\end{aligned}
$$

The Ricci tensor is obtained by a contraction, and is given by

$$
\mathrm{Ric} = -\frac{2}{9} \frac{c}{(u^0)^{\frac{2}{3}}} K_1 - \frac{2}{9} \frac{1}{(u^0)^2} K_2 + \frac{4}{9} \frac{c}{(u^0)^{\frac{2}{3}}} K_3
$$

where

$$
\begin{aligned}
K_1 &= \omega^1 \otimes \omega^1 + \omega^2 \otimes \omega^2 + \omega^3 \otimes \omega^3 \\
K_2 &= -3\omega^0 \otimes \omega^0 \\
K_3 &= 2K_1 \,,
\end{aligned}
$$

and so

$$
\mathrm{Ric} = \frac{2}{3} \frac{1}{(u^0)^2} \omega^0 \otimes \omega^0 + \frac{2}{3} \frac{c}{(u^0)^{\frac{2}{3}}} \left(\omega^1 \otimes \omega^1 + \omega^2 \otimes \omega^2 + \omega^3 \otimes \omega^3 \right)
$$

with the component matrix

$$\left(\mathrm{Ric}_{ik}\right) = \begin{pmatrix} \dfrac{2}{3}\dfrac{1}{(u^0)^2} & 0 & 0 & 0 \\[2ex] 0 & \dfrac{2}{3}\dfrac{c}{(u^0)^{\frac{2}{3}}} & 0 & 0 \\[2ex] 0 & 0 & \dfrac{2}{3}\dfrac{c}{(u^0)^{\frac{2}{3}}} & 0 \\[2ex] 0 & 0 & 0 & \dfrac{2}{3}\dfrac{c}{(u^0)^{\frac{2}{3}}} \end{pmatrix}.$$

The mixed Ricci tensor has the component matrix

$$\left(\mathrm{Ric}_{k}^{i}\right) = \begin{pmatrix} \dfrac{2}{3}\dfrac{1}{(u^0)^2} & 0 & 0 & 0 \\[2ex] 0 & -\dfrac{2}{3}\dfrac{1}{(u^0)^2} & 0 & 0 \\[2ex] 0 & 0 & -\dfrac{2}{3}\dfrac{1}{(u^0)^2} & 0 \\[2ex] 0 & 0 & 0 & -\dfrac{2}{3}\dfrac{1}{(u^0)^2} \end{pmatrix}$$

with the trace

$$S = -\frac{4}{3}\frac{1}{(u^0)^2}.$$

In the next chapter, we will introduce the Einstein tensor $G = \mathrm{Ric} - \frac{1}{2}Sg$, in order to formulate the Einstein field equations. For the Einstein-de Sitter spacetime its component matrix is given by

$$\left(G_{ik}\right) = \begin{pmatrix} \dfrac{4}{3}\dfrac{1}{(u^0)^2} & 0 & 0 & 0 \\[1ex] 0 & 0 & 0 & 0 \\[1ex] 0 & 0 & 0 & 0 \\[1ex] 0 & 0 & 0 & 0 \end{pmatrix}.$$

The connection forms ω_j^k in Definition 8.8 refer to the coordinate vector fields $\partial_1, ..., \partial_n$ and use the Christoffel symbols, which are sometimes cumbersome to calculate. The concept can also be used more generally and then refers to vector fields $X_1, ..., X_n$, which form bases in the tangent spaces.

Definition 8.10 Associated with a basis $X_1, ..., X_n$, the **connection forms** ω_j^k are defined by

$$\nabla_x X_j(P) = \langle x, \omega_j^k(P) \rangle X_k(P) \quad \text{for} \quad x \in M_P .$$

$\blacklozenge$

For a vector field X, we have

$$\nabla_X X_j = \langle X, \omega_j^k \rangle X_k .$$

The question now is how to calculate the connection forms without Christoffel symbols. We will see that this can be done quite comfortably for semi-orthonormal bases.

Definition 8.11 Vector fields $X_1, ..., X_n$ on an n-dimensional semi-Riemannian manifold are said to form a **semi-orthonormal basis**, if

$$g(X_i, X_j) = \begin{cases} 0 & \text{for } i \neq j \\ \varepsilon_i = \pm 1 & \text{for } i = j \end{cases}$$

$\blacklozenge$

The basis dual to $X_1, ..., X_n$ consists of covector fields $A^1, ..., A^n$ with

$$\langle X_i, A^k \rangle = \begin{cases} 0 & \text{for } i \neq k \\ 1 & \text{for } i = k \end{cases}$$

and

$$g(A^i, A^j) = \begin{cases} 0 & \text{for } i \neq j \\ \varepsilon_i & \text{for } i = j \end{cases} .$$

Theorem 8.20 *For the connection forms* ω_j^k *generated by a semi-orthonormal basis* $X_1, ..., X_n$ *with the dual basis* $A^1, ..., A^n$, *there holds*

(1) $\omega_j^k = -\varepsilon_j \varepsilon_k \omega_k^j$
(2) $dA^i = -\omega_k^i \wedge A^k$

Proof For any vector field X, we have

$$0 = Xg(X_i, X_j) = g(\nabla_X X_i, X_j) + g(X_i, \nabla_X X_j)$$

$$= \langle X, \omega_i^k \rangle g(X_k, X_j) + \langle X, \omega_j^l \rangle g(X_i, X_l) = \langle X, \omega_i^j \rangle \varepsilon_j + \langle X, \omega_j^i \rangle \varepsilon_i ,$$

which gives (1). We show (2) by verifying

$$dA^i(X_j, X_l) = -\omega_k^i \wedge A^k(X_j, X_l) .$$

The left hand side is

$$dA^i(X_j, X_l) = X_j\langle X_l, A^i\rangle - X_l\langle X_j, A^i\rangle - \langle[X_j, X_l], A^i\rangle = \langle[X_l, X_j], A^i\rangle$$

and the right hand side is also

$$-\omega_k^i \wedge A^k(X_j, X_l) = -\langle X_j, \omega_k^i\rangle\langle X_l, A^k\rangle + \langle X_l, \omega_k^i\rangle\langle X_j, A^k\rangle$$

$$= -\langle X_j, \omega_l^i\rangle + \langle X_l, \omega_j^i\rangle = -\langle\nabla_{X_j}X_l, A^i\rangle + \langle\nabla_{X_l}X_j, A^i\rangle = \langle[X_l, X_j], A^i\rangle .$$

Theorem 8.21 *The connection forms ω_j^i associated with a semi-orthonormal basis $X_1, ..., X_n$ with the dual basis $A^1, ..., A^n$ are calculated via*

$$2\langle X_l, \omega_j^i\rangle = dA^i(X_j, X_l) + \varepsilon_i\varepsilon_j dA^j(X_l, X_i) + \varepsilon_i\varepsilon_l dA^l(X_j, X_i)$$

$$= -\langle[X_j, X_l], A^i\rangle - \varepsilon_i\varepsilon_j\langle[X_l, X_i], A^j\rangle - \varepsilon_i\varepsilon_l\langle[X_j, X_i], A^l\rangle .$$

Proof According to Theorem 8.20(2), we have

$$dA^i(X_j, X_l) = -\langle X_j, \omega_k^i\rangle\langle X_l, A^k\rangle + \langle X_l, \omega_k^i\rangle\langle X_j, A^k\rangle = -\langle X_j, \omega_l^i\rangle + \langle X_l, \omega_j^i\rangle .$$

We exchange i, j, l cyclically and obtain by using Theorem 8.20(1) that

$$dA^j(X_l, X_i) = -\langle X_l, \omega_i^j\rangle + \langle X_i, \omega_l^j\rangle = \varepsilon_i\varepsilon_j\langle X_l, \omega_j^i\rangle + \langle X_i, \omega_l^j\rangle$$

and

$$dA^l(X_i, X_j) = -\langle X_i, \omega_j^l\rangle + \langle X_j, \omega_i^l\rangle = \varepsilon_j\varepsilon_l\langle X_i, \omega_l^j\rangle - \varepsilon_i\varepsilon_l\langle X_j, \omega_l^i\rangle .$$

The linear combination of these three equations with the coefficients 1, $\varepsilon_i\varepsilon_j$, $-\varepsilon_i\varepsilon_l$ is the claimed equation. The alternative formulation is based on the calculation rules for the outer derivative.

From the connection forms ω_j^k, the curvature forms are given, in accordance with Definition 8.9 by

$$\Omega_j^k = d\omega_k^j + \omega_l^j \wedge \omega_k^l .$$

The curvature tensor R can be described with reference to the semi-orthonormal basis $X_1, ..., X_n$ and its dual basis $A^1, ..., A^n$, by

$$R = X_i \otimes A^j \otimes \Omega_j^i$$

with the components

$$R_{ijk}^l = \langle X_u, A^l\rangle\langle X_i, A^v\rangle\Omega_v^u(X_j, X_k) = \Omega_i^l(X_j, X_k) .$$

The Ricci tensor has the components

$$\mathrm{Ric}_{ik} = \Omega_i^j(X_j, X_k) \ .$$

The computation of all the n^2 curvature forms is not necessary, since it follows from Theorem 8.20(1) that

$$\Omega_i^j = -\varepsilon_i \varepsilon_j \Omega_j^i \ .$$

Matter

9

9.1 Mass

We now clarify how mass distribution in spacetime is described. Whether the mass rests or flows depends on the observer. They register a flow velocity, which based on the point of view formulated in Sect. 5.1, must be a future-pointing, time-like tangent vector z with $g(z, z) = 1$. There is such a vector at every point, i.e., it is a vector field Z with $g(Z, Z) = 1$ and such that $Z(P)$ is future-pointing. An intensity in the sense of **density** and an isotropic **pressure**, described by scalar fields ϱ and p, must also be taken into account. Since the mass in particular depends on the observer, this is also to be expected from the density. The number $\varrho(P)$ should be the density that the observer moving in the flow Z registers as observer $z = Z(P)$. However, other interactions such as viscosity and temperature exchange should not play a role. Overall, these considerations motivate why one thinks of the triple $[Z, \varrho, p]$ as an **ideal flow**. One expects that there are further relationships between X, ϱ and p, analogous to the basic equations of hydrodynamics in Newtonian mechanics. We can formulate such laws comfortably only in Sect. 11.5.

As already mentioned, the density being the quotient of mass by volume, depends on the observer. How the mass is obtained has already been defined in Definition 5.6. The volume is influenced by contraction in length (see the introduction). The volume measurement in the tangent space, introduced below, takes this into account.

Definition 9.1 For an observer x, three space-like vectors v_1, v_2, v_3 span a parallelepiped having the **volume** $V(x, v_1, v_2, v_3)$, where V is the volume form (Definition 6.4) and the vectors v_1, v_2, v_3 are arranged to form a right-handed system. ♦

© The Author(s), under exclusive license to Springer Nature Switzerland AG 2023
R. Oloff, *The Geometry of Spacetime*, Graduate Texts in Physics,
https://doi.org/10.1007/978-3-031-16139-1_9

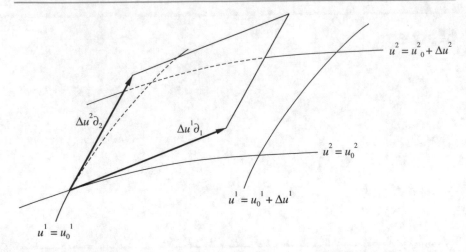

$u^2 = u^2_0 + \Delta u^2$

$\Delta u^2 \partial_2$ $\Delta u^1 \partial_1$

$u^2 = u^2_0$

$u^1 = u^1_0 + \Delta u^1$

$u^1 = u^1_0$

Fig. 9.1 Calculation of volume

If $v_i \in x^\perp$, then $V(x, v_1, v_2, v_3)$ is the Euclidean volume induced by the metric $-g$, because both $V(x, ., ., .)$ and the Euclidean volume depend trilinearly on the three edge vectors v_1, v_2, v_3 and give the value 1 for an orthonormal system.

   In order to calculate the density of the flow to be measured by an observer x, the three pairs of coordinate planes can be identified in the $x^\perp$-space with the corresponding parallelepiped in the tangent space. Figure 9.1 shows the analogue in one reduced dimension.

   The observer x registers a flow velocity v, and so

$$\frac{x + v}{\sqrt{1 + g(v, v)}} = z .$$

The linearly independent space-like vectors $v_1, v_2, v_3 \in x^\perp$ are arranged as a right-handed system. For z the parallelepiped spanned by these vectors has the volume

$$V(z, v_1, v_2, v_3) = V\left(\frac{x + v}{\sqrt{1 + g(v, v)}}, v_1, v_2, v_3\right) = \frac{V(x, v_1, v_2, v_3)}{\sqrt{1 + g(v, v)}} .$$

In there exists matter, which for z has the mass $\varrho V(z, v_1, v_2, v_3)$. For the observer x, the perceived mass, according to Definition 5.6, is given by

$$\frac{\varrho V(z, v_1, v_2, v_3)}{\sqrt{1 + g(v, v)}} = \frac{\varrho V(x, v_1, v_2, v_3)}{1 + g(v, v)} .$$

The observer x thus registers the density

$$\frac{\varrho}{1 + g(v, v)} = \varrho(g(x, z))^2 .$$

9.2   Energy and Momentum of a Flow

In non-relativistic hydrodynamics, a flow is described by the position- and time-dependent **flow vector** $v = (v^1, v^2, v^3)$. Here we use an orthonormal basis e_1, e_2, e_3. Also relevant is the **pressure** p, which also depends on the location and on time. The **density** ϱ is considered to be constant in the case of an incompressible liquid. ϱv is then the so-called **current density vector**. Here we ignore friction, heat flow and gravity.

In the general case of variable density, one has that in a region Ω

$$\iint_{\partial\Omega} \varrho v \, d\vec{o} = \iiint_{\Omega} \operatorname{div}(\varrho v) \, dx$$

and so the rate of change of mass content of Ω is given by

$$\iiint_{\Omega} \left(-\frac{\partial \varrho}{\partial t}\right) dx = -\frac{d}{dt} \iiint_{\Omega} \varrho \, dx = \iiint_{\Omega} \operatorname{div}(\varrho v) \, dx .$$

This gives rise to the **continuity equation**

$$\frac{\partial \varrho}{\partial t} + \operatorname{div}(\varrho v) = 0$$

and in particular for incompressible fluids, $\operatorname{div} v = 0$.

The force of a substance in a region Ω due to the pressure p is

$$-\iint_{\partial\Omega} p \, d\vec{o} = -\left(\iint_{\partial\Omega} p e_1 \, d\vec{o}, \iint_{\partial\Omega} p e_2 \, d\vec{o}, \iint_{\partial\Omega} p e_3 \, d\vec{o}\right)$$

$$= -\left(\iiint_{\Omega} \frac{\partial p}{\partial x^1} dx, \iiint_{\Omega} \frac{\partial p}{\partial x^2} dx, \iiint_{\Omega} \frac{\partial p}{\partial x^3} dx\right) = -\iiint_{\Omega} \operatorname{grad} p \, dx ,$$

which causes an acceleration determined by

$$\iiint_{\Omega} \varrho \frac{dv}{dt} dx = -\iiint_{\Omega} \operatorname{grad} p \, dx .$$

Here the derivative $\frac{dv}{dt}$ is understood in the sense

$$\frac{d}{dt} v\left(t, x^1(t), x^2(t), x^3(t)\right) = \frac{\partial v}{\partial t} + \operatorname{grad} v \cdot v$$

with

$$\operatorname{grad} v = \begin{pmatrix} \operatorname{grad} v^1 \\ \operatorname{grad} v^2 \\ \operatorname{grad} v^3 \end{pmatrix} = \begin{pmatrix} \dfrac{\partial v^1}{\partial x^1} & \dfrac{\partial v^1}{\partial x^2} & \dfrac{\partial v^1}{\partial x^3} \\[2ex] \dfrac{\partial v^2}{\partial x^1} & \dfrac{\partial v^2}{\partial x^2} & \dfrac{\partial v^2}{\partial x^3} \\[2ex] \dfrac{\partial v^3}{\partial x^1} & \dfrac{\partial v^3}{\partial x^2} & \dfrac{\partial v^3}{\partial x^3} \end{pmatrix} = \dfrac{\partial v}{\partial x} \; .$$

Equating the integrands yields the well-known **Euler equation**

$$\frac{\partial v}{\partial t} + \frac{\partial v}{\partial x} v + \frac{1}{\varrho} \operatorname{grad} p = 0 \; .$$

A flow has energy content, and is described by an energy density, which is composed of two summands. One component is $\varrho|v|^2/2$, the kinetic energy of the flowing substance, and the other summand $\varepsilon\varrho$ describes the internal energy, contained in the molecular movements. Here ε is the internal energy per unit mass.

The part of the energy current density determined by the summand $\varepsilon\varrho$ is obviously $\varepsilon\varrho v$. The part s, corresponding to $\varrho|v|^2/2$, of the energy current density is determined by the balance equation

$$-\frac{d}{dt} \iiint\limits_{\Omega} \frac{\varrho|v|^2}{2} \, dx = \iint\limits_{\partial\Omega} s \, d\vec{o}$$

and so is characterised by

$$-\frac{\partial}{\partial t} \frac{\varrho|v|^2}{2} = \operatorname{div} s \; .$$

To calculate s, we put in

$$-\frac{\partial}{\partial t} \frac{\varrho|v|^2}{2} = -\varrho v \cdot \frac{\partial v}{\partial t}$$

in the Euler equation and use the identity

$$\left(\frac{\partial v}{\partial x} \right)^T v = \operatorname{grad} \frac{|v|^2}{2} \; ,$$

resulting in the following representation in the case of an incompressible fluid

$$-\frac{\partial}{\partial t} \frac{\varrho|v|^2}{2} = \varrho v \cdot \left(\frac{\partial v}{\partial x} v + \frac{1}{\varrho} \operatorname{grad} p \right) = \varrho \left(\frac{\partial v}{\partial x} \right)^T v \cdot v + v \cdot \operatorname{grad} p$$

$$= \varrho v \cdot \operatorname{grad} \frac{|v|^2}{2} + \operatorname{div}(pv) = \operatorname{div} \left(\left(\frac{\varrho|v|^2}{2} + p \right) v \right) \; .$$

Thus we read off

$$s = \left(\frac{\varrho |v|^2}{2} + p \right) v.$$

All in all, the **energy density** $(|v|^2/2 + \varepsilon)\varrho$ has resulted in the **energy current density** $(\varrho |v|^2/2 + p + \varepsilon \varrho)v$.

This is so far the Newtonian view of energy density and energy current density. In the theory of relativity, mass is understood as energy (see Sect.5.2). Consequently, the number ϱ has to be added to the energy density. Accordingly, the summand ϱv has to be added to the energy current density. Finally, for large speeds v, the change in volume, as explained in the previous section, must also be taken into account. We will come back to this in the following section.

Now we turn to the Newtonian concepts of momentum density and momentum current density. Taking into account the contribution of each individual particle, the flow in the region Ω has the momentum

$$\iiint_\Omega \varrho v \, dx = \left(\iiint_\Omega \varrho v^1 \, dx, \iiint_\Omega \varrho v^2 \, dx, \iiint_\Omega \varrho v^3 \, dx \right),$$

and the **momentum density** is thus ϱv. Its rate of change with respect to time can be formulated as the following, using the Euler equation

$$\frac{d}{dt} \iiint_\Omega \varrho v \, dx = \iiint_\Omega \varrho \frac{\partial v}{\partial t} \, dx = - \iiint_\Omega \varrho \frac{\partial v}{\partial x} v \, dx - \iiint_\Omega \operatorname{grad} p \, dx.$$

Using the Gauss integral theorem, this gives for the ith component that

$$\frac{d}{dt} \iiint_\Omega \varrho v^i \, dx = -\varrho \iiint_\Omega v \cdot \operatorname{grad} v^i \, dx - \iiint_\Omega \frac{\partial p}{\partial x^i} \, dx$$

$$= -\varrho \iiint_\Omega \operatorname{div}(v^i v) \, dx - \iiint_\Omega \operatorname{div}(p e_i) \, dx = - \iint_{\partial \Omega} (\varrho v^i v + p e_i) \, d\vec{o},$$

and altogether, the balance equation gives

$$-\frac{d}{dt} \iiint_\Omega \varrho v \, dx = \left(\iint_{\partial \Omega} P_{1k} n^k \, do, \iint_{\partial \Omega} P_{2k} n^k \, do, \iint_{\partial \Omega} P_{3k} n^k \, do \right)$$

with the outward normal vector $n^k e_k$ and the component matrix

$$(P_{ik}) = \begin{pmatrix} \varrho v^1 v^1 + p & \varrho v^1 v^2 & \varrho v^1 v^3 \\ \varrho v^2 v^1 & \varrho v^2 v^2 + p & \varrho v^2 v^3 \\ \varrho v^3 v^1 & \varrho v^3 v^2 & \varrho v^3 v^3 + p \end{pmatrix}$$

of the so-called **momentum current density tensor**. If the momentum is interpreted as a linear form, as should be done in this context too, the balance equation takes the form

$$-\frac{d}{dt} \iiint_\Omega \varrho v \, dx \cdot y = \iint_{\partial\Omega} P(y, \vec{n}) \, do \,,$$

and the momentum current density tensor P is seen to be twice covariant. From the matrix, we also see that it is symmetric. P can be formulated in a coordinate-free manner as

$$P(x, y) = \varrho \, (v \cdot x) \, (v \cdot y) + p \, (x \cdot y).$$

The linear form $P(x, \,.)$ gives the momentum that flows per unit time and per unit area through a surface with the normal unit vector x.

9.3   The Energy-Momentum Tensor

The Newtonian concept of the momentum was combined with the energy via the energy-momentum-form in Definitions 5.5 and 5.6. Similarly, we now combine the momentum current density tensor, momentum density, energy current density and energy density to form the energy momentum tensor T. Like the momentum current density tensor P, T is symmetric and doubly covariant, but acts on the entire tangent space at the point in question. The following considerations motivate Definition 9.2 below.

Just like the Newtonian concepts of energy and momentum, derived from the momentum form, depend on the observer x specifies the aforementioned non-relativistic concepts from the energy-momentum tensor T, with the subspace $x^\perp$ again playing the role of the Euclidean space $\mathbb{R}^3$. Modelled on the momentum current density tensor, for $u \in x^\perp$, the linear form $T(u, \,.)$ is the momentum form, which flows through a surface with the normal vector u per unit of time. In particular, the restriction of $-T(u, \,.)$ to $x^\perp$ is the (Newtonian) momentum that flows through this surface, and $T(u, x)$ is the energy that flows through this surface. The restriction of $T(\,.\,, x)$ to $x^\perp$ is thus the energy current density, simultaneously also the momentum density, and $T(x, x)$ is then the corresponding energy density.

The flow vector z can itself be an observer. No energy flows for this observer, the energy density is ϱ and the momentum current density tensor is the p-fold dot product of the two vectors involved. According to the above interpretations, we have

$$T(x, z) = 0 \quad \text{f"ur } x \in z^{\perp}$$
$$T(z, z) = \varrho$$
$$T(x, y) = p(-g(x, y)) \quad \text{f"ur } x, y \in z^{\perp} .$$

Theorem 9.1 *Let T be a symmetric doubly covariant tensor on a Lorentz space, and for the time-like vector z satisfies $g(z, z) = 1$. Then T is uniquely determined by the relations*

$$T(z, z) = \varrho$$
$$T(x, z) = 0 \quad \text{f"ur } x \in z^{\perp}$$
$$T(x, y) = -pg(x, y) \quad \text{f"ur } x, y \in z^{\perp}$$

and has the representation

$$T(x, y) = (\varrho + p)g(x, z)g(y, z) - pg(x, y) .$$

Proof Every vector x in the Lorentz space can be decomposed into the sum of a multiple of z and an element of the orthogonal subspace $z^{\perp}$. The decomposition coefficients can be determined and we have

$$x = g(x, z)z + (x - g(x, z)z) .$$

Thus for the tensor T, we obtain

$$\begin{aligned} T(x, y) &= T(g(x, z)z + (x - g(x, z)z), \; g(y, z)z + (y - g(y, z)z)) \\ &= g(x, z)g(y, z)T(z, z) + T(x - g(x, z)z, \; y - g(y, z)z) \\ &= g(x, z)g(y, z)\varrho - pg(x - g(x, z)z, \; y - g(y, z)z) \\ &= \varrho g(x, z)g(y, z) - pg(x, y) + pg(x, z)g(y, z) . \end{aligned}$$

Definition 9.2 The **energy-momentum tensor** T of an ideal flow $[Z, \varrho, p]$ is the tensor field

$$T(P)(x, y) = (\varrho(P) + p(P))g(x, Z(P))g(y, Z(P)) - p(P)g(x, y)$$

for $x, y \in M_P$. For an observer x at point P, the flow has the **energy density** $T(P)(x, x)$. The **energy current density** and **momentum density** are the restriction of $-T(P)(x, \,)$ to $x^{\perp}$, and the restriction of the bilinear form $T(P)$ to $x^{\perp}$ is the **momentum current density**.   ◆

An observer $x \in M_P$, as already mentioned in Sect. 9.1, attributes to a flow the flow velocity v, characterised by

$$\frac{x+v}{\sqrt{1+g(v,v)}} = Z(P) .$$

Consequently, he observes an energy density

$$T(x,x) = (\varrho + p) \left(g \left(x, \frac{x+v}{\sqrt{1+g(v,v)}} \right) \right)^2 - pg(x,x)$$

$$= \frac{\varrho + p}{1+g(v,v)} - p = \frac{\varrho}{1+g(v,v)} - \frac{pg(v,v)}{1+g(v,v)} .$$

When considering the Newtonian limiting case of low speeds, the part containing the pressure p is negligible. It should also be noted that the ϱ used here is not the same ϱ from the previous section. If one substitutes $\varrho = (1+\varepsilon)\sigma$ and by Taylor expanding, one obtains

$$\frac{1}{\sqrt{1+g(v,v)}} = 1 - \frac{1}{2}g(v,v) + \cdots$$

which gives

$$\frac{\varrho}{1+g(v,v)} \approx \frac{\varepsilon\sigma}{1+g(v,v)} + \frac{\sigma(1-g(v,v)/2)}{\sqrt{1+g(v,v)}} = \frac{\varepsilon\sigma}{1+g(v,v)} + \frac{\sigma}{\sqrt{1+g(v,v)}} + \frac{\sigma v^2/2}{\sqrt{1+g(v,v)}} ,$$

and thus σ plays the role of ϱ from the previous section, but where the volume is converted for the observer x.

To characterise the restriction of $-T(x, .)$ to $x^\perp$ is to represent the number $T(x,y)$ for $y \in x^\perp$. We have

$$T(x,y) = (\varrho + p) \, g \left(x, \frac{x+v}{\sqrt{1+g(v,v)}} \right) g \left(y, \frac{x+v}{\sqrt{1+g(v,v)}} \right) - pg(x,y)$$

$$= \frac{\varrho + p}{\sqrt{1+g(v,v)}} \frac{g(y,v)}{\sqrt{1+g(v,v)}} = g \left(y, \frac{(\varrho + p)v}{1+g(v,v)} \right) .$$

For comparison with the Newtonian energy current density we substitute $\varrho = (1+\varepsilon)\sigma$ again, and read that the restriction of the linear form $-T(x, .)$ to $x^\perp$ is represented by the scalar product $-g$ with the vector

$$\frac{((1+\varepsilon)\sigma + p)v}{1+g(v,v)} = \frac{\sigma v}{1+g(v,v)} + \frac{\varepsilon\sigma v + pv}{1+g(v,v)} \approx \left(\frac{\sigma + \sigma v^2/2}{\sqrt{1+g(v,v)}} + \frac{\varepsilon\sigma + p}{1+g(v,v)} \right) v .$$

The momentum current density is the bilinear form restricted to $x^\perp$ given by

$$T(y_1, y_2) = (\varrho + p)\, g\left(y_1, \frac{x+v}{\sqrt{1 + g(v,v)}}\right) g\left(y_2, \frac{x+v}{\sqrt{1 + g(v,v)}}\right) - p\, g(y_1, y_2)$$

$$= \frac{\varrho + p}{1 + g(v,v)}\, g(y_1, v)\, g(y_2, v) - p\, g(y_1, y_2)$$

$$= \varrho\, \frac{g(y_1, v)\, g(y_2, v)}{1 + g(v,v)} + p\, \frac{g(y_1, v)\, g(y_2, v) - g(y_1, y_2) - g(y_1, y_2)\, g(v,v)}{1 + g(v,v)}$$

$$\approx \frac{\varrho\, g(y_1, v)\, g(y_2, v) - p\, g(y_1, y_2)}{1 + g(v,v)}\,,$$

and the last approximation essentially corresponds to the formula given in the previous section.

Finally, the component representation of the energy-momentum tensor should be mentioned. From Definition 9.2 we have

$$T_{ik} = (\varrho + p)g_{ij}g_{kl}z^j z^l - p g_{ik}\,.$$

For a Lorentz basis, we have

$$
\begin{aligned}
T_{00} &= (\varrho + p)z^0 z^0 - p \\
T_{0k} &= T_{k0} = -(\varrho + p)z^0 z^k \qquad \text{for } k = 1, 2, 3 \\
T_{kk} &= (\varrho + p)z^k z^k + p \qquad \text{for } k = 1, 2, 3 \\
T_{ik} &= (\varrho + p)z^i z^k \qquad\qquad \text{for } i, k = 1, 2, 3,\ i \neq k\,.
\end{aligned}
$$

For an observer x, giving a Lorentz-basis x, e_1, e_2, e_3, we have

$$z^0 = g(x, z) = g\left(x, \frac{x+v}{\sqrt{1 + g(v,v)}}\right) = \frac{1}{\sqrt{1 - \sum_{i=1}^{3}(v^i)^2}}$$

and

$$z^k = g\left(e_k, \frac{x+v}{\sqrt{1 + g(v,v)}}\right) = \frac{v^k}{\sqrt{1 - \sum_{i=1}^{3}(v^i)^2}} \qquad \text{for } k = 1, 2, 3\,,$$

where v^1, v^2, v^3 are the components of the flow vector v as perceived by the observer x.

9.4 Charge

Definition 9.3 Electric current $[J, \sigma]$ is a pointwise future directed vector field J with $g(J, J) = 1$ and a scalar field σ. The covector field $S = -\sigma g(., J)$ is called the **charge-current-form**. For an observer x, $L = -S(x) = \sigma g(x, J)$ is the **charge density** and the vector $I \in x^{\perp}$, which represents, with respect to g, the restriction of S to $x^{\perp}$, is the **current density** I. ♦

A comoving observer $x = J$ registers a charge density $L = -S(J) = \sigma$. The scalar field σ is thus to be interpreted as the rest charge density. Charge density L and current density I depend on the observer and must be converted when changing to another observer.

Theorem 9.2 *An observer x registers a charge density L and a current density with the components I^1, I^2, I^3 with respect to the orthonormal basis $e_1, e_2, e_3 \in x^{\perp}$. Another observer x', whom the first observer sees at a relative velocity βe_1, then measures a charge density*

$$L' = \frac{L - \beta I^1}{\sqrt{1 - \beta^2}}$$

and a current density with the components

$$(I^1)' = \frac{I^1 - \beta L}{\sqrt{1 - \beta^2}}, \quad (I^2)' = I^2, \quad (I^3)' = I^3$$

with respect to the basis $(\beta x + e_1)/\sqrt{1 - \beta^2}, e_2, e_3 \in (x')^{\perp}$.

Proof The observer x sees the other observer

$$x' = \frac{x + \beta e_1}{\sqrt{1 - \beta^2}} .$$

The latter measures a charge density

$$L' = -S(x') = \sigma g\left(\frac{x + \beta e_1}{\sqrt{1 - \beta^2}}, J\right) = \frac{\sigma}{\sqrt{1 - \beta^2}}(g(x, J) + \beta g(e_1, J)) = \frac{L - \beta I^1}{\sqrt{1 - \beta^2}} .$$

The first component of the current density transforms as follows

$$(I^1)' = g\left(\frac{\beta x + e_1}{\sqrt{1 - \beta^2}}, I\right) = -\sigma g\left(\frac{\beta x + e_1}{\sqrt{1 - \beta^2}}, J\right)$$

$$= \frac{-\sigma \beta g(x, J) - \sigma g(e_1, J)}{\sqrt{1 - \beta^2}} = \frac{-\beta L + g(e_1, I)}{\sqrt{1 - \beta^2}} = \frac{I^1 - \beta L}{\sqrt{1 - \beta^2}} .$$

The calculation also chows that the other two components do not change.

In Sect. 6.5 we formulated the classical Maxwell equations in vacuum using differential forms. Now that we have clarified how to describe the charge, we can now deal with the general case. The equations $\operatorname{div} B = 0$ and $\operatorname{rot} E + \frac{\partial B}{\partial t} = 0$ also apply in general and are summarised according to Theorem 6.15 as $dF = 0$. The other two Maxwell equations are known to be $\operatorname{div} E = 4\pi L$ and $\operatorname{rot} B - \frac{\partial E}{\partial t} = 4\pi I$ with the charge density L and the current density I.

Theorem 9.3 *The equations* $\operatorname{div} E = 4\pi L$ *and* $\operatorname{rot} B - \frac{\partial E}{\partial t} = 4\pi I$ *are equivalent to* $d*F = *4\pi S$.

Proof As with the proof of Theorem 6.13, we again choose a chart, whose coordinate vector fields $\partial_0, \partial_1, \partial_2, \partial_3$ form a pointwise positively oriented Lorentz-basis. The observer ∂_0 registers a charge density $L = -S(\partial_0) = -S_0$ and a current density $I \in \partial_0^\perp$, characterised by $S(y) = g(y, I)$ for $y \in \partial_0^\perp$. The components S_1, S_2, S_3 of S are thus calculated from the components of I as follows

$$S_i = S(\partial_i) = I^i \quad \text{for } i = 1, 2, 3 \,.$$

The charge-current-form S hence has the representation

$$S = -L du^0 + I^1 du^1 + I^2 du^2 + I^3 du^3 \,.$$

When calculating the Hodge-star-operator in the sense of Definition 6.5, $J\partial_0 = du^0$ and $J\partial_i = -du^i$ for $i = 1, 2, 3$ must be taken into account. We have

$$
\begin{aligned}
*du^0 &= (-1)^3 (-du^1) \wedge (-du^2) \wedge (-du^3) \\
*(-du^1) &= -(-1)^2 du^0 \wedge (-du^2) \wedge (-du^3) \\
*(-du^2) &= (-1)^2 du^0 \wedge (-du^1) \wedge (-du^3) \\
*(-du^3) &= -(-1)^2 du^0 \wedge (-du^1) \wedge (-du^2)
\end{aligned}
$$

and altogether

$$*S = -L du^1 \wedge du^2 \wedge du^3 + I^1 du^0 \wedge du^2 \wedge du^3 - I^2 du^0 \wedge du^1 \wedge du^3 + I^3 du^0 \wedge du^1 \wedge du^2.$$

To represent the differential of the Maxwell tensor, we replace, in the representation of dF given in the proof of Theorem 6.15, the electric field strength E by $-B$ and the magnetic field strength B by E, and obtain

$$d * F = \left(-\frac{\partial E^1}{\partial u^1} - \frac{\partial E^2}{\partial u^2} - \frac{\partial E^3}{\partial u^3} \right) du^1 \wedge du^2 \wedge du^3$$

$$+ \left(\frac{\partial B^3}{\partial u^2} - \frac{\partial B^2}{\partial u^3} - \frac{\partial E^1}{\partial u^0} \right) du^0 \wedge du^2 \wedge du^3$$

$$+ \left(\frac{\partial B^3}{\partial u^1} - \frac{\partial B^1}{\partial u^3} + \frac{\partial E^2}{\partial u^0} \right) du^0 \wedge du^1 \wedge du^3$$

$$+ \left(\frac{\partial B^2}{\partial u^1} - \frac{\partial B^1}{\partial u^2} - \frac{\partial E^3}{\partial u^0} \right) du^0 \wedge du^1 \wedge du^2 \,.$$

The equivalence to be shown can now be read off.

9.5   Energy and Momentum in the Electromagnetic Field

In addition to mass, the charge also makes a contribution to the energy-momentum tensor. As preparation for its calculation, we first formulate and prove the **energy theorem** (Theorem 9.4) and the **momentum theorem** (Theorem 9.5) from classical electrodynamics.

Theorem 9.4 *If the time-dependent vector fields* E, B, I *are coupled via* $\operatorname{rot} E + \frac{\partial B}{\partial t} = 0$ *and* $\operatorname{rot} B = \frac{\partial E}{\partial t} + 4\pi I$, *then for each integration domain* Ω, *there holds*

$$\iiint_{\Omega} I \cdot E \, dx + \frac{d}{dt} \iiint_{\Omega} \frac{E \cdot E + B \cdot B}{8\pi} \, dx + \iint_{\partial\Omega} \frac{E \times B}{4\pi} \, d\vec{o} = 0 \,.$$

Proof From the given equations and the identity

$$\operatorname{div}(E \times B) = B \cdot \operatorname{rot} E - E \cdot \operatorname{rot} B$$

one obtains, using the Gauss integral theorem, that

$$\frac{d}{dt} \iiint_{\Omega} \frac{E \cdot E + B \cdot B}{8\pi} \, dx = \iiint_{\Omega} \frac{E \cdot \frac{\partial E}{\partial t} + B \cdot \frac{\partial B}{\partial t}}{4\pi} \, dx$$

$$= \iiint_{\Omega} \frac{E \cdot \operatorname{rot} B - B \cdot \operatorname{rot} E}{4\pi} \, dx - \iiint_{\Omega} I \cdot E \, dx$$

$$= -\iint_{\partial\Omega} \frac{E \times B}{4\pi} \, d\vec{o} - \iiint_{\Omega} I \cdot E \, dx.$$

Of course, the vector fields E, B and I in Theorem 9.4 are to be interpreted as electrical and magnetic field strength and current density, and the two given equations are two of the for Maxwell equations. The first integral in the equation formulated in the conclusion of the energy theorem is the change in the kinetic energy of the charge in the region Ω. This can be justified as follows: A particle with charge e moving in the field with velocity v experiences a change in its momentum p according to the **Lorentz-force**

$$\frac{dp}{dt} = e(E + v \times B) .$$

This results in a change in its kinetic energy, determined by

$$\frac{d}{dt}\left(\frac{mv^2}{2}\right) = mvv' = vp' = evE .$$

If there is an electric current with density I instead of the individual particle, ev is to be replaced by I and integrated with the integrand $I \cdot E$ over the region Ω.

Given the equation formulated in the conclusion of the energy theorem, it makes sense to think of the scalar field $(E \cdot E + B \cdot B)/8\pi$ as the **energy density** and the vector field $(E \times B)/4\pi$, called the **Pointing-vector**, as the **energy current density** of the field. Then the equation describes the conservation of energy. An increase in the kinetic energy of the particles in the region Ω, an outflow of energy from the region, and an increase in the amount of energy in the region determined by the field strengths, all add up to zero in total.

The following momentum theorem uses volume integrals of vector-valued functions and a surface integral of the first kind of a vector function or second kind of an integrand that is linear-mapping valued.

Theorem 9.5 *Under the same situation of Theorem 9.4, there holds*

$$\iiint_{\Omega} (LE + I \times B)\, dx + \frac{d}{dt} \iiint_{\Omega} \frac{E \times B}{4\pi}\, dx + \iint_{\partial\Omega} M\, d\vec{o} = 0 ,$$

where the linear mapping M, with respect to a positively oriented orthonormal basis, has the matrix representation

$$M_{ii} = \frac{E \cdot E + B \cdot B}{8\pi} - \frac{E^i E^i + B^i B^i}{4\pi}$$

and

$$M_{ik} - -\frac{E^i E^k + B^i B^k}{4\pi}$$

Proof According to the assumed Maxwell equations

$$\frac{d}{dt}(E \times B) = \frac{dE}{dt} \times B + E \times \frac{dB}{dt} = \text{rot}B \times B + \text{rot}E \times E - 4\pi I \times B \ .$$

In general, for vector fields U and V, there holds, with respect to a positively oriented orthonormal basis, that

$$\text{rot}U \times V = \begin{pmatrix} 0 & \partial_2 U^1 - \partial_1 U^2 & \partial_3 U^1 - \partial_1 U^3 \\ \partial_1 U^2 - \partial_2 U^1 & 0 & \partial_3 U^2 - \partial_2 U^3 \\ \partial_1 U^3 - \partial_3 U^1 & \partial_2 U^3 - \partial_3 U^2 & 0 \end{pmatrix} \begin{pmatrix} v^1 \\ v^2 \\ v^3 \end{pmatrix}$$

$$= \begin{pmatrix} \partial_1 U^1 & \partial_2 U^1 & \partial_3 U^1 \\ \partial_1 U^2 & \partial_2 U^2 & \partial_3 U^2 \\ \partial_1 U^3 & \partial_2 U^3 & \partial_3 U^3 \end{pmatrix} \begin{pmatrix} v^1 \\ v^2 \\ v^3 \end{pmatrix} - \begin{pmatrix} \partial_1 U^1 & \partial_1 U^2 & \partial_1 U^3 \\ \partial_2 U^1 & \partial_2 U^2 & \partial_2 U^3 \\ \partial_3 U^1 & \partial_3 U^2 & \partial_3 U^3 \end{pmatrix} \begin{pmatrix} v^1 \\ v^2 \\ v^3 \end{pmatrix} \ ,$$

and so

$$(\text{rot}\, U \times V)^i = V\,\text{grad}\, U^i - V\partial_i U \ .$$

In particular, thanks to the Maxwell equations,

$$(\text{rot}\, B \times B)^i = B \cdot \text{grad}\, B^i - B \cdot \partial_i B = \text{div}(B^i B) - \frac{1}{2}\partial_i (B \cdot B)$$

and analogously

$$(\text{rot}E \times E)^i = \text{div}(E^i E) - 4\pi L E^i - \frac{1}{2}\partial_i (E \cdot E) \ .$$

Upon integrating, we obtain

$$\frac{d}{dt} \iiint_\Omega (E \times B)^i \, dx = \iiint_\Omega \text{div}(E^i E + B^i B) \, dx$$

$$-4\pi \iiint_\Omega (LE + I \times B)^i \, dx - \frac{1}{2} \iiint_\Omega \partial_i (E \cdot E + B \cdot B) \, dx$$

$$= \iint_{\partial\Omega} (E^i E + B^i B - \frac{e_i}{2}(E \cdot E + B \cdot B)) \, d\vec{o} - 4\pi \iiint_\Omega (LE + I \times B)^i \, dx \ .$$

This is the ith component of the claimed equation. The matrix M can be read off.

The momentum theorem is interpreted in a similar manner to the energy theorem. The momentum p of a particle with charge e changes due to the influence of an electromagnetic field, according to the Lorentz force law

$$\frac{dp}{dt} = e(E + v \times B).$$

The rate of change of the momentum of the charge contained in a region Ω with the charge density L and current density I is thus given by

$$\iiint\limits_{\Omega} (LE + I \times B)\, dx.$$

This is the first integral in the momentum theorem. The second integral describes the change in the momentum inherent in the electromagnetic field and the third the momentum flowing through the surface $\partial\Omega$. So the Poynting vector $(E \times B)/4\pi$ can be understood as a momentum density as well as an energy current density. The matrix M then describes the momentum current density. If the momentum is interpreted as a linear form, it is a symmetric bilinear form. The momentum $M(d\vec{o}, .)$ flows across a small planar surface with the normal vector $d\vec{o}$ per unit of time.

Energy density, energy current density and momentum current density of the electromagnetic field are now combined according to the format described in Definition 9.2 to obtain the electromagnetic energy-momentum tensor.

Theorem 9.6 *The components of the energy-momentum tensor T of the electromagnetic fields are calculated from the components of the Faraday tensor by the formula*

$$T_{ik} = \frac{1}{4\pi}\left(F_{ij}F_k^j + \frac{1}{4}g_{ik}F_{lm}F^{lm}\right).$$

Proof With respect to a Lorentz basis, the component matrix of the energy-momentum tensor is given by

$$(T_{ik}) = \frac{1}{4\pi}
\begin{pmatrix}
0 & E^1 & E^2 & E^3 \\
-E^1 & 0 & -B^3 & B^2 \\
-E^2 & B^3 & 0 & -B^1 \\
-E^3 & -B^2 & B^1 & 0
\end{pmatrix}
\begin{pmatrix}
0 & E^1 & E^2 & E^3 \\
E^1 & 0 & B^3 & -B^2 \\
E^2 & -B^3 & 0 & B^1 \\
E^3 & B^2 & -B^1 & 0
\end{pmatrix}$$

$$+ \frac{B \cdot B - E \cdot E}{8\pi}
\begin{pmatrix}
1 & 0 & 0 & 0 \\
0 & -1 & 0 & 0 \\
0 & 0 & -1 & 0 \\
0 & 0 & 0 & -1
\end{pmatrix}.$$

So in the top left corner

$$\frac{E \cdot E}{4\pi} + \frac{B \cdot B - E \cdot E}{8\pi} = \frac{E \cdot E + B \cdot B}{8\pi}$$

is the energy density. In the remaining upper row and left column, we have the Poynting vector $(E \times B)/4\pi$, as wanted. In the diagonal (i, i) position, for $i = 1, 2, 3$, we have

$$\frac{-E^i E^i + (B \cdot B - B^i B^i)}{4\pi} - \frac{B \cdot B - E \cdot E}{8\pi} = \frac{E \cdot E + B \cdot B}{8\pi} - \frac{E^i E^i + B^i B^i}{4\pi},$$

and outside the diagonal, for (i, k), $i, k = 1, 2, 3$, $i \neq k$, we have

$$\frac{-E^i E^k - B^k B^i}{4\pi}.$$

The submatrix for the position $(0, 0)$ is thus the matrix M formulated in Theorem 9.5, describing the momentum current density.

Now that the matrix representation for the energy-momentum tensor has been obtained for a basis, we only need to show that this representation transforms as a doubly covariant tensor when changing to another basis. Let (α_k^i) and (β_k^i) be two matrices that are inverses of each other, describing the coupling between two bases (see Theorem 3.5). Then for the first expression we have

$$\bar{F}_{ij} \bar{F}_k^j = \alpha_i^l \alpha_j^r F_{lr} \alpha_k^s F_s^t \beta_t^j = \alpha_i^l F_{lr} \alpha_k^s F_s^r$$

and for the second one we have

$$\bar{g}_{ik} \bar{F}_{lm} \bar{F}^{lm} = \alpha_i^n \alpha_k^p g_{np} \alpha_l^q \alpha_m^r F_{qr} F^{st} \beta_s^l \beta_t^m = \alpha_i^n \alpha_k^p g_{np} F_{qr} F^{qr}.$$

This is the required transformation behavior, and so the matrix representation is verified for all bases.

9.6   The Einstein Field Equation

In Newtonian mechanics, gravitation means the fact that a point mass m is drawn to another point mass M located at a distance r with a force of the magnitude mM/r^2. The fact that there is usually a factor (gravitational constant) in this formula is due to the units of measurement commonly used. More generally it can be said in Newtonian physics that a mass distribution, described by a mass density, determines a gravitational field.

In the general theory of relativity, this law of gravity is replaced by the following point of view: *Matter, described by the energy-momentum tensor T, generates a curvature of spacetime via the* **Einstein field equation**

$$\mathrm{Ric} - \tfrac{1}{2} S g = 8\pi T,$$

and particles then move along geodesics. The next chapter explains in detail what geodesics are and how the movement along geodesics implies Newton's law of gravitation results in the non-relativistic limiting case. Although Einstein's field equation is postulated as an axiom in general relativity and therefore requires no proof, the considerations that led to its discovery are of interest. The starting point was Einstein's idea that matter bends spacetime. The equation that relates matter to curvature must of course be of a tensorial nature. The energy-momentum tensor is available to describe the matter. Since this is doubly covariant, a $(0, 2)$-tensor must also be used to describe the curvature. The Ricci tensor is the first choice. However, in contrast to the energy-momentum tensor, this is not divergence-free (see Sect. 11.5). This deficiency can be remedied by subtracting an appropriate multiple of Sg. The Einstein tensor $G = \mathrm{Ric} - \tfrac{1}{2} S g$ already introduced at the end of Chap. 8, proves to be divergence-free and can be matched with the energy-momentum tensor. This motivates the field equation $G = \mu T$. Finally, a comparison with Newtonian mechanics results in $\mu = 8\pi$ (Example 2 in Sect. 10.4).

The tensor $\mathrm{Ric} - \tfrac{1}{2} S g - \Lambda g$ is also divergence-free. In fact, for a long time, Einstein favored the field equation

$$\mathrm{Ric} - \tfrac{1}{2} S g - \Lambda g = 8\pi T,$$

with Λ a very small constant, called the **cosmological constant**, because the version of the field equation with $\Lambda = 0$ does not allow for static world models (see Sect. 15.3), and a changing universe was considered absurd at the time. After HUBBLE discovered in 1929 that the galaxies move away from us at an escape velocity proportional to their distance from us, EINSTEIN dropped this more general field equation. Today one is again inclined to believe that a field equation with a constant Λ applies. However, Λ is so small that it only affects the calculations in cosmological orders of magnitude; Λ can be neglected for the considerations in the solar system.

An energy-momentum tensor is generated by an ideal flow (Definition 9.2) or by an electromagnetic field (Theorem 9.6) or by both. In the latter case, both parts should be added. If the (total) energy-momentum tensor is known in a situation, then the Ricci tensor can be calculated. The following version of Einstein's field equation is more suitable than the original one.

Theorem 9.7 *The Einstein field equation* $\mathrm{Ric} - \tfrac{1}{2} S g = 8\pi T$ *is equivalent to the equation*

$$\mathrm{Ric} = 8\pi \left(T - \tfrac{1}{2} C T g \right)$$

where $C T = g^{ik} T_{ik}$.

Proof From the component representation of the original field equation

$$\text{Ric}_{ik} - \tfrac{1}{2}Sg_{ik} = 8\pi T_{ik}$$

it follows, by index pulling and contraction, that

$$S - \tfrac{1}{2}S \cdot 4 = 8\pi CT$$

and thus $S = -8\pi CT$. Inserting this into the original field equation, the new formulation results. Analogously, the old formulation can be obtained from the new formulation.

9.7   Spherically Symmetric Solutions

The spacetime that is generated by a single spherically symmetric celestial body is obviously also spherically symmetric. Spherical symmetry of the celestial body means in particular that the body does not rotate. In applying the consequences from the spherically symmetric model to the Sun or the Earth, we neglect their rotation.

Thanks to the spherical symmetry, a two-parameter family of spherical surfaces can be used as the basic set M for spacetime. Each of the spherical surfaces can be identified with the surface S_2 of the unit sphere and equipped with an atlas using the usual spherical coordinates ϑ and φ. The metric g on the four-dimensional tangent space then should be such that the restriction of $-g$ to the two-dimensional subspaces spanned by ∂_ϑ and ∂_φ is the one on the tangent spaces of spherical surfaces. Thus there should hold that

$$g(\partial_\vartheta, \partial_\vartheta) = -r^2 , \quad g(\partial_\varphi, \partial_\varphi) = -r^2 \sin^2\vartheta , \quad g(\partial_\vartheta, \partial_\varphi) = 0.$$

One of the other two parameters has already been mentioned: The positive parameter r is called the **Schwarzschild radius**. Except for the sign, this is the same metric as the one on a spherical surface in $\mathbb{R}^3$, the radius of which is the Schwarzschild radius. As the tangent spaces of the manifold to be constructed are four-dimensional, another parameter t must be introduced, which can take any real value, and is called the **Schwarzschild time**. It will be shown that, up to an additive constant, the Schwarzschild time is the proper time of an infinitely distant observer at rest.

Thus we take the basic point set as the Cartesian product $M = \mathbb{R} \times (0, \infty) \times S_2$. The fact that a chart $Q \to (\vartheta, \varphi)$ on S_2 gives rise to a chart $\varphi(t, r, Q) = (t, r, \vartheta, \varphi)$ on $\mathbb{R} \times (0, \infty) \times S_2$ was already mention in Sect. 1.1 (Example 4). The coordinate vector fields $\partial_t, \partial_r, \partial_\vartheta, \partial_\varphi$ are induced by this chart. The metric g has already been specified on the two-dimensional subspaces spanned by ∂_ϑ and ∂_φ. Owing to symmetry, a reorientation of the angles ϑ and φ ought not to alter the component matrix of g. Thus $g(\partial_t, -\partial_\vartheta) = g(\partial_t, \partial_\vartheta)$ and thus $g(\partial_t, \partial_\vartheta) = 0$ and analogously

$g(\partial_t, \partial_\varphi) = g(\partial_r, \partial_\vartheta) = g(\partial_r, \partial_\varphi) = 0$. Furthermore, the remaining matrix element must be independet of ϑ, φ and t. Hence the component matrix has the form

$$(g_{ik}) = \begin{pmatrix} a(r) & c(r) & 0 & 0 \\ c(r) & b(r) & 0 & 0 \\ 0 & 0 & -r^2 & 0 \\ 0 & 0 & 0 & -r^2\sin^2\vartheta \end{pmatrix},$$

whereby it itself and thus also the $(2, 2)$-submatrix formed by the functions a, b, c must be indefinite. The matrix can be diagonalised by means of a coordinate transformation of the form $\bar{t} := t + \alpha(r)$ with a suitable function α. The new chart is then

$$\bar{\varphi}(t, r, Q) := (t + \alpha(r), r, \vartheta, \varphi),$$

and the induced coordinate vector fields are denoted by $\bar\partial_{\bar t}, \bar\partial_r, \bar\partial_\vartheta, \bar\partial_\varphi$. For a C^∞-function f on M, we have

$$\bar\partial_r f = \frac{\partial}{\partial_r}(f \circ \bar\varphi^{-1})(\bar t, r, \vartheta, \varphi) = \frac{\partial}{\partial_r}(f \circ \varphi^{-1})(\bar t - \alpha(r), r, \vartheta, \varphi) = \partial_r f - \alpha' \partial_t f,$$

and so $\bar\partial_r = \partial_r - \alpha' \partial_t$. The other coordinate vector fields are not affected by the transformation, i.e., we have $\bar\partial_{\bar t} = \partial_t$, $\bar\partial_\vartheta = \partial_\vartheta$ and $\bar\partial_\varphi = \partial_\varphi$. By choosing the function α such that

$$0 = g(\bar\partial_{\bar t}, \bar\partial_r) = g(\partial_t, \partial_r - \alpha' \partial_t) = c - \alpha' a$$

α must be a primitive of c/a. The diagonal element $g(\partial_r, \partial_r)$ is transformed to

$$g(\bar\partial_r, \bar\partial_r) = g(\partial_r, \partial_r) - 2\alpha' g(\partial_t, \partial_r) + (\alpha')^2 g(\partial_t, \partial_t) = b - 2c^2/a + c^2/a = b - c^2/a.$$

Since the original component matrix was indefinite, so is

$$g(\bar\partial_{\bar t}, \bar\partial_{\bar t}) g(\bar\partial_r, \bar\partial_r) = ab - c^2 < 0.$$

According to the interpretation of t, respective $\bar t$ as time, the vector $\bar\partial_{\bar t}$ must be timelike. Thus $A := a$ and $B := c^2/a - b$ must be positive functions. Finally, we remove the bars again and get the following result: Every point M corresponds to a real number t, a positive number r and a point Q of S_2. If a suitable subset of M is used, and the use is made of spherical coordinates on S_2, then the point Q can be assigned the angular coordinates ϑ and φ. With respect to this chart and the induced coordinate vector fields $\partial_t, \partial_r, \partial_\vartheta, \partial_\varphi$, the metric has the component matrix given by

$$(g_{ik}) = \mathrm{diag}(A(r), -B(r), -r^2, -r^2\sin^2\vartheta)$$

with positive function A and B. This is the so-called **standard metric** on this manifold M.

In the special case $A(r) = 1 - 2\mu/r$ and $B(r) = 1/A(r)$, we had already calculated the covariant derivatives of the coordinate vector fields in Sect. 7.5. In the general case, similarly one obtains

$$\nabla_{\partial_t}\partial_t = \frac{A'}{2B}\,\partial_r\,, \qquad \nabla_{\partial_r}\partial_r = \frac{B'}{2B}\,\partial_r\,, \qquad \nabla_{\partial_\vartheta}\partial_\vartheta = -\frac{r}{B}\,\partial_r\,,$$

$$\nabla_{\partial_\varphi}\partial_\varphi = -\frac{r\sin^2\vartheta}{B}\,\partial_r - \sin\vartheta\cos\vartheta\,\partial_\vartheta\,, \qquad \nabla_{\partial_t}\partial_r = \frac{A'}{2A}\,\partial_t\,,$$

$$\nabla_{\partial_t}\partial_\vartheta = 0 = \nabla_{\partial_t}\partial_\varphi\,, \qquad \nabla_{\partial_r}\partial_\vartheta = \frac{1}{r}\,\partial_\vartheta\,, \qquad \nabla_{\partial_r}\partial_\varphi = \frac{1}{r}\,\partial_\varphi\,, \qquad \nabla_{\partial_\vartheta}\partial_\varphi = \cot\vartheta\,\partial_\varphi\,.$$

Correspondingly, the results from Sect. 8.5 are generalised to

$$R^0_{101} = \frac{1}{2A}\left(\frac{A'B'}{2B} - A'' + \frac{A'A'}{2A}\right) = -R^0_{110}$$

$$R^1_{001} = \frac{1}{2B}\left(\frac{A'A'}{2A} - A'' + \frac{A'B'}{2B}\right) = -R^1_{010}$$

$$R^0_{202} = -\frac{A'r}{2AB} = -R^0_{220}$$

$$R^2_{002} = -\frac{A'}{2Br} = -R^2_{020}$$

$$R^0_{303} = -\frac{A'r\sin^2\vartheta}{2AB} = -R^0_{330}$$

$$R^3_{003} = -\frac{A'}{2Br} = -R^3_{030}$$

$$R^1_{212} = -\frac{B'r}{2B^2} = -R^1_{221}$$

$$R^2_{112} = -\frac{B'}{2Br} = -R^2_{121}$$

$$R^1_{313} = \frac{B'r\sin^2\vartheta}{2B^2} = -R^1_{331}$$

$$R^3_{113} = -\frac{B'}{2Br} = -R^3_{131}$$

$$R^2_{323} = \left(1 - \frac{1}{B}\right)\sin^2\vartheta = -R^2_{332}$$

$$R^3_{223} = \frac{1}{B} - 1 = -R^3_{232}\,.$$

All other components of the curvature tensor are again zero. For the Ricci tensor, the diagonal elements of the component matrix are

$$\mathrm{Ric}_{00} = \frac{A''}{2B} - \frac{A'A'}{4AB} - \frac{A'B'}{4B^2} + \frac{A'}{Br}$$

$$\mathrm{Ric}_{11} = \frac{A'B'}{4AB} - \frac{A''}{2A} + \frac{A'A'}{4A^2} + \frac{B'}{Br}$$

$$\mathrm{Ric}_{22} = -\frac{A'r}{2AB} + \frac{B'r}{2B^2} + 1 - \frac{1}{B}$$

$$\mathrm{Ric}_{33} = \sin^2\vartheta \, \mathrm{Ric}_{22}$$

and the others are zero.

9.8 Outer and Inner Schwarzschild Metric

We now determine the spacetime generated by a single fixed spherically symmetric star. Thus we need to determine the functions A and B occuring in the standard metric of the previous section. As there is no matter outside the fixed star, the energy-momentum tensor is zero there, and the Einstein field equation in the version described in Theorem 9.7 becomes $\mathrm{Ric} = 0$. From $\mathrm{Ric}_{00} = \mathrm{Ric}_{11} = 0$, it follows that

$$0 = \frac{\mathrm{Ric}_{00}}{A} + \frac{\mathrm{Ric}_{11}}{B} = \frac{A'}{ABr} + \frac{B'}{B^2 r} = \frac{1}{rB}\left(\frac{A'}{A} + \frac{B'}{B}\right),$$

and so

$$0 = \frac{A'}{A} + \frac{B'}{B} = \frac{A'B + AB'}{AB} = (\log(AB))'.$$

Consequently, the product function AB must be a constant. It is physically plausible that the metric must look like the Minkowski metric far away from the fixed star. In particular,

$$\lim_{r\to\infty} A(r) = \lim_{r\to\infty} B(r) = 1$$

and so $AB = 1$, i.e., $B = 1/A$. In light of this, it follows from $(\mathrm{Ric})_{22} = 0$, that

$$-\frac{A'r}{2} - \frac{A'r}{2} + 1 - A = 0,$$

which can be rewritten as $(rA(r))' = 1$. Integrating this results in $rA(r) = r - 2\mu$, with an appropriately designated constant, and so $A(r) = 1 - 2\mu/r$. Hence the sought-for metric is given by

$$(g_{ik}) = \mathrm{diag}\left(1 - \frac{2\mu}{r}, \left(1 - \frac{2\mu}{r}\right)^{-1}, -r^2, -r^2\sin^2\vartheta\right).$$

This is the Schwarzschild metric introduced in Sect. 4.5, more precisely the **outer Schwarzschild metric**. In most textbooks it is given as a Schwarzschild line element

$$ds^2 = (1 - 2\mu/r)\,dt^2 - (1 - 2\mu/r)^{-1}dr^2 - r^2(d\vartheta^2 + \sin^2\varphi\,d\varphi^2).$$

In the next chapter, after Theorem 10.6, we will explain why the positive number μ is to be interpreted as the mass (in geometric units) of the fixed star. The outer Schwarzschild metric can only be used for $r > 2\mu$. It is only needed outside the fixed star anyway, and its radius is usually much bigger than 2μ. For example for our sun, $2\mu = 2,954$ km and for the Earth, $8,899$ mm. The radius can only be smaller than 2μ for stars with an unimaginably high density. Then this is a so-called **black hole**.

In contrast to the exterior of a fixed star, the relativistic effects in its interior are not accessible to experimental observation. If we still want to determine the metric inside, then we must do so using the energy-momentum tensor and Einstein's field equation. The starting point is again the standard metric, where the functions $A(r)$ and $B(r)$ must be determined. The energy-momentum tensor has the components

$$T_{ik} = (\rho + p)g_{ij}Z^j g_{kl}Z^l - pg_{ik}$$

based on a flow Z with $Z^1 = Z^2 = Z^3 = 0$. Since

$$1 = g(Z, Z) = AZ^0 Z^0,$$

we have $Z^0 = 1/\sqrt{A}$. Inserting this gives the diagonal entries of the component matrix as

$$T_{00} = (\rho + p)A \cdot \frac{1}{\sqrt{A}} \cdot A \cdot \frac{1}{\sqrt{A}} - pA = \rho A, \quad T_{11} = pB, \quad T_{22} = pr^2, \quad T_{33} = pr^2 \sin^2\vartheta.$$

With the contraction $CT = \rho - 3p$ the matrix arising on the right hand side of the field equation is given by

$$8\pi \operatorname{diag}\left((\rho + 3p)A/2, \ (\rho - p)B/2, \ (\rho - p)r^2/2, \ (\rho - p)r^2(\sin\vartheta)^2/2\right).$$

Comparing with the diagonal elements of the Ricci tensor given at the end of the previous section, we arrive at the three equations

$$\frac{A''}{2B} - \frac{A'A'}{4AB} - \frac{A'B'}{4B^2} + \frac{A'}{Br} = 4\pi(\rho + 3p)A$$

$$\frac{A'B'}{4AB} - \frac{A''}{2A} + \frac{A'A'}{4A^2} + \frac{B'}{Br} = 4\pi(\rho - p)B$$

$$-\frac{A'r}{2AB} + \frac{B'r}{2B^2} + 1 - \frac{1}{B} = 4\pi(\rho - p)r^2.$$

Their linear combination with the coefficients $r^2/(2A)$, $r^2/(2B)$ and 1, respectively yields

$$\frac{B'r}{B^2} + 1 - \frac{1}{B} = 8\pi\rho r^2,$$

which can be rewritten as

$$1 - 8\pi\rho r^2 = \frac{B - B'r}{B^2} = (r/B)' .$$

Upon integrating, this gives

$$\frac{r}{B(r)} = \int_0^r (1 - 8\pi\rho(s)s^2)ds = r - 2\mu(r)$$

with

$$\mu(r) = 4\pi \int_0^r \rho(s)s^2 ds .$$

Solving for B, we obtain

$$B(r) = \left(1 - \frac{2\mu(r)}{r}\right)^{-1} .$$

According to the physical situation, the density function ρ can be continued beyond the radius R of the fixed star to $\rho(r) = 0$ for $r > R$. This implies that for $r > R$ $\mu(r) = \mu(R)$, which is the total mass μ, and thus $B(r)$ for $r > R$ is also the corresponding coefficient in the outer Scharzschild metric, and in particular, there is a continuous transition at $r = R$.

To determine $A(r)$, we put in $B(r)$ and

$$B'(r) = B^2(r) \left(8\pi\rho(r)r - \frac{2\mu(r)}{r^2}\right)$$

into the third equation of the three equations obtained from the field equation, and get

$$-\frac{A'(r)r}{2A(r)} \left(1 - \frac{2\mu(r)}{r}\right) + 4\pi\rho(r)r^2 + \frac{\mu(r)}{r} = 4\pi(\rho(r) - p(r))r^2$$

and so

$$\frac{A'(r)}{A(r)} = \left(8\pi p(r)r + \frac{2\mu(r)}{r^2}\right)\left(1 - \frac{2\mu(r)}{r}\right)^{-1} .$$

As $\ln A$ is a primitive for the function on the left hand side, it is also a primitive for

$$f(r) = \left(8\pi p(r)r + \frac{2\mu(r)}{r^2}\right)\left(1 - \frac{?\mu(r)}{r}\right)^{-1} .$$

Consequently, $\ln A$ can be determined up to an additive constant as follows: As $p(r) = 0$ and $\mu(r) = \mu$ for $r > R$, we have that f can be continued for $r > R$ as

$$f(r) = \frac{2\mu}{r^2}\left(1 - \frac{2\mu}{r}\right)$$

. Hence f has a primitive for $F(r) = \ln(1 - 2\mu/r)$, and so

$$\ln A(r) = \ln\left(1 - \frac{2\mu}{r}\right).$$

For $r > R$, $A(r)$ corresponds to the corresponding coefficient of the outer Schwarzschild metric. The above primitive can be written as the improper integral

$$\ln\left(1 - \frac{2\mu}{r}\right) = -\int_r^\infty \frac{2\mu}{s^2}\left(1 - \frac{2\mu}{s}\right)^{-1} ds$$

.This formulation can also be used for $r < R$, and this results in

$$A(r) = \exp\left[-\int_r^\infty \left(8\pi p(s)s + \frac{2\mu(s)}{s}\right)\left(1 - \frac{2\mu(s)}{s}\right)^{-1} ds\right].$$

Together with $B(r) = (1 - 2\mu(r)/r)^{-1}$, this describes the **inner Schwarzschild metric** for $r < R$, and for $r > R$, they are the previously determined coefficients of the outer Schwarzschild metric.

We now return to the outer Schwarzschild metric. We had derived it based on the assumption of time-independence and isotropy. We will now show that isotropy implies time-independence, so that the latter is not needed as an assumption. The starting point is again the metric

$$(g_{ik}) = \mathrm{diag}(A, -B, -r^2, -r^2\sin^2\vartheta),$$

where the coefficients A and B can now also depend on t in addition to r. Of the 10 formulas characterising the covariant derivative, three have to be generalised:

$$\nabla_{\partial_t}\partial_t = \frac{1}{2A}\frac{\partial A}{\partial t}\partial_t + \frac{1}{2B}\frac{\partial A}{\partial r}\partial_r$$

$$\nabla_{\partial_r}\partial_r = \frac{1}{2A}\frac{\partial B}{\partial t}\partial_t + \frac{1}{2B}\frac{\partial B}{\partial r}\partial_r$$

$$\nabla_{\partial_t}\partial_r = \frac{1}{2A}\frac{\partial A}{\partial r}\partial_t + \frac{1}{2B}\frac{\partial B}{\partial t}\partial_r,$$

while the others remain unchanged. The usual calculation steps lead to

$$\text{Ric}_{00} = \frac{A''}{2B} - \frac{A'}{4B}\left(\frac{A'}{A} + \frac{B'}{B}\right) + \frac{A'}{Br} - \frac{\ddot{B}}{2B} + \frac{\dot{B}}{4B}\left(\frac{\dot{A}}{A} + \frac{\dot{B}}{B}\right)$$

$$\text{Ric}_{11} = \frac{\ddot{B} - A''}{2A} + \frac{(A')^2 - \dot{A}\dot{B}}{4A^2} + \frac{A'B' - (\dot{B})^2}{4AB} + \frac{B'}{Br}$$

$$\text{Ric}_{22} = \frac{r}{2B}\left(\frac{B'}{B} - \frac{A'}{A}\right) + 1 - \frac{1}{B}$$

$$\text{Ric}_{01} = \frac{\dot{B}}{Br} \, ,$$

where $\dot{}$ means partial differentiation with respect to t, and $'$ means partial differentiation with respect to r. From $\text{Ric}_{01} = 0$ follows the time-independence of B, and the formulas for Ric_{ii} are reduced to their original form. So the previous calculation gives in particular the time-independence of $A = 1/B$. This again results in the time-independent Schwarzschild metric.

Geodesics

10.1 Time

It is a basic fact of the theory of relativity that the time calculation depends on the respective observer. According to Definition 5.1, an observer in a spacetime M is an M-valued function γ of the real variable t with $g(\gamma'(t), \gamma'(t)) = 1$ and $\gamma'(t)$ future-pointing. The variable t is the **time** given by a watch carried by the observer γ, determined of course only up to an additive constant. The observer γ sees the passage of time as a change in what is happening. Scala fields, vector fields, and more generally, tensor fields, change their functional value. What the observer registers at time t_0 as a change $\frac{\partial}{\partial t} f$ of a scalar field f, or a change $\frac{\partial}{\partial t} X$ of a vector field X, is the number $\gamma'(t_0) f$, respectively the tangent vector $\nabla_{\gamma'(t_0)} X$, that is, the application of the tangent vector $\gamma'(t_0)$ to f or the covariant derivative with respect to $\gamma'(t_0)$ of X.

The image points $\gamma(t)$ of an observer γ form his **world line**. When constructing an observer, it sometimes makes sense to initially ignore the requirement $g(\gamma', \gamma') = 1$. For an M-valued function β of a real variable s with a future-pointing derivative $\beta'(s)$, an observer γ can be found by means of a variable transformation, the world line of which is the range of β. To this end, we fix two numbers s_0 and t_0 and determine the number $t = \tau(s)$ corresponding to the number s using the equation

$$t - t_0 = \int_{s_0}^{s} \sqrt{g\left(\beta'(\sigma), \beta'(\sigma)\right)}\, d\sigma .$$

This gives

$$\tau'(s) = \sqrt{g\left(\beta'(s), \beta'(s)\right)} .$$

© The Author(s), under exclusive license to Springer Nature Switzerland AG 2023
R. Oloff, *The Geometry of Spacetime*, Graduate Texts in Physics,
https://doi.org/10.1007/978-3-031-16139-1_10

The sought-after observer is $\gamma = \beta \circ \tau^{-1}$, since now, in contrast to β, there holds that

$$g(\gamma'(t),\, \gamma'(t)) = g\left(\frac{\beta'(\tau^{-1}(t))}{\tau'(\tau^{-1}(t))},\, \frac{\beta'(\tau^{-1}(t))}{\tau'(\tau^{-1}(t))}\right)$$

$$= g\left(\frac{\beta'(s)}{\sqrt{g(\beta'(s), \beta'(s))}},\, \frac{\beta'(s)}{\sqrt{g(\beta'(s), \beta'(s))}}\right) = 1.$$

We record the formula for calculating the time.

Theorem 10.1 *An observer, who moves along the world line described by β:* $[s_1, s_2] \longrightarrow M$, *needs the time*

$$\int_{s_1}^{s_2} \sqrt{g(\beta'(\sigma),\, \beta'(\sigma))}\, d\sigma$$

for this.

A consequence of this is a spectacular basic fact in the theory of relativity, known as the **twin paradox**: *One twin leaves the other at a constant speed, turns back after a certain time, and moves back in the same way. When he meets the other twin again, it turns out that they are no longer the same age. Less time has passed for the travelling twin than for the one who stayed home.*

To confirm and quantify this effect, we use the **Minkowski space** $\mathbb{R}^4$, equipped with the Lorentz metric

$$g\left((\xi^0, \xi^1, \xi^2, \xi^3),\, (\eta^0, \eta^1, \eta^2, \eta^3)\right) = \xi^0 \eta^0 - \xi^1 \eta^1 - \xi^2 \eta^2 - \xi^3 \eta^3.$$

This is permissible because a tangent space of a spacetime is always isometrically isomorphic to this four-dimensional space, and with the problem at hand, we can restrict ourselves to small periods of time, so that we can neglect the dependence of the tangential spaces $T(P)$ on the point P. The world lines

$$\alpha(s) = (s, 0, 0, 0). \qquad 0 \le s \le 2T$$

and

$$\beta(s) = \begin{cases} (s, vs, 0, 0) & \text{for} \quad 0 \le s \le T \\ (s, 2vT - vs, 0, 0) & \text{for} \quad T \le s \le 2T \end{cases}$$

shown in Fig. 10.1 can be compared. While the time

$$\int_0^{2T} \sqrt{1}\, ds = 2T$$

Fig. 10.1 Twin paradox

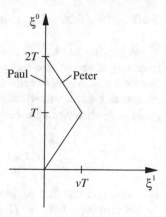

elapses for the dormant twin until the reunion, for the travelling twin, the reunion already takes place in time

$$\int_0^T \sqrt{1-v^2}\,ds + \int_T^{2T} \sqrt{1-(-v)^2}\,ds = 2T\sqrt{1-v^2}.$$

Here v is the magnitude of the travelling twin's speed as registered by the resting twin. The ratio $1/\sqrt{1-v^2}$ of the time periods represents the **time dilation** (see also the interpretation of Theorem 5.4).

In this particular case, we encounter a phenomenon that is postulated generally in the theory of relativity: *The world line γ of a force-free movement from P to Q is characterised by the fact that the time integral*

$$\int_{s_1}^{s_2} \sqrt{g(\beta'(s),\,\beta'(s))}\,ds$$

is maximal for γ among all world lines from P to Q that are close to γ. Such world lines are called **geodesics**. The name comes from the classical extremal problem on the surface of the Earth: What is the shortest curve joining two points that are not too far apart? This is of course a so-called great circle, the intersection curve of the surface of the sphere with a plane passing through the center of the sphere. In the natural parameter representation, i.e., with $|\gamma'(s)| = 1$, a great circle can be characterised by the fact that the „acceleration vector" $\gamma''(s)$ is perpendicular to the respective tangential plane. This criterion also applies in the case of a general curved surface in $\mathbb{R}^3$. The equivalent formulation $\nabla_{\gamma'(s)}\gamma'(s) = 0$ has the advantage that it only makes use of the intrinsic geometry of the surface and is thus useful and valid in a Riemannian manifold. We will show below that the condition $\nabla_{\gamma'(s)}\gamma'(s) = 0$ also applies to a geodesic γ in spacetime with $|\gamma'(s)| = 1$. The fact that the minimum in the case of a Riemannian manifold now corresponds to a maximum in the case of spacetimes is due to the signature of the metric. The method developed below does not anyway distinguish between maxima and minima and is equally applicable in both scenarios.

10.2   The Euler-Lagrange Equation

The following problem is considered in the calculus of variations: Given a twice continuously differentiable function $L(t, x_1, \ldots, x_n, y_1, \ldots, y_n)$ of $2n + 1$ variables,
numbers $a < b$ and n-tuples $(\alpha_1, \ldots, \alpha_n)$, $(\beta_1, \ldots, \beta_n) \in \mathbb{R}^n$, find all continuously
differentiable real-valued functions $x_1, \ldots, x_n \in C^2 [a, b]$ such that $x_i(a) = \alpha_i$ and
$x_i(b) = \beta_i$ that minimise the integral

$$\int_a^b L\Big(t, x_1(t), \ldots, x_n(t), x_1'(t), \ldots, x_n'(t)\Big)\, dt\,.$$

It is assumed that there exists a relative minimiser of the integral in the so-called
radial topology. Let $C = C^1([a, b], \mathbb{R}^n)$ be the vector space of all continuously differentiable $\mathbb{R}^n$-valued functions x on $[a, b]$ and C_0 be the subspace of functions
$x \in C$ such that $x(a) = x(b) = 0$. If the functional

$$\varphi(x) = \int_a^b L(t, x(t), x'(t))\, dt$$

has a relative minimiser at $x^0 = (x_1^0, \ldots, x_n^0)$, then for every function $z \in C_0$, the
real-valued function $g(\eta) = \varphi(x^0 + \eta z)$ has a relative minimum at $\eta = 0$. This
implies

$$0 = g'(0) = \frac{d}{d\eta}\varphi(x^0 + \eta z)\Big|_{\eta=0}$$

$$= \int_a^b \frac{d}{d\eta} L\Big(t, x^0(t) + \eta z(t), x^{0'}(t) + \eta z'(t)\Big)\Big|_{\eta=0} dt$$

$$= \int_a^b \left[\sum_{i=1}^n \frac{\partial L}{\partial x_i}(t, x^0(t), x^{0'}(t)) z_i(t) + \sum_{i=1}^n \frac{\partial L}{\partial y_i}(t, x^0(t), x^{0'}(t)) z_i'(t)\right] dt\,.$$

Using integration by parts on the second summand, we obtain

$$\int_a^b \sum_{i=1}^n \frac{\partial L}{\partial y_i}(t, x^0(t), x^{0'}(t)) z_i'(t)\, dt$$

$$= \sum_{i=1}^n \frac{\partial L}{\partial y_i}(t, x^0(t), x^{0'}(t)) z_i(t)\Big|_a^b - \int_a^b \sum_{i=1}^n \frac{d}{dt}\frac{\partial L}{\partial y_i}(t, x^0(t), x^{0'}(t)) z_i(t)\, dt\,.$$

Since $z(a) = z(b) = 0$, we obtain

$$\sum_{i=1}^n \int_a^b \left[\frac{\partial L}{\partial x_i}(t, x^0(t), x^{0'}(t)) - \frac{d}{dt}\frac{\partial L}{\partial y_i}(t, x^0(t), x^{0'}(t))\right] z_i(t)\, dt = 0$$

for all continuously differentiable real-valued functions $z_1, \ldots, z_n$ satisfying $z_i(a) = z_i(b) = 0$. That this fact implies

$$\frac{\partial L}{\partial x_i}(t, x^0(t), x^{0'}(t)) - \frac{d}{dt}\frac{\partial L}{\partial y_i}(t, x^0(t), x^{0'}(t)) = 0$$

for $a \le t \le b$ and $i = 1, \ldots, n$, is the so-called **fundamental lemma of the calculus of variations**. This can easily be shown by assuming the contrary, and constructing appropriate functions z_i that lead to a contradiction. So we have the following result:

Theorem 10.2 *Suppose that the real-valued function $L(t, x_1, \ldots, x_n, y_1, \ldots, y_n)$ is twice continuously differentiable. If the continuously differentiable $\mathbb{R}^n$-valued function $x(t)$ is a local extremiser of the extremum problem*

$$min\left\{\int_a^b L(t, x(t), x'(t)) \, dt \ : \ x(a) = \alpha, \ x(b) = \beta\right\},$$

then it satisfies the **Euler-Lagrange equations**

$$\frac{\partial L}{\partial x_i}(t, x(t), x'(t)) = \frac{d}{dt}\frac{\partial L}{\partial y_i}(t, x(t), x'(t)), \quad i = 1, \ldots, n.$$

10.3   The Geodesic Equation

We now return to the extremisation problem of force-free movement as formulated in Sect. 10.1. γ is a particle on which no forces act, and β is another parameterisation of this world line with a future-pointing derivative $\beta'(s)$. For points P and Q on this world line, the integral

$$\int_a^b \sqrt{g(\beta'(s), \beta'(s))} \, ds$$

is locally maximised among curves which assume points P and A at a and b. In particular, this also applies to points P and Q that are close enough to lie in a common chart φ. Then the problem becomes

$$max\left\{\int_a^b \sqrt{g_{ik}(\xi(s))\,(\xi^i)'(s)\,(\xi^k)'(s)} \, ds, \quad \xi(a) = \alpha, \ \xi(b) = \beta\right\}.$$

Theorem 10.3 *For a particle γ let $\xi = \varphi \circ \gamma$ be the chart representation of the curve in $\mathbb{R}^n$. For ξ, the Euler-Lagrange equations of the geodesic problem are equivalent to the* **geodesic equations**

$$(\xi^j)'' + \Gamma_{ik}^j(\varphi^{-1}(\xi))\,(\xi^i)'(\xi^k)' = 0, \quad j = 1, \ldots, n.$$

Proof With

$$L(t, x^1, \ldots, x^n, y^1, \ldots, y^n) = \sqrt{g_{lk}(x^1, \ldots, x^n)\, y^l y^k}$$

the Euler-Lagrange equations are given by

$$\frac{\partial L}{\partial x^i}(t, \xi(t), \xi'(t))$$

$$= \frac{\partial^2 L}{\partial t \partial y^i}(t, \xi(t), \xi'(t)) + \frac{\partial^2 L}{\partial x^j \partial y^i}(t, \xi(t), \xi'(t))\,(\xi^j)'(t) + \frac{\partial^2 L}{\partial y^j \partial y^i}(t, \xi(t), \xi'(t))\,(\xi^j)''(t).$$

First, for variables $t, x^1, \ldots, x^n, y^1, \ldots, y^n$ with

$$g_{lk}(x^1, \ldots, x^n)\, y^l y^k = 1$$

the needed partial derivatives must be calculated. We have

$$\frac{\partial L}{\partial x^i} = \frac{1}{2}\left(\frac{\partial}{\partial x^i} g_{lk}\right) y^l y^k$$

$$\frac{\partial^2 L}{\partial t \partial y^i} = \frac{\partial}{\partial y^i}\frac{\partial L}{\partial t} = 0$$

$$\frac{\partial L}{\partial y^i} = g_{ik}\, y^k$$

$$\frac{\partial^2 L}{\partial x^j \partial y^i} = \left(\frac{\partial}{\partial x^j} g_{ik}\right) y^k$$

$$\frac{\partial^2 L}{\partial y^j \partial y^i} = g_{ij}.$$

The Euler-Lagrange equations in the present case are

$$\frac{1}{2}\left(\frac{\partial}{\partial x^i} g_{jk}\right)(\xi^j)'(\xi^k)' = \left(\frac{\partial}{\partial x^j} g_{ik}\right)(\xi^j)'(\xi^k)' + g_{ij}\,(\xi^j)''.$$

By symmetrising the coefficients of the first sum on the right hand side and rearranging, we obtain

$$g_{ij}\,(\xi^j)'' = \frac{1}{2}\left(\frac{\partial}{\partial x^i} g_{jk} - \frac{\partial}{\partial x^j} g_{ik} - \frac{\partial}{\partial x^k} g_{ij}\right)(\xi^j)'(\xi^k)'.$$

By using the contravariant metric, these equations can be used to solve for the second derivatives. On the one hand

$$g^{il} g_{ij}\,(\xi^j)'' = (\xi^l)''$$

and on the other hand,

$$\frac{1}{2}g^{il}\left(\frac{\partial}{\partial x^i}g_{jk} - \frac{\partial}{\partial x^j}g_{ik} - \frac{\partial}{\partial x^k}g_{ij}\right)(\xi^j)'(\xi^k)' = -\Gamma^l_{jk}(\xi^j)'(\xi^k)',$$

and so

$$(\xi^l)'' = -\Gamma^l_{jk}(\xi^j)'(\xi^k)'.$$

The idea of a geodesic on a curved surface is that it is as straight as possible. With γ in natural parameter representation, that is, $|\gamma'| = 1$, this means that the tangential component of the acceleration γ'' formed in the ambient surrounding space R^3 vanishes. This can be expressed as $\nabla_{\gamma'}\gamma' = 0$. In fact, in the general case, this condition is equivalent to the geodesic equations formulated in Theorem 10.3, just as in $\mathbb{R}^n$.

Theorem 10.4 *For every particle (observer), there holds*

$$\nabla_{\gamma'}\gamma' = 0.$$

Proof After the conversion to $\xi = \varphi \circ \gamma$ using a map φ, the tangent vector γ' has the representation

$$\gamma' = (\xi^i)'\partial_i$$

and can be applied on the real-valued functions $(\xi^k)'$ in the sense

$$\gamma'(\xi^k)' = (\xi^k)''.$$

The following holds for the covariant derivative to be determined

$$\nabla_{\gamma'}\gamma' = \nabla_{(\xi^i)'\partial_i}(\xi^k)'\partial_k = (\xi^k)''\partial_k + (\xi^i)'(\xi^k)'\Gamma^j_{ik}\partial_j.$$

Thus the geodesic equations say that all the components of $\nabla_{\gamma'}\gamma'$ are zero.

In Newtonian mechanics, the motion of a particle in a gravitational field is clearly determined by the initial position and initial speed. So it can be expected that this also holds in spacetime. In fact, when the geodesic problem is expressed by using a chart, then Theorem 10.3 amounts to a system of differential equations for the functions $\xi^1, \ldots, \xi^n, (\xi^1)', \ldots, (\xi^n)'$, and from the structure of the system, it can be concluded that the corresponding initial value problem is (locally) clearly solvable. *For $P \in M$ and a future-pointing tangent vector $x \in M_P$ with $g(x, x) = 1$, there exists an interval I with $0 \in I$ and a geodesic $\gamma : I \to M$ with $\gamma(0) = P$ and $\gamma'(0) = x$.*

We now write the geodesic equations for the important case of Schwarzschild spacetime. The Christoffel symbols are given at the end of Sect. 7.5 and serve as coefficients in these equations. The additional condition $g(\gamma', \gamma') = 1$ must also be noted.

Theorem 10.5 *In the (outer) Schwarzschild spacetime, for a particle under the constraint*

$$h(r)(t')^2 - \frac{1}{h(r)}(r')^2 - r^2(\vartheta')^2 - r^2\sin^2\vartheta\,(\varphi')^2 = 1$$

the following **geodesic equations** *hold:*

$$t'' = -\frac{2\mu}{r^2 h(r)}t'r'$$

$$r'' = -\frac{h(r)\mu}{r^2}(t')^2 + \frac{\mu}{r^2 h(r)}(r')^2 + rh(r)(\vartheta')^2 + rh(r)\sin^2\vartheta\,(\varphi')^2$$

$$\vartheta'' = \sin\vartheta\cos\vartheta\,(\varphi')^2 - \frac{2}{r}r'\vartheta'$$

$$\varphi'' = -\frac{2}{r}r'\varphi' - 2\cot\vartheta\,\vartheta'\varphi'$$

with $h(r) = 1 - \frac{2\mu}{r}$.

Of particular interest is the description of radially moving particles. To do this, the equations given in Theorem 10.5 must be specialised for the case of constant angular coordinates ϑ and φ.

Theorem 10.6 *For a particle* $\gamma(\tau) = (t(\tau), r(\tau), Q)$ *in the Schwarzschild spacetime, there holds,*

$$r'' = -\frac{\mu}{r^2}.$$

Proof In the equation

$$r'' = -\frac{\mu}{r^2}\left(h(r)(t')^2 - \frac{1}{h(r)}(r')^2\right)$$

we use the constraint equation

$$h(r)(t')^2 - \frac{1}{h(r)}(r')^2 = 1$$

to obtain the desired conclusion.

In Newtonian physics, a point mass moving radially from a celestial body of mass μ experiences a radial acceleration $r'' = -G\mu/r^2$. The fact that the grvitational constant G does not appear as a factor, or is one, in Theorem 10.6, is a consequence of having set the speed of light to one (geometric units). By comparison, the constant μ from the Schwarzschild spacetime metric, which occurred when it was motivated in Sect. 9.7 for purely mathematical reasons, is identified as the mass of the celestial body that generates spacetime. When interpreting the formula in Theorem 10.6, it should be noted that r'' describes the radial acceleration only approximately for small speeds r' and large distances r. For the stationary observer $\partial_t/\sqrt{1-2\mu/r}$, the tangent vector $\gamma' = t'\partial_t + r'\partial_r$ has the relative speed

$$
v = \frac{t'\partial_t + r'\partial_r}{g\left(\partial_t/\sqrt{1-\frac{2\mu}{r}},\, t'\partial_t + r'\partial_r\right)} - \frac{\partial_t}{\sqrt{1-\frac{2\mu}{r}}} = \frac{r'}{t'\sqrt{1-\frac{2\mu}{r}}}\partial_r
$$

with the absolute value (speed)

$$
\sqrt{-g(v,v)} = \frac{r'}{t'(1-\frac{2\mu}{r})}.
$$

The derivative of this expression would be associated with the observed acceleration. For $r \gg 2\mu$, $r' \ll 1$ and therefore $t' \approx 1$ that would mean

$$
\left(\frac{r'}{t'\left(1-\frac{2\mu}{r}\right)}\right)' \approx r''.
$$

10.4   The Geodesic Deviation

If two geodesics γ_1 and γ_2 in $\mathbb{R}^n$ intersect at a point $P = \gamma_1(0) = \gamma_2(0)$, then the distance between the points $\gamma_1(t)$ and $\gamma_2(t)$ increases in proportion to the parameter t. If this is not the case for geodesics on a semi-Riemannian manifold, then this is a symptom of the curvature. To clarify this fact for a spacetime, we consider a chart φ with the following property: The coordinate vector fields ∂_1, ∂_2, ∂_3 are space-like, and for each assignment of coordinates u^1, u^2, u^3, the curve $\gamma(t) = \varphi^{-1}(t, u^1, u^2, u^3)$ is a geodesic with the future-pointing vector $\partial_0 = \gamma'$ and $g(\partial_0, \partial_0) = 1$. Such coordinates are called **geodesic coordinates**. From the physical point of view, it is plausible, at least in the case of the two examples discussed below, that such charts exist.

The distance between the points $\varphi^{-1}(t, u^1, u^2, u^3)$ and $\varphi^{-1}(t, u^1 + \Delta u, u^2, u^3)$ is given approximately by the length $\sqrt{-g(\Delta u\,\partial_1, \Delta u\,\partial_1)}$ of the vector $\Delta u\,\partial_1$. The vector ∂_1 at the position $\varphi^{-1}(t, u^1, u^2, u^3)$ describes, roughly speaking, for the displacement of $\varphi^{-1}(t, u^1, u^2, u^3)$ to $\varphi^{-1}(t, u^1 + 1, u^2, u^3)$. The covariant derivative $\nabla_{\partial_0}\partial_1$ describes the change in this displacement vector with increasing parameter t, and would be constant in flat space. The double covariant derivative $\nabla_{\partial_0}\nabla_{\partial_0}\partial_1$, which would be zero in flat space, is called the **geodesic deviation** with respect to the coordinate u^1. The geodesic deviation with respect to u^2 and u^3 are defined analogously.

Theorem 10.7 *The geodesic deviation with respect to u^i is described in terms of the curvature operator via the formula*

$$\nabla_{\partial_0} \nabla_{\partial_0} \partial_i = R(\partial_0, \partial_i)\partial_0 .$$

Proof According to Definition 8.1, Theorems 7.2 (D5) and 10.4, we have

$$R(\partial_0, \partial_i)\partial_0 = \nabla_{\partial_0} \nabla_{\partial_i} \partial_0 - \nabla_{\partial_i} \nabla_{\partial_0} \partial_0 = \nabla_{\partial_0} \nabla_{\partial_0} \partial_i .$$

Example 1. We approximately determine the components of the curvature tensor of the Schwarzschild spacetime by calculating the geodesic deviations with respect to the coordinates r, ϑ, φ while neglecting relativistic effects. A particle that starts with the intial velocity zero, and falls towards the center due to gravity, obeys the equation $r''(t) = -\mu/(r(t))^2$. Let d be a positive number, which should be interpreted to be relatively small. While a particle starting at a point P_0 with the coordinates r_0, ϑ_0, φ_0 has the initial acceleration $-\mu/r_0^2$, a particle starting at point P_1 with the coordinates $r_0 + d$, ϑ_0, φ_0 (see Fig. 10.2) starts with the initial acceleration

$$-\frac{\mu}{(r_0 + d)^2} = -\mu \left(\frac{1}{r_0^2} + \frac{-2}{r_0^3} d + \cdots \right) .$$

Hence, the displacement vector from the particle starting at P_0 to the particle starting at P_1 has the second derivative with respect to time given approximately by $d(2\mu/r_0^3)\partial_r$. It can be seen that the geodesic deviation with respect to the coordinate r at the point P_0 is approximately $(2\mu/r_0^3)\partial_r$. By Theorem 10.7, $(2\mu/r_0^3)\partial_r$ is an approximation of the tangent vector

$$R(\partial_t, \partial_r)\partial_t = R_{001}^0 \partial_t + R_{001}^1 \partial_r + R_{001}^2 \partial_\vartheta + R_{001}^3 \partial_\varphi ,$$

and a comparison of coefficients results, in particular,

$$R_{001}^1 \approx 2\mu/r_0^3 , \quad R_{001}^2 \approx 0 , \quad R_{001}^3 \approx 0 .$$

A particle starting at point P_2 with the coordinates r_0, $\vartheta_0 + d$, φ_0 also moves towards the center again with the acceleration $-\mu/r_0^2$ (see Fig. 10.2 again). The displacement vector from the particle starting at P_0 to the particle starting at P_2 has a second derivative approximately equal to

$$(\tan d) \left(-\frac{\mu}{r_0^2} \right) \frac{\partial_\vartheta}{r_0} \approx d \left(-\frac{\mu}{r_0^3} \right) \partial_\vartheta .$$

The geodesic deviation with respect to ϑ is thus approximately $-(\mu/r_0^3)\partial_\vartheta$, and this vector approximates $R(\partial_t, \partial_\vartheta)\partial_t$, implying that

$$R_{002}^1 \approx 0, \quad R_{002}^2 \approx -\mu/r_0^3 , \quad R_{002}^3 \approx 0 .$$

Fig. 10.2 Radially falling
particles, starting at P_0, P_1,
and P_2

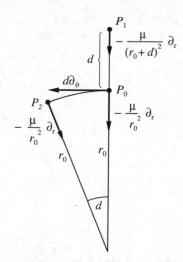

Analogously, a comparison of a particle starting at point P_3 with the coordinates $r_0, \vartheta_0, \varphi_0 + d$ with the particle starting at P_0 results in the geodesic deviation with respect to φ being given approximately by $-(\mu/r_0^3)\partial_\varphi$, and this implies for the corresponding components of the curvature tensor that

$$R_{003}^1 \approx 0\,, \quad R_{003}^2 \approx 0\,, \quad R_{003}^3 \approx -\mu/r_0^3\,.$$

Summarising, at the point with the coordinates t, r, ϑ, φ, we have

$$\begin{pmatrix} R_{001}^1 & R_{001}^2 & R_{001}^3 \\ R_{002}^1 & R_{002}^2 & R_{002}^3 \\ R_{003}^1 & R_{003}^2 & R_{003}^3 \end{pmatrix} \approx \begin{pmatrix} \dfrac{2\mu}{r^3} & 0 & 0 \\ 0 & -\dfrac{\mu}{r^3} & 0 \\ 0 & 0 & -\dfrac{\mu}{r^3} \end{pmatrix}.$$

In comparison, the precise result can be read from the matrices given in Sect. 8.5 for the curvature operators:

$$\begin{pmatrix} R_{001}^1 & R_{001}^2 & R_{001}^3 \\ R_{002}^1 & R_{002}^2 & R_{002}^3 \\ R_{003}^1 & R_{003}^2 & R_{003}^3 \end{pmatrix} = \begin{pmatrix} \dfrac{2\mu}{r^3}\left(1 - \dfrac{2\mu}{r}\right) & 0 & 0 \\ 0 & -\dfrac{\mu}{r^3}\left(1 - \dfrac{2\mu}{r}\right) & 0 \\ 0 & 0 & -\dfrac{\mu}{r^3}\left(1 - \dfrac{2\mu}{r}\right) \end{pmatrix}.$$

Example 2. We now calculate curvature components inside a spherically symmetric homogeneous celestial body of constant density ρ. We imagine that a straight shaft is drilled through the celestial body, which runs through the center (Fig. 10.3). If a point mass m is located in a shaft at a distance r from the center, then a gravitational force of magnitude $m(4/3)\pi\rho r^3/r^2$ acts on it, giving the particle an acceleration

Fig. 10.3 Free fall in radial tubes

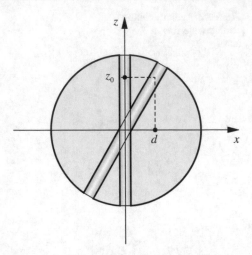

$r'' = -(4/3)\pi\rho r$. The resulting motion is periodic, which is described when the initial speed is zero by $r(t) = r_0 \cos(2\sqrt{\pi\rho/3}\, t)$ (here r is supposed to be signed). We can now use Cartesian coordinates with the z-axis along the shaft and the coordinate origin at the center. The particle dropped from z_0 at time instant $t = 0$ moves according to $z(t) = z_0 \cos(2\sqrt{\pi\rho/3}t)$. This displacement vector from this particle to a particle starting at z_0+d has the second derivative $-(4/3)\pi\rho d\partial_z$. The geodesic deviation with respect to the coordinate z is therefore

$$R(\partial_t, \partial_z)\partial_t = -(4/3)\pi\rho\, \partial_z\ ,$$

and this implies $R^1_{030} = R^2_{030} = 0$ und $R^3_{030} = (4/3)\pi\rho$. A particle is dropped through another drilled shaft from the position with the Cartesian coordinates $(d, 0, z_0)$; see Fig. 10.3. For the geodesic deviation with respect to x, we clearly have

$$R(\partial_t, \partial_x) = -(4/3)\pi\rho\partial_x\ .$$

The geodesic deviation with respect to y can be calculated analogously. Altogether, we have

$$\begin{pmatrix} R^1_{010} & R^2_{010} & R^3_{010} \\ R^1_{020} & R^2_{020} & R^3_{020} \\ R^1_{030} & R^2_{030} & R^3_{030} \end{pmatrix} = \frac{4\pi\rho}{3} \begin{pmatrix} 1\ 0\ 0 \\ 0\ 1\ 0 \\ 0\ 0\ 1 \end{pmatrix}\ .$$

The second example is of fundamental importance because it provides the motivation of one of the constants in Einstein's field equation. An equation

$$\mathrm{Ric} - (1/2)Sg = \mu T$$

with an unknown constant μ is equivalent to

$$\mathrm{Ric} = \mu(T - (1/2)CTg)$$

(see Theorem 9.7). In the present case, we take approximately

$$(g_{ik}) = \mathrm{diag}(1, -1, -1, -1)$$

and also neglect the pressure. The energy-momentum tensor T then has $T_{00} = \rho$ in the top left corner, and zeros elsewhere. On the right hand side of the modified field equation, we have

$$\mu(T - (1/2)CTg) = \mu\,\mathrm{diag}(\rho/2\,,\ \rho/2\,,\ \rho/2\,,\ \rho/2)\,.$$

On the other hand, the calculation of the geodesic deviations resulted in the curvature components $R^i_{0i0} = (4/3)\pi\rho$ for $i = 1, 2, 3$. Together with $R^0_{000} = 0$ this gives the number $4\pi\rho$ in the top left corner of the matrix for Ric. Finally, the equation $4\pi\rho = \mu\rho/2$ identifies the unknown constant μ as $\mu = 8\pi$.

According to classical Newtonian mechanics, different forces act on the individual mass points of an extended solid body that is in an inhomogeneous gravitational field, which lead to tensions in this body. In the theory of relativity, this phenomenon is explained by the fact that the mass points follow different geodesics, and this effect can be quantified using the concept of geodesic deviation. In the situation described at the beginning of this section, the geodesic deviation $\nabla_{\partial_0}\nabla_{\partial_0}\partial_1$ stands for the acceleration that a mass point at the point with coordinates $t,\ u^1+1,\ u^2,\ u^3$ experiences relative to the observer at the point with the coordinates $t,\ u^1,\ u^2,\ u^3$. For the observer, what happens is perceived as if a force is acting in the neighboring point, and this is the so-called **tidal force**. For the points along the u^i-coordinate line, this tidal force per distance is $\nabla_{\partial_0}\nabla_{\partial_0}\partial_i$. According to Theorem 10.7, this is the tangent vector $R(\partial_0, \partial_i)\partial_0$.

Definition 10.1 For an observer $z \in M_P$, the **tidal force** F_z in $z^\perp$ is defined by

$$F_z x = R(z, x)z\,.$$

♦

The linearity of F_z is clear. That $F_z x \in z^\perp$ follows from

$$g(z, F_z x) = g(z, R(z, x)z) = R(z, z, z, x) = 0$$

where we have used the skew-symmetry of the covariant curvature tensor with respect to the first two vector variables (see Theorem 8.5).

Theorem 10.8 *The tidal force operator F_z in the Euclidean space $z^\perp$, with the scalar product $-g$, is self adjoint and has the trace*

$$\mathrm{tr}\,F_z = \mathrm{Ric}(z, z)\,.$$

Proof By Theorem 8.5,

$$g(F_z x, y) = g(R(z, x)z, y) = R(y, z, z, x) = R(z, x, y, z) = R(x, z, z, y) = g(x, F_z y) .$$

With an orthonormal basis e_1, e_2, e_3 in $z^\perp$ and the corresponding linear forms $a^i = -g(e_i, .)$, we obtain

$$\langle F_z e_i, a^i \rangle = \langle R(z, e_i)z, a^i \rangle = -R(a^i, z, z, e_i) = R(a^i, z, e_i, z) = \mathrm{Ric}(z, z) .$$

Example. We calculate the matrix of the tidal force operator in the Schwarzschild spacetime for the observer $z = (1/\sqrt{h(r)}) \, \partial_t$ with respect to the orthonormal basis $e_1 = \sqrt{h(r)} \, \partial_r$, $e_2 = (1/r) \, \partial_\vartheta$ and $e_3 = (1/(r \sin \vartheta)) \, \partial_\varphi$. From the matrices given in Sect. 8.4, we can conclude that

$$F_z e_1 = R \left(\frac{1}{\sqrt{h(r)}} \, \partial_t , \sqrt{h(r)} \partial_r \right) \frac{1}{\sqrt{h(r)}} \, \partial_t = \frac{1}{\sqrt{h(r)}} \frac{2\mu h(r)}{r^3} \, \partial_r = \frac{2\mu}{r^3} \, e_1$$

and analogously

$$F_z e_2 = -\frac{\mu}{r^3} e_2 \quad \text{und} \quad F_z e_3 = -\frac{\mu}{r^3} e_3 .$$

Consequently, the matrix of F_z with respect to this basis is given by

$$((F_z)_i^k) = \mathrm{diag} \left(\frac{2\mu}{r^3}, -\frac{\mu}{r^3}, -\frac{\mu}{r^3} \right)$$

and has zero trace.

10.5   Perihelion Precession

Among other things, Kepler's laws state that the planets move on elliptical orbits, with the Sun at one of the focal points. Since the planets also influence each other through gravity, the point (**perihelion**) of the ellipse closest to the Sun can shift over time. However, decades of systematic observations of Mercury have shown that its perihelion, minus the influence of the other planets, also rotates the Sun at an angular velocity of $43, 1 \pm 0, 5$ arc s per century. This effect could only be explained with the general theory of relativity.

In order to be able to compare the relativistic result with the classical one, we first recapitulate the considerations that led to the formulation of the corresponding Kepler's law. The situation is taking place in a plane equipped with polar coordinates r and φ, where the Sun is placed at the origin. The motion of the planet is described by

$$x(t) = r(t) \cos \varphi(t)$$
$$y(t) = r(t) \sin \varphi(t) .$$

Its acceleration

$$\ddot{x} = (\ddot{r} - r\dot{\varphi}^2)\cos\varphi - (2\dot{r}\dot{\varphi} + r\ddot{\varphi})\sin\varphi$$
$$\ddot{y} = (\ddot{r} - r\dot{\varphi}^2)\sin\varphi + (2\dot{r}\dot{\varphi} + r\ddot{\varphi})\cos\varphi$$

has a radial component $\ddot{r} - r\dot{\varphi}^2$ and a tangential component $2\dot{r}\dot{\varphi} + r\ddot{\varphi}$. Since the Sun's attraction is only radial, we have

$$2\dot{r}\dot{\varphi} + r\ddot{\varphi} = 0 \, .$$

As

$$(r^2\dot{\varphi})^{\cdot} = 2r\dot{r}\dot{\varphi} + r^2\ddot{\varphi} = 0$$

the quantity $r^2\dot{\varphi}$ is constant (conservation of angular momentum), say L. According to Newton's law of gravitation, we have for the radial acceleration, that

$$\ddot{r} - r\dot{\varphi}^2 = -\frac{\mu}{r^2} \, .$$

We want to represent r as a function of φ. However, it turns out to be easier to formulate the reciprocal value $u = \frac{1}{r}$ in terms of φ. We have

$$\dot{r} = \frac{d}{dt}\frac{1}{u} = -\frac{1}{u^2}\frac{du}{d\varphi}\dot{\varphi} = -(r^2\dot{\varphi})\frac{du}{d\varphi} = -L\frac{du}{d\varphi}$$

and

$$\ddot{r} = -L\frac{d^2u}{d\varphi^2}\dot{\varphi} = -Lu^2\,(r^2\dot{\varphi})\frac{d^2u}{d\varphi^2} = -L^2\,u^2\frac{d^2u}{d\varphi^2} \, .$$

Inserting the above for radial acceleration into the equation from earlier yields

$$-\frac{\mu}{r^2} = \ddot{r} - r\dot{\varphi}^2 = -L^2u^2\frac{d^2u}{d\varphi^2} - L^2u^3 \, .$$

This is the **Binet differential equation**

$$\frac{d^2u}{d\varphi^2} + u = \frac{\mu}{L^2}$$

having the general solution

$$u(\varphi) = \lambda\cos(\varphi - \varphi_0) + \mu/L^2 \, .$$

Since the coordinate system can be rotated if necessary, we can assume that $\varphi_0 = \pi$. This results in the representation

$$r(\varphi) = \frac{1}{\mu/L^2 - \lambda\cos\varphi} = \frac{L^2/\mu}{1 - (\lambda\,L^2/\mu)\cos\varphi} \, .$$

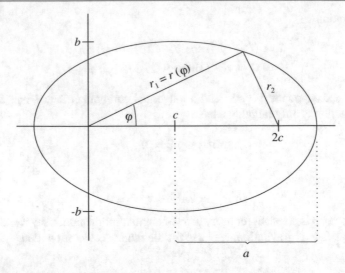

Fig. 10.4 Polar representation of the ellipse

One can see that for $|\lambda L^2/\mu| < 1$, this is an ellipse: We consider an ellipse with the focal points at $(0, 0)$ and $(2c, 0)$. Let a be the major axis and b be the minor axis. For a point (x, y) on the ellipse, let r_1 be the distance to $(0, 0)$ and r_2 be the distance to $(2c, 0)$ (see Fig. 10.4). Then

$$r_1 + r_2 = 2a$$

and

$$(r_1 + r_2)(r_1 - r_2) = r_1^2 - r_2^2 = x^2 + y^2 - [(x - 2c)^2 + y^2] = 4cx - 4c^2 ,$$

so that

$$r_1 - r_2 = 2\frac{c}{a}x - 2\frac{c^2}{a} .$$

Adding these, we obtain

$$r = r_1 = \frac{c}{a}x - \frac{c^2}{a} + a = \frac{c}{a}r\cos\varphi + \frac{b^2}{a}$$

and finally the polar representation

$$r = \frac{b^2/a}{1 - \varepsilon\cos\varphi} \qquad \text{with the \textbf{eccentricity} } \varepsilon = c/a .$$

A comparison shows that the planetary orbits in Newtonian physics are actually ellipses.

We will now work on determining the planetary orbit from the relativistic point of view. In the geodesic equations of the Schwarzschild spacetime, we set $\vartheta = \frac{\pi}{2}$. Then the geodesic equation gives

$$\varphi'' = -\frac{2}{r}r'\varphi'$$

and so

$$(r^2\varphi')' = r^2\varphi'' + 2rr'\varphi' = 0$$

showing that $r^2\varphi' = L$ is a constant. From the geodesic equation

$$r'' = -\frac{\mu}{r^2}\left(h(r)(t')^2 - \frac{1}{h(r)}(r')^2\right) + rh(r)(\varphi')^2$$

and the constraint

$$h(r)(t')^2 - \frac{1}{h(r)}(r')^2 - r^2(\varphi')^2 = 1$$

it follows that

$$r'' = -\frac{\mu}{r^2}(1 + r^2(\varphi')^2) + rh(r)(\varphi')^2 = -\frac{\mu}{r^2} + (r - 3\mu)(\varphi')^2 \ .$$

For $u = 1/r$, we have $u' = -r'/r^2$ and

$$u'' = -\frac{r''r^2 - 2r(r')^2}{r^4} = 2\frac{(r')^2}{r^3} - \frac{r''}{r^2} \ .$$

If u is thought of as a function of φ, then we have $\frac{du}{d\varphi} = u'/\varphi'$ and

$$\frac{d^2u}{d\varphi^2} = \left(\frac{u'}{\varphi'}\right)'\frac{1}{\varphi'} = \frac{u''\varphi' - u'\varphi''}{(\varphi')^3} = 2\frac{(r')^2}{r^3(\varphi')^2} - \frac{r''}{r^2(\varphi')^2} + \frac{r'\varphi''}{r^2(\varphi')^3}$$

$$= 2\frac{(r')^2}{r^3(\varphi')^2} + \frac{\mu}{r^4(\varphi')^2} - \frac{r - 3\mu}{r^2} - \frac{2(r')^2}{r^3(\varphi')^2} = \frac{\mu}{L^2} - u + 3\mu u^2 \ .$$

Theorem 10.9 *A particle moved is Schwarzschild spacetime with $\vartheta = \frac{\pi}{2}$. The function $\varphi \to u = 1/r$ satisfies the differential equation*

$$\frac{d^2u}{d\varphi^2} + u = \frac{\mu}{L^2} + 3\mu u^2$$

with $L = r^2\varphi'$.

Because of the additional summand $3\mu u^2$ in contrast to Binet's differential equation, the differential equation in Theorem 10.9 can no longer be solved in an elementary manner. But since r is very large in comparison to μ, μu^2 is therefore very small, which means that the additional summand has little influence, and the classical Keplerian solution

$$u = \frac{\mu}{L^2}(1 + \varepsilon \cos \varphi)$$

(with perihelion at $\varphi = 0$) to the Binet equation, can be considered as a first approximation. This by itself does not make much progress compared to the classical result, but at least motivates one to solve the differential equation

$$\frac{d^2 u}{d\varphi^2} + u = \frac{\mu}{L^2} + 3\mu \left(\frac{\mu}{L^2}\right)^2 (1 + \varepsilon \cos \varphi)^2 = \frac{\mu}{L^2} + \frac{3\mu^3}{L^4}\left(1 + \frac{\varepsilon^2}{2} + 2\varepsilon \cos \varphi + \frac{\varepsilon^2}{2}\cos 2\varphi\right)$$

instead of the differential equation given in Theorem 10.9, in order to get an even better approximation. This is again a second order linear differential equation. Using the method of undetermined coefficients, the particular solution

$$\frac{\mu}{L^2} + \frac{3\mu^3}{L^4} + \frac{3\mu^3 \varepsilon^2}{2L^4} + \frac{3\mu^3 \varepsilon}{L^4}\varphi \sin \varphi - \frac{\mu^3 \varepsilon^2}{2L^4}\cos 2\varphi$$

can be determined. So the function

$$u(\varphi) = \frac{\mu}{L^2}(1 + \varepsilon \cos \varphi) + \frac{3\mu^3}{L^4}\left(1 + \frac{\varepsilon^2}{2} - \frac{\varepsilon^2}{6}\cos 2\varphi + \varepsilon\varphi \sin \varphi\right)$$

is also a solution. There holds for it that $u'(0) = 0$ and $u(0) \approx (\mu/L^2)(1 + \varepsilon)$. This function is considered to be a better approximation of the solution to the differential equation from Theorem 10.9, and it is now used for further arguments. A perihelion is a local maximum of the function $u(\varphi)$. The derivative expression

$$u'(\varphi) = -\frac{\mu\varepsilon}{L^2}\sin \varphi + \frac{3\mu^3 \varepsilon}{L^4}\left(\frac{\varepsilon}{3}\sin 2\varphi + \sin \varphi + \varphi \cos \varphi\right)$$

shows that the perihelion at $\varphi = 0$ is followed by the next perihelion at $2\pi + \delta$ for a small positive number δ, which satisfies approximately that

$$0 \approx -\frac{\mu\varepsilon}{L^2}\sin \delta + \frac{3\mu^3 \varepsilon}{L^4}(2\pi + \delta) \cos \delta \,,$$

and so

$$\delta \approx \tan \delta \approx \frac{3\mu^2}{L^2}(2\pi + \delta) \approx \frac{6\pi\mu^2}{L^2} \,.$$

For the planet Mercury, this is a value that causes the perihelion to rotate about 43 arc s over the course of a century (Fig. 10.5).

Fig. 10.5 Perihelion rotation
of a planetary orbit

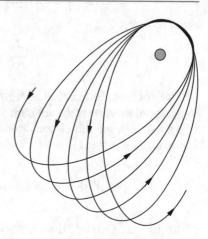

10.6   Light Deflection

In Newtonian physics, one takes as a fact that light travels in a straight line. In the theory of relativity it is obvious to expect to replace the straight line by geodesics. *Light rays are geodesics whose tangent vectors are light-like, i.e., geodesics γ for which $g(\gamma', \gamma') = 0$.* The term geodesic here of course is not meant as one which has extremal arc length, but that $\nabla_{\gamma'}\gamma' = 0$. One also says that the curve γ describes a photon.

The relativistic point of view has the consequence that a light beam that passes close to a fixed star is slightly deflected (Fig. 10.6). To prove and quantify this effect, we can once again use the geodesic equations of the Schwarzschild spacetime from Theorem 10.5, but constraint is now

$$h(r)(t')^2 - \frac{1}{h(r)}(r')^2 - r^2(\vartheta')^2 - r^2 \sin^2 \vartheta (\varphi')^2 = 0 \,.$$

If we modify the calculation done in Theorem 10.9 appropriatly, then we obtain the following result.

Theorem 10.10 *For a light beam $\gamma(\tau) = (t(\tau), r(\tau), Q(\tau))$, where $Q(\tau)$ has the angular coordinates $\pi/2$ and $\varphi(\tau)$, the function $\varphi \to u = 1/r$ satisifies the differ-*

Fig. 10.6 Deflection of a
beam of light

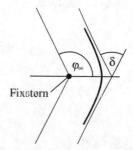

Fixstern

ential equation

$$\frac{d^2u}{d\varphi^2} + u = 3\mu u^2 .$$

This nonlinear differential equation can only be solved approximately. The first rough approximation is the solution $\tilde{u}(\varphi) = (1/r_0) \cos \varphi$ of the differential equation $\tilde{u}'' + \tilde{u} = 0$ with the initial conditions $\tilde{u}(0) = 1/r_0$ and $\tilde{u}'(0) = 0$. A much better approximation is then the solution of the inhomogeneous linear differential equation

$$u'' + u = 3\mu \left(\frac{1}{r_0} \cos \varphi \right)^2 = \frac{3\mu}{2r_0^2}(1 + \cos 2\varphi)$$

with $u(0) = 1/r_0$ and $u'(0) = 0$, too. Using the method of undetermined coefficients, one can find the particular solution

$$\varphi \to \frac{3\mu}{2r_0^2} - \frac{\mu}{2r_0^2} \cos 2\varphi .$$

Using the intial conditions then gives the function

$$u(\varphi) = \frac{\mu}{2r_0^2}(3 - \cos 2\varphi) + \frac{r_0 - \mu}{r_0^2} \cos \varphi ,$$

and this serves as a useful approximation to the solution of the differential equation given in Theorem 10.10 with the present initial conditions. The angle φ_∞ marked in Fig. 10.6 is a zero of this solution function. From

$$0 = u(\varphi_\infty) = \frac{\mu}{r_0^2}(2 - \cos^2 \varphi_\infty) + \frac{r_0 - \mu}{r_0^2} \cos \varphi_\infty$$

we obtain the quadratic equation

$$(\cos \varphi_\infty)^2 + \left(1 - \frac{r_0}{\mu} \right) \cos \varphi_\infty - 2 = 0$$

with the solution

$$\cos \varphi_\infty = \frac{r_0}{2\mu} - \frac{1}{2} - \sqrt{\left(\frac{r_0}{2\mu} - \frac{1}{2} \right)^2 + 2} .$$

The assumption that $r_0 \gg 2\mu$ leads to the approximation

$$\cos \varphi_\infty \approx \frac{r_0}{2\mu} - \frac{1}{2} - \left(\sqrt{\left(\frac{r_0}{2\mu} - \frac{1}{2}\right)^2 + 2 \frac{1/2}{\sqrt{\left(\frac{r_0}{2\mu} - \frac{1}{2}\right)^2}}} \right)$$

$$= \frac{-1}{\frac{r_0}{2\mu} - \frac{1}{2}} = -\frac{2\mu}{r_0 - \mu} \approx -2\mu/r_0.$$

This results in the deflection angle

$$\delta = 2\left(\varphi_\infty - \frac{\pi}{2}\right) \approx 2\sin\left(\varphi_\infty - \frac{\pi}{2}\right) = -2\cos\varphi_\infty \approx 4\mu/r_0 .$$

For our Sun, with $\mu \approx 1,477\,\mathrm{km}$ and with the solar radius $r_0 = 6,96 \cdot 10^5\,\mathrm{km}$, the deflection angle becomes $\delta = 8,489 \cdot 10^{-6}$, that is, 1.751 arc s. In principle, this effect can be observed during a solar eclipse. However, the measurement errors are still considerable with the current state of art, and so the observation result is an interval of approximately 1.5–2 arc s.

10.7   Red Shift

In addition to perihelion rotation and light deflection, the red shift is one of the classic tests of general relativity. What is meant here is the **gravitational red shift**, the **cosmological red shift** will be discussed only in Chap. 15.

The gravitational red shift means the effect that the light emitted from the surface of a star is received by a distant observer at a reduced frequency. The spectral lines shift towards the red end of the spectrum. This phenomenon is explained by the following two facts. First, for the function $t(r)$, which described the trajectory of a photon emitted radially outwards in the $r-t$-plane (r being the Schwarzschild radius and t the Schwarzschild time), we have, thanks to

$$h(r)(t')^2 - \frac{1}{h(r)}(r') = 0$$

the differential equation

$$\frac{dt}{dr} = \frac{t'}{r'} = \frac{1}{h(r)} .$$

The right hand side depends only on r, and so two photons emitted one after the other in the Schwarzschild-time-distance Δt keep this distance in the sense of Schwarzschild time. Secondly, the Schwarzschild time is not the proper time of an observer resting at a point with a Schwarzschild radius r, and the scale of this proper time depends on r.

Theorem 10.11 *Radial light signals that are emitted outwards from the Schwarzschild radius r_1 with the frequency ν_1, only have the frequency*

$$\nu_2 = \sqrt{\left(1 - \frac{2\mu}{r_1}\right) \Big/ \left(1 - \frac{2\mu}{r_1}\right)} \, \nu_1$$

at $r = r_2$.

Proof At the point $r = r_1$, N signals (or N wave crests) are sent radially outwards at regular intervals in periods Δt in Schwarzschild time. This process takes proper time there equal to

$$\int_t^{t+\Delta t} \sqrt{g(\partial_t, \partial_t)} \, d\tau = \sqrt{h(r_1)} \, \Delta t \; .$$

Consequently, the frequency for the transmitter is $\nu_1 = N/(\sqrt{h(r_1)} \, \Delta t)$. The Schwarzschild time Δt also passes for the receiver, but the proper time is $\sqrt{h(r_2)} \, \Delta t$ there, and hence the frequency

$$\nu_2 = \frac{N}{\sqrt{h(r_2)} \, \Delta t} = \sqrt{\frac{1 - 2\mu/r_1}{1 - 2\mu/r_2}} \, \nu_1 \; .$$

Covariant Differentiation of Tensor Fields

11.1 Parallel Transport of Vectors

For a semi-Riemannian manifold, the concept of covariant differentiation of vector fields enables the construction of isomorphism between tangent spaces via parallel transport. This then results in a characterisation of the covariant derivative of vector fields, which can be generalised to a definition of the covariant derivative of tensor fields.

Definition 11.1 A vector field X on a semi-Riemannian manifold M is said to be **parallel along the curve** γ, if for all t, $\nabla_{\gamma'(t)} X = 0$. ♦

Examples are depicted in Figs. 11.1 and 11.2. If M is a plane, the vector field X is parallel along an arbitrary chosen curve if and only if the image vectors $X(\gamma(t))$ are identical except for the usual parallel translation in linear space. The situation is more complicated on curved surfaces. If the curve is a geodesic, then the angle between the vectors $X(\gamma(t))$ and $\gamma'(t)$ must stay constant.

The covariant derivative mentioned in Definition 11.1 can be calculated in a chart φ via the formula

$$\nabla_{\gamma'(t)} X = (\varphi^i \circ \gamma)'(t) \left(\frac{\partial(X^k \circ \varphi^{-1})}{\partial x^i} (\varphi(\gamma(t))) \, \partial_k + X^k(\gamma(t)) \, \Gamma^j_{ik}(\gamma(t)) \, \partial_j \right)$$

$$= \left((X^j \circ \gamma)'(t) + \Gamma^j_{ik}(\gamma(t)) (\varphi^i \circ \gamma)'(t) (X^k \circ \gamma)(t) \right) \partial_j .$$

The covariant derivative is therefore the zero vector if and only if for $j = 1, \ldots, n$, we have

$$(X^j \circ \gamma)'(t) = -\Gamma^j_{ik}(\gamma(t)) (\varphi^i \circ \gamma)'(t) (X^k \circ \gamma)(t) .$$

© The Author(s), under exclusive license to Springer Nature Switzerland AG 2023
R. Oloff, *The Geometry of Spacetime*, Graduate Texts in Physics,
https://doi.org/10.1007/978-3-031-16139-1_11

Fig. 11.1 Parallel translation
in the plane

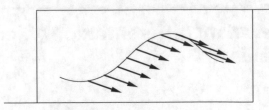

Fig. 11.2 Parallel translation
along the equator

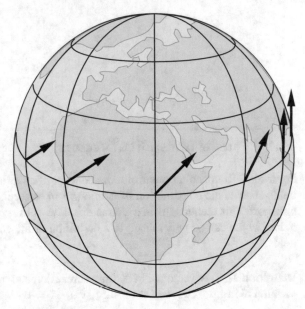

This is a homogeneous system of linear differential equations for the real functions
$X^j \circ \gamma$. As is well-known the initial value problem for such a system can be solved
uniquely. This establishes the following uniqueness theorem.

Theorem 11.1 *If the vector fields X and Y are parallel along the curve γ and
coincide at the point $\gamma(t_0)$, then for all t, $X(\gamma(t)) = Y(\gamma(t))$.*

This theorem gives rise to isomorphisms between the tangent spaces $M_{\gamma(t)}$.

Definition 11.2 For two real numbers s and t from the defining interval of the curve
γ, the linear map $\tau_{s,t}$ from $M_{\gamma(t)}$ to $M_{\gamma(s)}$ that sends the tangent vector $x \in M_{\gamma(t)}$
to the tangent vector $X(t) \in M_{\gamma(s)}$, where X is the vector field parallel along γ with
$X(s) = x$, is called **parallel transport along** γ.   ◆

The maps $\tau_{s,t}$ are obviously linear and injective, and because of the equality of
the dimensions of the tangent space, they are thus surjective, that is they are overall
bijective. It is important to point out the influence of the curve γ on parallel transport.
Parallel transport along different curves generally results in different outcomes. An
illustrative counterexample is shown in Fig. 11.3. A vector is transported from the

Fig. 11.3 Parallel transport
along various curves

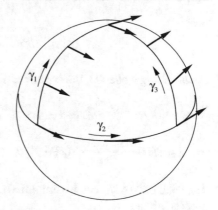

point P on the equator along the great circle γ_1 parallelly to the north pole N, and on
the other hand, for comparison, is first transported along a quarter γ_2 of the equator to
Q and then along the great circle γ_3 to N. The resulting vectors differ by a right angle.
This counterexample can also be rephrased as follows: The vector is rotated through
the angle $\pi/2$ by parallel transport along the closed curve $(-\gamma_1) \cup \gamma_2 \cup \gamma_3$. This
example thus illustrates the following theorem of classical differential geometry,
closely related to the **integral theorem of Gauss-Bonnet**: *The angle by which a
tangent vector is rotated by parallel transport along a simple closed curve is equal
to the integral over the enclosed surface of the Gaussian curvature.*

For parallel transport on a curved surface, the lengths of the vector and the angles
which they form with each other remain unchanged. This also holds in general.

Theorem 11.2 *For $x, y \in M_{\gamma(t)}$ and for all parameter values s, $g(\tau_{s,t}x, \tau_{s,t}y) =
g(x, y)$.*

Proof We choose two vector fields X and Y that are parallel along γ with $X(\gamma(t)) =
x$ and $Y(\gamma(t)) = y$. Then

$$\frac{d}{ds}g(\tau_{s,t}x, \tau_{s,t}y) = \frac{d}{ds}g\Big(X\big(\gamma(s)\big),\, Y\big(\gamma(s)\big)\Big) = \gamma'(s)g(X, Y)$$

$$= g(\nabla_{\gamma'(s)}X,\, Y) + g(X,\, \nabla_{\gamma'(s)}Y) = g(0, Y) + g(X, 0) = 0.$$

In Sect. 7.1, we had seen why it is initially not possible to introduce the covariant
derivative of a vector field as the limit of difference quotients. Using the parallel
transport of vectors, it is now possible to interpret the covariant derivative as the
limit of suitably formulated difference quotients.

Theorem 11.3 *The covariant derivative of the vector field Y with respect to the
tangent vector $\gamma'(t)$ of a curve γ can be expressed as*

$$\nabla_{\gamma'(t)}Y = \lim_{h \longrightarrow 0} \frac{1}{h}\Big(\tau_{t,t+h}Y(t+h) - Y(t)\Big) = v'(t)$$

with

$$v(s) = \tau_{t,s} Y\big(\gamma(s)\big).$$

Proof Using a chart φ, the covariant derivative becomes

$$\nabla_{\gamma'(t)} Y(\gamma(t)) = \nabla_{(\varphi^i \circ \gamma)'(t)\, \partial_i (\gamma(t))} Y^k(\gamma(t))\, \partial_k(\gamma(t))$$

$$= (\varphi^i \circ \gamma)'(t) \frac{\partial(Y^k \circ \varphi^{-1})}{\partial u^i} (\varphi(\gamma(t)))\, \partial_k(\gamma(t)) + (\varphi^i \circ \gamma)'(t)\, Y^k(\gamma(t))\, \Gamma^j_{ik}(\gamma(t))\, \partial_j(\gamma(t)).$$

The coefficient of ∂_k can be simplified even further, because the classical Chain Rule applies, giving

$$(Y^k \circ \gamma)'(t) = ((Y^k \circ \varphi^{-1}) \circ (\varphi \circ \gamma))'(t) = \frac{\partial(Y^k \circ \varphi^{-1})}{\partial u^i} (\varphi(\gamma(t)))\, (\varphi^i \circ \gamma)'(t).$$

Altogether, the covariant derivative is given by

$$\nabla_{\gamma'(t)} Y(\gamma(t)) = \Big[(Y^j \circ \gamma)'(t) + (\varphi^i \circ \gamma)'(t)\, Y^k(\gamma(t))\, \Gamma^j_{ik}(\gamma(t))\Big] \partial_j(\gamma(t)).$$

We now calculate $v'(t)$. The equation

$$Y(\gamma(s)) = \tau_{s,t} v(s)$$

means, in terms of components, that

$$Y^k(\gamma(s)) = (\tau_{s,t})^k_i v^i(s).$$

By differentiation with respect to the parameters, we obtain

$$(Y^k \circ \gamma)'(s) = \frac{\partial}{\partial s} (\tau_{s,t})^k_i v^i(s) + (\tau_{s,t})^k_i (v^i)'(s).$$

Since for any tangent vector $w = w^i \partial_i \in M_{\gamma(t)}$, the vector function

$$\tau_{s,t} w = (\tau_{s,t})^k_i w^i \partial_k$$

is parallel along γ, the following equation holds for its kth component:

$$\frac{\partial}{\partial s} (\tau_{s,t})^k_i w^i = -\Gamma^k_{lr}(\gamma(s))\, (\varphi^l \circ \gamma)'(s)\, (\tau_{s,t})^r_i w^i,$$

and in particular for $w = v(s)$. So we obtain

$$(Y^k \circ \gamma)'(s) = -\Gamma^k_{lr}(\gamma(s))\, (\varphi^l \circ \gamma)'(s)\, (\tau_{s,t})^r_i v^i(s) + (\tau_{s,t})^k_i (v^i)'(s).$$

For $s = t$, this gives

$$(v^k)'(t) = (Y^k \circ \gamma)'(t) + \Gamma^k_{lr}(\gamma(t)) \, (\varphi^l \circ \gamma)'(t) \, Y^r(\gamma(t)) \, .$$

This is the kth component of $\nabla_{\gamma'(t)} Y(\gamma(t))$.

In order to use the just proven relationship for the covariant differentiation of vector fields as an access point towards the covariant differentiation of tensor fields, we must first generalise the concept of parallel transport to also apply to tensors.

11.2 Parallel Transport of Tensors

So far we have formulated parallel transport $\tau_{s,t}$ along a curve γ as a map from $M_{\gamma(t)}$ to $M_{\gamma(s)}$. The associated dual map $\tau^*_{s,t}$ maps $M^*_{\gamma(s)}$ to $M^*_{\gamma(t)}$. In order to obtain a map from $M^*_{\gamma(t)}$ to $M^*_{\gamma(s)}$, again denoted by $\tau_{s,t}$, one could use $(\tau^{-1}_{s,t})^* = \tau^*_{t,s}$.

Definition 11.3 For $a \in M^*_{\gamma(t)}$, $\tau_{s,t}a \in M^*_{\gamma(s)}$ is defined by

$$\langle y, \tau_{s,t}a \rangle = \langle \tau_{t,s}y, a \rangle \qquad \text{for} \quad y \in M_{\gamma(s)} \, .$$

The thus defined linear map $\tau_{s,t}$ from $M^*_{\gamma(t)}$ to $M^*_{\gamma(s)}$ is also called **parallel transport along** γ. ◆

The parallel transport $\tau_{s,t}$ of covectors is thus the dual map $\tau^*_{t,s}$ of parallel transport $\tau_{t,s}$ of vectors. Hence every parallel transport of covectors is an isomorphism. Although the same notation is used, it will be clear from context whether parallel transport of vectors or foe covectors is meant. This also applies to parallel transport of general tensors.

Definition 11.4 For $T \in (M_{\gamma(t)})^p_q$, $\tau_{s,t}T \in (M_{\gamma(s)})^p_q$ is defined by

$$\tau_{s,t}T(b^1, \ldots, b^p, y_1, \ldots, y_q) = T(\tau_{t,s}b^1, \ldots, \tau_{t,s}b^p, \tau_{t,s}y_1, \ldots, \tau_{t,s}y_q)$$

for $b^1, \ldots, b^p \in M^*_{\gamma(s)}$ and $y_1, \ldots, y_q \in M_{\gamma(s)}$. ◆

Definition 11.4 generalises Definitions 11.2 and 11.3, because when $p = 0$ and $q = 1$, Definition 11.4 means the same as Definition 11.3. Parallel transport of a vector x is the same as parallel transport of the one-fold contravariant tensor x, because according to Definitions 11.4 and 11.3, we have

$$\tau_{s,t}x(b) = x(\tau_{t,s}b) = \langle x, \tau_{t,s}b \rangle = \langle \tau_{s,t}x, b \rangle \, ,$$

and so the parallely transported $(1, 0)$-tensor $\tau_{s,t}x$ can be represented by the vector $\tau_{s,t}x$.

The characterisation of parallel transport in the form

$$\tau_{s,t} T(\tau_{s,t} a^1, \ldots, \tau_{s,t} a^p, \tau_{s,t} x_1, \ldots, \tau_{s,t} x_q) = T(a^1, \ldots, a^p, x_1, \ldots, x_q)$$

is sometimes useful. In particular,

$$\langle \tau_{s,t} x, \tau_{s,t} a \rangle = \langle x, a \rangle .$$

Furthermore, obviously

$$\tau_{s,t}(S \otimes T) = (\tau_{s,t} S) \otimes (\tau_{s,t} T) .$$

However, the fact that parallel transport commutes with every contraction needs a mathematical proof.

Theorem 11.4 *For $u \in \{1, \ldots, p\}$, $v \in \{1, \ldots, q\}$ and $T \in (M_{\gamma(t)})_p^q$, we have*

$$C_v^u \tau_{s,t} T = \tau_{s,t} C_v^u T .$$

Proof Let $x_1, \ldots, x_n$ be a basis for $M_{\gamma(t)}$ and $a^1, \ldots, a^n$ be the corresponding dual basis. Since the parallel transport of vectors is an isomorphism, the vectors $\tau_{s,t} x_i$ form again a basis for $M_{\gamma(s)}$, and because

$$\langle \tau_{s,t} x_i, \tau_{s,t} a^k \rangle = \langle x_i, a^k \rangle$$

the covectors $\tau_{s,t} a^k$ are the corresponding dual vectors. For $y_1, \ldots, y_{v-1}$, $y_{v+1}, \ldots, y_q \in M_{\gamma(s)}$ and $b^1, \ldots, b^{u-1}, b^{u+1}, \ldots, b^p \in M_{\gamma(s)}^*$, we have

$$C_v^u \tau_{s,t} T\left(\ldots, b^{u-1}, b^{u+1}, \ldots, y_{v-1}, y_{v+1}, \ldots \right)$$

$$= \tau_{s,t} T\left(\ldots, b^{u-1} \tau_{s,t} a^i, b^{u+1}, \ldots, y_{v-1}, \tau_{s,t} x_i, y_{v+1}, \ldots \right)$$

$$= T\left(\ldots, \tau_{t,s} b^{u-1}, a^i, \tau_{t,s} b^{u+1}, \ldots, \tau_{t,s} y_{v-1}, x_i, \tau_{t,s} y_{v+1}, \ldots \right)$$

as well as

$$\tau_{s,t} C_v^u T\left(\ldots, b^{u-1}, b^{u+1}, \ldots, y_{v-1}, y_{v+1}, \ldots \right)$$

$$= C_v^u T\left(\ldots, \tau_{t,s} b^{u-1}, \tau_{t,s} b^{u+1}, \ldots, \tau_{t,s} y_{v-1}, \tau_{t,s} y_{v+1}, \ldots \right)$$

$$= T\left(\ldots, \tau_{t,s} b^{u-1}, a^i, \tau_{t,s} b^{u+1}, \ldots, \tau_{t,s} y_{v-1}, x_i, \tau_{t,s} y_{v+1}, \ldots \right) .$$

As mentioned earlier, now analogous to Theorem 11.3, we introduce the covariant differentiation of tensor fields.

Definition 11.5 The **covariant derivative of the tensor field** S **with respect to the tangent vector** $\gamma'(t)$ generated by a curve γ is defined by

$$\nabla_{\gamma'(t)}S = \lim_{h \to 0} \frac{1}{h}\Big(\tau_{t,t+h}S(t+h) - S(t)\Big) = v'(t)$$

with

$$v(s) = \tau_{t,s}S(\gamma(s)) .$$

$\blacklozenge$

The covariant differentiation with respect to a vector $x \in M_P$ turns a tensor field $S \in T_q^p(P)$ to a tensor $\nabla_x S \in (M_P)_q^p$. Of course, a vector field can also be used for the derivation instead of a single vector. Then a vector field $X \in \mathcal{X}(P)$ and a tensor field $S \in T_q^p(P)$ combine to form the tensor field $\nabla_X S \in T_q^p(P)$, defined by

$$\nabla_X S(P) = \nabla_{X(P)}S = \nabla_{\gamma'(0)}S$$

with $\gamma(0) = P$ and $\gamma'(0) = X(P)$.

11.3 Calculation Rules and Component Representation

Theorem 11.5 *For any vector field* X *and tensor fields* S *and* T, *we have*

$$\nabla_X(S \otimes T) = \nabla_X S \otimes T + S \otimes \nabla_X T .$$

Proof As with the proof of the classical product rule for differentiating a product of two functions of a real variable, we split the difference quotient into two summands and obtain

$$\nabla_{\gamma'(0)}(S \otimes T)$$

$$= \lim_{h \to 0} \frac{\tau_{0,h}\Big(S(\gamma(h)) \otimes T(\gamma(h))\Big) - S(\gamma(0)) \otimes T(\gamma(0))}{h}$$

$$- \lim_{h \to 0} \frac{\tau_{0,h}S(\gamma(h)) \otimes \tau_{0,h}T(\gamma(h)) - S(\gamma(0)) \otimes T(\gamma(0))}{h}$$

$$= \lim_{h \to 0} \frac{\tau_{0,h}S(\gamma(h)) - S(\gamma(0))}{h} \otimes \tau_{0,h}T(\gamma(h)) +$$

$$+ \lim_{h \to 0} S(\gamma(0)) \otimes \frac{\tau_{0,h}T(\gamma(h)) - T(\gamma(0))}{h}$$

$$= \nabla_{\gamma'(0)}S \otimes T(\gamma(0)) + S(\gamma(0)) \otimes \nabla_{\gamma'(0)}T .$$

Theorem 11.6 *For any vector field X, and tensor field S and any contraction C, we have*

$$\nabla_X CS = C\nabla_X S.$$

Proof By Theorem 11.4, we have

$$\nabla_{\gamma'(0)} CS$$

$$= \lim_{h \to 0} \frac{\tau_{0,h} CS(\gamma(h)) - CS(\gamma(0))}{h}$$

$$= \lim_{h \to 0} C\frac{\tau_{0,h} S(\gamma(h)) - S(\gamma(0))}{h} = C\nabla_{\gamma'(0)} S.$$

A covector field is primarily intended to be applied to a vector field. With this in mind, the question arises as to what effect the covector field $\nabla_X A$ has on a vector field Y. The following Theorem 11.7 gives in the special case $p=0$ and $q=1$ a characterisation of the scalar field $\langle Y, \nabla_X A\rangle$. Arranged somewhat more suggestively, the equation is

$$X\langle Y, A\rangle = \langle \nabla_X Y, A\rangle + \langle Y, \nabla_X A\rangle,$$

and his follows directly from Theorem 11.5 giving

$$\nabla_X (Y \otimes A) = \nabla_X Y \otimes A + Y \otimes \nabla_X A$$

via contraction.

Theorem 11.7 *For any (p, q)-tensor field S, vector fields $X, Y_1, \ldots, Y_q$ and covector fields $A^1, \ldots, A^p$, there holds*

$$(\nabla_X S)(A^1, \ldots, Y_q)$$

$$= X(S(A^1, \ldots, Y_q)) - S(\nabla_X A^1, A^2, \ldots, Y_q) - \cdots - S(A^1, \ldots, Y_{q-1}, \nabla_X Y_q).$$

Proof We first determine the "complete" contraction of a (p, q)-tensor s with a (q, p)-tensor of the form $x_1 \otimes \cdots \otimes x_q \otimes a^1 \otimes \cdots \otimes a^p$, where the corresponding inputs are offset in pairs. With $y_1, \ldots, y_n$ a basis for the underlying space and the associated dual basis $b^1, \ldots, b^n$, we have

$$(s \otimes x_1 \otimes \ldots \otimes x_q \otimes a^1 \otimes \ldots \otimes a^p)(b^{i_1}, \ldots, b^{i_p}, b^{i_{p+1}}, \ldots, b^{i_{p+q}}, y_{i_{p+1}}, \ldots, y_{i_{p+q}}, y_{i_1}, \ldots, y_{i_p})$$

$$= s(b^{i_1}, \ldots, b^{i_p}, y_{i_{p+1}}, \ldots, y_{i_{p+q}})\langle x_1, b^{i_{p+1}}\rangle \cdots \langle x_q, b^{i_{p+q}}\rangle\langle y_{i_1}, a^1\rangle \cdots \langle y_{i_p}, a^p\rangle$$

$$= s(\ldots, \langle y_{i_p}, a^p\rangle b^{i_p}, \langle x_1, b^{i_{p+1}}\rangle y_{i_{p+1}}, \ldots) = s(a^1, \ldots, a^p, x_1, \ldots, x_q).$$

Contraction of the tensor s with the simple tensor is thus the insertion of its component parts into s. This result naturally also applies to (p, q)-tensor fields, vector fields and covector fields. By Theorem 11.5, we obtain

$$\nabla_X(S \otimes Y_1 \otimes \cdots \otimes Y_q \otimes A^1 \otimes \cdots \otimes A^p)$$
$$= (\nabla_X S) \otimes Y_1 \otimes \cdots \otimes A^p + S \otimes (\nabla_X Y_1) \otimes \cdots \otimes A^p + \ldots + S \otimes Y_1 \otimes \cdots \otimes (\nabla_X A^p).$$

Complete contraction provides, by the calculation rule that has just been proven, that

$$X(S(A^1, \ldots, A^p, Y_1, \ldots, Y_q))$$
$$= (\nabla_X S)(A^1, \ldots, Y_q) + S(\nabla_X A^1, \ldots, Y_q) + \cdots + S(A^1, \ldots, \nabla_X Y_q).$$

According to Definition 11.5, the covariant derivative $\nabla_X S(P)$ of the tensor field S with the vector field X at the point P uses from the vector field just its value $X(P)$, and the tensor $\nabla_X S(P)$ depends on this vector in a linear manner.

Definition 11.6 The covariant derivative of the (p, q)-tensor field S is the $(p, q+1)$ tensor field ∇S, which at P maps the tangent vector $x \in M_P$ to the (p, q)-tensor $\nabla_x S$. ♦

In other words, at the point P, the covectors $a^1, \ldots, a^p \in M_P^*$ and the vectors $x_0, x_1, \ldots, x_q \in M_P$ are multilinearly mapped to the number

$$(\nabla S)(P)(a^1, \ldots, a^p, x_0, x_1, \ldots, x_q) = \nabla_{x_0} S(a^1, \ldots, a^p, x_1, \ldots, x_q).$$

Example The covariant derivative of the metric g is zero, because of the property (D4) in Theorem 7.2:

$$(\nabla_X g)(Y, Z) = Xg(Y, Z) - g(\nabla_X Y, Z) - g(Y, \nabla_X Z) = 0.$$

We now investigate how the components of the covariant derivative ∇S of a (p, q)-tensor field S are calculated from its components in a chart. In the special case $p=1$ and $q=0$, Theorem 7.3 says

$$(\nabla Y)_j^i = \partial_j Y^i + \Gamma_{jk}^i Y^k.$$

As a generalisation, we obtain the following result.

Theorem 11.8 *The components of the covariant derivative ∇S of the (p, q)-tensor field S are calculated from its components as follows:*

$$(\nabla S)_{j_0 j_1 \cdots j_q}^{i_1 \cdots i_p} = \partial_{j_0} S_{j_1 \cdots j_q}^{i_1 \cdots i_p} +$$

$$+ \Gamma_{j_0 k_1}^{i_1} S_{j_1 \cdots j_q}^{k_1 i_2 \cdots i_p} + \cdots + \Gamma_{j_0 k_p}^{i_p} S_{j_1 \cdots j_q}^{i_1 \cdots i_{p-1} k_p} - \Gamma_{j_0 j_1}^{l_1} S_{l_1 j_2 \cdots j_q}^{i_1 \cdots i_p} - \cdots - \Gamma_{j_0 j_q}^{l_q} S_{j_1 \cdots j_{q-1} l_q}^{i_1 \cdots i_p}.$$

Proof The coordinate vector fields $\partial_1, \ldots, \partial_n$ and the dual covector fields $du^1, \ldots, du^n$ are to be inserted in the expression for $\nabla_{\partial_{j_0}} S$ given in Theorem 11.7. This results in components of the form

$$\partial_{j_0}\Big(S(du^{i_1}, \ldots, du^{i_p}, \partial_{j_1}, \ldots, \partial_{j_q})\Big) = \partial_{j_0} S^{i_1 \cdots i_p}_{j_1 \cdots j_q}$$

and

$$-S(du^{i_1}, \ldots, du^{i_p}, \partial_{j_1}, \ldots, \nabla_{\partial_{j_0}} \partial_{j_r}, \ldots, \partial_{j_q}) = -\Gamma^{l_r}_{j_0 j_r} S^{i_1 \cdots i_p}_{j_1 \cdots l_r \cdots j_q}.$$

The calculation rule

$$\langle Z, \nabla_X A \rangle = X \langle Z, A \rangle - \langle \nabla_X Z, A \rangle$$

should be used to calculate the derivatives of covector fields. In particular,

$$\langle \partial_{k_s}, \nabla_{\partial_{j_0}} du^{i_s} \rangle = -\langle \nabla_{\partial_{j_0}} \partial_{k_s}, du^{i_s} \rangle = -\Gamma^{i_s}_{j_0 k_s},$$

and so

$$\nabla_{\partial_{j_0}} du^{i_s} = -\Gamma^{i_s}_{j_0 k_s} du^{k_s}.$$

This leads to the remaining components

$$-S(du^{i_1}, \ldots, \nabla_{\partial_{j_0}} du^{i_s}, \ldots, du^{i_p}, \partial_{j_1}, \ldots, \partial_{j_q}) = \Gamma^{i_s}_{j_0 k_s} S^{i_1 \cdots k_s \cdots i_p}_{j_1 \cdots j_q}.$$

Finally, we mention the notation widely used in the physics literature,

$$S^{i_1 \cdots i_p}_{j_1 \cdots j_q, j} = \partial_j S^{i_1 \cdots i_p}_{j_1 \cdots j_q}$$

and

$$S^{i_1 \cdots i_p}_{j_1 \cdots j_q; j} = (\nabla_{\partial_j} S)^{i_1 \cdots i_p}_{j_1 \cdots j_q}.$$

11.4 The Second Bianchi Identity

While the first Bianchi identity (Theorem 8.3(2)) describes a cyclic property of the curvature tensor, the second Bianchi identity (Theorem 11.9) deals with such a property of the covariant derivative of the curvature tensor. The curvature tensor was originally introduced as a trilinear mapping (Definition 8.2). In the sense of a basis-independent isomorphism, this corresponds to a mapping that assigns a number to a linear form and three vectors quarti-linearly. The curvature tensor as a tensor field thus assigns a scalar field to a covector field and three vector fields quarti-linearly and $\mathcal{F}$-homogeneously. Its covariant derivative in the sense of Definition 11.6 and Theorem 11.7 is a multilinear mapping from $\mathcal{K}(M) \times \mathcal{X}(M) \times \mathcal{X}(M) \times \mathcal{X}(M) \times \mathcal{X}(M)$ to

$\mathcal{F}(M)$, which can ultimately also be understood as a quarti-linear map ∇R, which assigns to four vector fields X, Y, Z, V a vector field $(\nabla_V R)(Y, Z)X$, such that for each covector field A, there holds

$$\langle (\nabla_V R)(Y, Z)X, A) \rangle = (\nabla_V R)(A, X, Y, Z)$$

$$= V R(A, X, Y, Z) - R(\nabla_V A, X, Y, Z) - R(A, \nabla_V X, Y, Z) -$$
$$- R(A, X, \nabla_V, Y, Z) - R(A, X, Y, \nabla_V Z)$$

$$= V \langle R(Y, Z)X, A \rangle - \langle R(Y, Z)X, \nabla_V A \rangle -$$
$$- \langle R(Y, Z)\nabla_V X + R(\nabla_V Y, Z)X + R(Y, \nabla_V Z)X, A \rangle$$

$$= \langle \nabla_V(R(Y, Z)X) - R(Y, Z)\nabla_V X - R(\nabla_V Y, Z)X - R(Y, \nabla_V Z)X, A \rangle.$$

The covariant derivative ∇R of the curvature tensor R is therefore also characterised by

$$(\nabla_V R)(Y, Z)X = \nabla_V(R(Y, Z)X) - R(Y, Z)\nabla_V X - R(\nabla_V Y, Z)X - R(Y, \nabla_V Z)X.$$

Theorem 11.9 *The covariant derivative ∇R of the curvature tensor R of a semi-Riemannian manifold fulfills the* **second Bianchi identity**

$$(\nabla_Z R)(X, Y) + (\nabla_X R)(Y, Z) + (\nabla_Y R)(Z, X) = 0$$

for vector fields X, Y, Z. In terms of components of the curvature tensor, we have

$$R^s_{ijk;l} + R^s_{ikl;j} + R^s_{ilj;k} = 0.$$

Proof We need to show

$$(\nabla_Z R)(X, Y)V + (\nabla_X R)(Y, Z)V + (\nabla_Y R)(Z, X)V = 0$$

for an additional vector field V. To do this, we use the representation formula for ∇R, established just before Theorem 11.9, the skew-symmetry Theorem 8.3(1), the multilinearity of R, the axiom (D5) for the Levi-Civita connection, and finally Definition 8.1. The result is a sum in which the summands cancel each other out in pairs, except for

$$-\nabla_{[[X,Z],Y]}V - \nabla_{[[Y,X],Z]}V - \nabla_{[[Z,Y],X]}V = \nabla_{[[Z,X],Y]+[[X,Y],Z+[[Y,Z],X]}V = \nabla_0 V = 0$$

(Jacobi-identity, Theorem 2.11). The formula in terms of the components of the curvature tensor describes the sth component of

$$(\nabla_{\partial_i} R)(\partial_j, \partial_k)\partial_i + (\nabla_{\partial_j} R)(\partial_k, \partial_l)\partial_i + (\nabla_{\partial_k} R)(\partial_l, \partial_j)\partial_i = 0.$$

11.5   Divergence

Definition 11.7 The **divergence** of a symmetric $(0, q)$-tensor field T is the $(0, q-1)$-tensor field $\mathrm{div}T$, determined as follows: Index pulling is applied to the covariant derivative ∇T with respect to the new vector variable and the resulting $(1, q)$-tensor field is contracted.   ◆

In components, this is expressed as follows: The components of $\mathrm{div}T$ are calculated from the components of T by

$$(\mathrm{div}T)_{i_1\cdots i_{q-1}} = g^{ji_q} T_{i_1\cdots i_q;j} \ .$$

In the case of $q=1$ and $M=\mathbb{R}^n$, the above definition corresponds to the classical notation of divergence. Here we are particularly interested in the case of $q=2$.

Example The divergence of the metric is zero. This is due to the fact that the covariant derivative of the metric is already zero (see example following Definition 11.6). The following theorem gives one more example.

Theorem 11.10 *In any semi-Riemannian manifold, $\mathrm{div}\mathrm{Ric} = \frac{1}{2}dS$.*

Proof Using the second Bianchi identity and the skew-symmetry properties of the curvature tensor, the components of the outer derivative of the curvature scalar S can be expressed in terms of the components of the divergence of the Ricci tensor. We have

$$S_{;i} = (g^{jk}(\mathrm{Ric})_{jk})_{;i} = g^{jk}(\mathrm{Ric})_{jk;i} = g^{jk} R^r_{jrk;i} = g^{jk}\left(R^r_{jri;k} - R^r_{jki;r}\right)$$
$$= g^{jk}(\mathrm{Ric})_{ji;k} - g^{jk}g^{sr}R_{sjki;r} = (\mathrm{div}\mathrm{Ric})_i + g^{sr}R^k_{ski;r} = 2(\mathrm{div}\mathrm{Ric})_i \ .$$

The mapping div from the subspace of symmetric members of $\mathcal{T}_q^0(M)$ to $\mathcal{T}_{q-1}^0(M)$ is clearly linear, that is, for symmetric covariant tensor fields S and T and numbers λ and μ, we have

$$\mathrm{div}(\lambda S + \mu T) = \lambda \, \mathrm{div}S + \mu \, \mathrm{div}T \ .$$

If there are scalar fields instead of numbers, then things are a bit more complicated. We are only interested in the equation $\mathrm{div}(Sg) = dS$, for the metric g and the curvature scalar S, which is easily verifiable by comparing components. Together with Theorem 11.10, this gives for the Einstein tensor $G = \mathrm{Ric} - \frac{1}{2}Sg$ that $\mathrm{div}G = 0$. This motivates the coefficient $\frac{1}{2}$ in the Einstein field equation, since the energy-momentum tensor must be divergence-free on physical grounds. To clarify this, the following two concepts are useful at least as abbreviations.

Definition 11.8 The **divergence** of a vector field is the contraction of its covariant derivative, i.e., $\mathrm{div}X = X^i_{;i}$.   ◆

Definition 11.9 The **gradient** of a scalar field f is the vector field $\operatorname{grad} f$, characterised by

$$g(\operatorname{grad} f, X) = df(X) = Xf$$

for every vector field X, i.e., $(\operatorname{grad} f)^i = g^{ij} \partial_j f$. ◆

Theorem 11.11 *For the energy-momentum tensor of an ideal flow $[Z, \rho, p]$, the statement $\operatorname{div} T = 0$ is equivalent to the* **energy equation** *(E) $Z\rho = -(\rho + p)\operatorname{div} Z$ and the* **force equation** *(K) $(\rho + p)\nabla_Z Z = \operatorname{grad}_\perp p$, where $\operatorname{grad}_\perp p$ is the part of $\operatorname{grad} p$ orthogonal to Z.*

Proof In terms of components of T, the statement $g^{jk} T_{ik;j} = 0$ is, thanks to the invertibility fo the matrix (g^{il}), equivalent to

$$0 = g^{il} g^{jk} T_{ik;j} = (g^{il} g^{jk} T_{ik})_{;j} = ((\rho + p) Z^l Z^j - p g^{lj})_{;j}$$

$$= (\partial_j \rho + \partial_j p) Z^l Z^j + (\rho + p)(Z^l_{;j} Z^j + Z^l Z^j_{;j}) = \partial_j p \, g^{lj}$$

that is

$$0 = Z(\rho + p)Z + (\rho + p)\nabla_Z Z + (\rho + p)(\operatorname{div} Z)Z - \operatorname{grad} p .$$

Since $g(Z, Z) = 1$, $\nabla_Z Z$ is orthogonal to Z. The part of the right hand side of the last vectorial equation that is orthogonal to Z is thus given by $(\rho + p)\nabla_Z Z - \operatorname{grad}_\perp p$. This proves (K). (E) results from inserting the right hand side in $g(Z, .)$. Conversely, (E) and (K) obviously also yield $g^{il} g^{jk} T_{ik;j} = 0$ and thus $\operatorname{div} T = 0$.

The equations (E) and (K) are reasonable from a physical point of view, and therefore $\operatorname{div} T = 0$ should also apply to an ideal flow. This is the requirement which was announced in Sect. 9.1.

The Lie Derivative

12

12.1 The Flow and Its Tangents

As already mentioned in Sect. 2.3, a vector field is to be interpreted as a flow. It is then natural to investigate where this flow transports a particle over a certain period of time (Fig. 12.1). A particle subject to the flow drifts along an integral curve γ, which is characterised by $\gamma'(t) = X(\gamma(t))$, that is,

$$(f \circ \gamma)'(t) = X(\gamma(t))f \quad \text{for } f \in \mathcal{F}(M).$$

By employing a chart φ, and with the notation

$$\varphi(\gamma(t)) = (\gamma^1(t), \dots, \gamma^n(t))$$

the left hand side becomes

$$(f \circ \gamma)'(t) = (f \circ \varphi^{-1} \circ \varphi \circ \gamma)(t) = \frac{\partial(f \circ \varphi^{-1})}{\partial u^k}(\varphi(\gamma(t)))\frac{d\gamma^k}{dt}(t)$$

and for the right hand side we obtain

$$X(\gamma(t))f = X^k(\gamma(t))\,\partial_k f = (X^k \circ \varphi^{-1})(\varphi(\gamma(t)))\frac{\partial(f \circ \varphi^{-1})}{\partial u^k}(\varphi(\gamma(t)))\,.$$

Comparing coefficients, we arrive at the differential equation system

$$\frac{d}{dt}\gamma^k(t) = (X^k \circ \varphi^{-1})(\gamma^1(t), \dots, \gamma^n(t))$$

for the coefficient functions $\gamma^1, \dots, \gamma^n$, characterising the curve γ. The right hand side of the system can be differentiated as often as required, which is why the initial

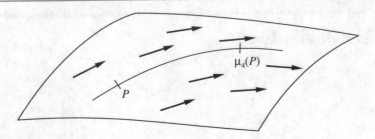

Fig. 12.1 The flow of a vector field

value problem is clearly solvable, i.e., the curve γ is uniquely determined locally by the initial position $\gamma(0)$.

Definition 12.1 For a vector field X and a real number t, let $\mu_t(P) = \gamma(t)$ with $\gamma' = X \circ \gamma$ and $\gamma(0) = P$. The family of maps μ_t is called the **flow of** X. ♦

As the solution of a system of differential equations with a right hand side that is infinitely many times differentiable depends in a C^∞ manner on the initial condition, it follows that the maps μ_t are C^∞ in every chart. Obviously there also holds $\mu_s \circ \mu_t = \mu_{s+t}$, and in particular, $\mu_{-t} = \mu_t^{-1}$. Thus the maps are also bijective, making them C^∞-diffeomorphisms.

Every C^∞-diffeomorphism μ on M sends a tangent vector $x \in M_P$ to a tangent vector $y \in M_{\mu(P)}$ via

$$yg = x(g \circ \mu) \qquad \text{for } g \in \mathcal{F}(\mu(P)).$$

The map $y : g \longrightarrow x(g \circ \mu)$ indeed has the properties of a tangent vector. For numbers α and β and $f, g \in \mathcal{F}(\mu(P))$

$$y(\alpha f + \beta g) = x((\alpha f + \beta g) \circ \mu) = \alpha x(f \circ \mu) + \beta x(g \circ \mu) = \alpha yf + \beta yg$$

and

$$y(f \cdot g) = x((f \cdot g) \circ \mu) = x((f \circ \mu) \cdot (g \circ \mu))$$

$$= x(f \circ \mu)(g \circ \mu)(P) + (f \circ \mu)(P)x(g \circ \mu) = yf\, g(\mu(P)) + f(\mu(P))\, yg.$$

Definition 12.2 The map T_μ, which maps the tangent vector $x \in M_P$ to the tangent vector $T_\mu x \in M_{\mu(P)}$ via

$$T_\mu x\, g = x(g \circ \mu) \qquad \text{for } g \in \mathcal{F}(\mu(P))$$

is called the **tangent of** μ. ♦

Example For a tangent vector $\gamma'(t_0)$ generated by a curve γ, $T_\mu \gamma'(t_0)$ is the vector $(\mu \circ \gamma)'(t_0)$ generated by the image curve $\mu \circ \gamma$, since for all $f \in \mathcal{F}$, we have

$$(\mu \circ \gamma)'(t_0) f = (f \circ \mu \circ \gamma)'(t_0) = \gamma'(t_0)(f \circ \mu) = T_\mu \gamma'(t_0) f .$$

Theorem 12.1 *The tangent T_μ of the diffeomorphism μ maps every tangent space M_P bijectively to $M_{\mu(P)}$. For diffeomorphisms μ and ν, there holds $T_{\mu \circ \nu} = T_\mu \circ T_\nu$.*

Proof For a linear combination $\alpha x + \beta y \in M_P$, one has

$$T_\mu(\alpha x + \beta y) g = (\alpha x + \beta y)(g \circ \mu) = \alpha x(g \circ \mu) + \beta y(g \circ \mu)$$

$$= \alpha T_\mu x \, g + \beta T_\mu y \, g = (\alpha T_\mu x + \beta T_\mu y) \, g ,$$

and so the restriction of T_μ to M_P is also linear. The tangent of $\mu \circ \nu$ satisfies

$$T_{\mu \circ \nu} x \, h = x(h \circ (\mu \circ \nu)) = x((h \circ \mu) \circ \nu) = (T_\nu x)(h \circ \mu) = (T_\mu(T_\nu x)) \, h$$

and so $T_{\mu \circ \nu} = T_\mu \circ T_\nu$. In particular, we have $T_\mu \circ T_{\mu^{-1}} = I$, and so $T_{\mu^{-1}}$ is the inverse of T_μ, therefore showing that $T_\mu : M_P \longrightarrow M_{\mu(P)}$ is bijective.

12.2   Pull-Back and Push-Forward

A bijective linear map between finite-dimensional vector spaces induces linear maps between the corresponding tensor spaces. In addition to the linear mapping itself, this generalises the concept of the dual mapping.

Definition 12.3 Let T be an invertible linear mapping from the finite-dimensional vector space E to the vector space F. The dual map T^* is defined as follows: for $g \in F_q^p$, $T^*g \in E_q^p$ is given by

$$T^*g(a^1, \ldots, a^p, x_1, \ldots, x_q) = g((T^*)^{-1}a^1, \ldots, (T^*)^{-1}a^p, Tx_1, \ldots, Tx_q)$$

for $x_1, \ldots, x_q \in E$ and $a^1, \ldots, a^p \in E^*$. For $f \in E_q^p$, $T_* f \in F_q^p$ is defined by

$$T_* f(b^1, \ldots, b^p, y_1, \ldots, y_q) = f(T^*b^1, \ldots, T^*b^p, T^{-1}y_1, \ldots, T^{-1}y_q)$$

for $y_1, \ldots, y_q \in F$ and $b^1, \ldots, b^p \in F^*$.   ◆

The maps T^* from F_q^p to E_q^p and T_* from E_q^p to F_q^p are clearly linear. For a covariant tensor g, in the definition of T^*g, one of course does not require the invertibility of T, and the same applies for the definition of T_*f in case of a contravariant tensor f.

Example 1 For a linear form $b \in F^*$, Definition 12.3 says

$$T^*b(x) = b(Tx) = \langle Tx, b \rangle = \langle x, T^*b \rangle = T^*b(x),$$

where at the end T^* means the dual mapping of T. For linear forms, thus the name for the new concept is compatible with the original connotation of T^*.

Example 2 A vector $x \in E$ is a one-fold contravariant tensor. According to Definition 12.3, we have

$$T_*x(b) = x(T^*b) = \langle x, T^*b \rangle = \langle Tx, b \rangle = Tx(b),$$

and so $T_*x = Tx$.

Example 3 For $y \in F$, we have

$$T^*y(a) = y((T^*)^{-1}a) = y((T^{-1})^*a) = \langle y, (T^{-1})^*a \rangle = \langle T^{-1}y, a \rangle = T^{-1}y(a),$$

and so $T^*y = T^{-1}y$.

Theorem 12.2

(1) For T from E to F, the maps T^* from F_q^p to E_q^p and T_* from E_q^p to F_q^p are inverses of each other.

(2) For T from E to F and tensors $f \in F_q^p$ and $g \in F_s^r$, we have

$$T^*(f \otimes g) = T^*f \otimes T^*g,$$

and for $f \in E_q^p$ and $g \in E_s^r$ we have

$$T_*(f \otimes g) = T_*f \otimes T_*g.$$

(3) For T from E to F and S from F to G, we have

$$(S \circ T)^* = T^* \circ S^*$$

and

$$(S \circ T)_* = S_* \circ T_*.$$

Proof (1) and (2) can be seen directly from Definition 12.3. (3) is based on the corresponding rules for the dual mapping and the inverse mapping. We have

$$(S \circ T)^* g(a^1, \ldots, x_q) = g(((S \circ T)^*)^{-1} a^1, \ldots, (S \circ T)x_q)$$

$$= g((S^*)^{-1}((T^*)^{-1} a^1), \ldots, S(Tx_q)) = S^* g((T^*)^{-1} a^1, \ldots, Tx_q) = T^*(S^* g)(a^1, \ldots, x_q)$$

and

$$(S \circ T)_* f(b^1, \ldots y_q) = f((S \circ T)^* b^1, \ldots, (S \circ T)^{-1} y_q)$$

$$= f(T^*(S^* b^1), \ldots, T^{-1}(S^{-1} y_q)) = T_* f(S^* b^1, \ldots, S^{-1} y_q) = S_*(T_* f)(b^1, \ldots, y_q) \,.$$

A real-valued function f of a real variable can without any difficulties be shifted to $g(x) = f(x+a)$. More generally, a new function g arises from a real-valued function f on a set M if one has a map μ via $g(x) = f(\mu(x))$. This is not so simple for a tensor field S on a manifold, because $S(P)$ must be a tensor on the tangent space M_P and therefore cannot be used as the function value at some other location Q. For the necessary conversion of the function values, the maps formulated in Definition 12.3 can be used.

Definition 12.4 For a C^∞-diffeomorphism μ on the manifold M, the map μ^* in $\mathcal{T}_q^p(M)$, defined by

$$\mu^* S = (T_\mu)^* \circ S \circ \mu \qquad \text{for } S \in \mathcal{T}_q^p(M) \,,$$

is called the **pull-back**, and the map μ_*, defined by

$$\mu_* S = (T_\mu)_* \circ S \circ \mu^{-1} \,,$$

the **push-forward**.   ◆

For $x_1, \ldots, x_q \in M_P$ and $a^1, \ldots, a^p \in M_P^*$, we therefore have

$$\mu^* S(P)(a^1, \ldots, x_q) = S(\mu(P))(((T_\mu)^*)^{-1} a^1, \ldots, T_\mu x_q)$$

and

$$\mu_* S(P)(a^1, \ldots, x_q) = S(\mu^{-1}(P))((T_\mu)^* a^1, \ldots, (T_\mu)^{-1} x_q) \,.$$

According to Example 3 for Definition 12.3, $(T_\mu)^*$ on a vector is the inverse mapping $T_\mu^{-1} = T_{\mu^{-1}}$, and the pull-back of a vector field is therefore

$$\mu^* X = T_{\mu^{-1}} \circ X \circ \mu \,,$$

and according to Example 2, $(T_\mu)_*$ for a vector is the map T_μ itself, and thus the push-forward of a vector field is given by

$$\mu_* X = T_\mu \circ X \circ \mu^{-1}.$$

With a scalar field f there is no reason to convert the function values; the pull-back is

$$\mu^* f = f \circ \mu$$

and the push-forward is

$$\mu_* f = f \circ \mu^{-1}.$$

This is inbuilt in Definition 12.4, if one views $(T_\mu)^*$ and $(T_\mu)_*$ as a number representing the identity map.

For fixed numbers p and q, the maps μ^* and μ_* in $\mathcal{T}_q^p(M)$ are clearly linear. Further properties are summarised in the following result.

Theorem 12.3

(1) For a C^∞-diffeomorphism μ on M, the maps μ^* and μ_* in $\mathcal{T}_q^p(M)$ are inverses of each other.
(2) There holds

$$\mu^*(S \otimes T) = \mu^* S \otimes \mu^* T$$

and

$$\mu_*(S \otimes T) = \mu_* S \otimes \mu_* T.$$

(3) For C^∞-diffeomorphisms μ and ν on M, we have

$$(\mu \circ \nu)^* = \nu^* \circ \mu^*$$

and

$$(\mu \circ \nu)_* = \mu_* \circ \nu_*.$$

Proof We have

$$\mu^*(\mu_* S) = (T_\mu)^* \circ ((T_\mu)_* \circ S \circ \mu^{-1}) \circ \mu = S$$

and analogously $\mu_*(\mu^* S) = S$, showing (1). (2) is obvious. (3) is essentially based on $T_{\mu \circ \nu} = T_\mu \circ T_\nu$ (Theorem 12.1). We have

$$(\mu \circ \nu)^* S = (T_{\mu \circ \nu})^* \circ S \circ (\mu \circ \nu) = (T_\nu)^* \circ (T_\mu)^* \circ S \circ \mu \circ \nu = \nu^*(\mu^* S)$$

and

$$(\mu \circ \nu)_* S = (T_{\mu \circ \nu})_* \circ S \circ (\mu \circ \nu)^{-1} = (T_\mu)_* \circ (T_\nu)_* \circ S \circ \nu^{-1} \circ \mu^{-1} = \mu_*(\nu_* S).$$

Finally, we determine the scalar field that is created by inserting vector fields and covector fields into a tensor field that is subject to pull-back or push-forward. We have

$$(\mu^* S)(B^1, \ldots, Y_q) = \mu^*(S(\mu_* B^1, \ldots, \mu_* Y_q))$$

and

$$(\mu_* S)(B^1, \ldots, Y_q) = \mu_*(S(\mu^* B^1, \ldots, \mu^* Y_q)),$$

in particular

$$\langle Y, \mu^* A \rangle = \mu^* \langle \mu_* Y, A \rangle$$

and

$$\langle Y, \mu_* A \rangle = \mu_* \langle \mu^* Y, A \rangle.$$

The more visually memorable versions are

$$\mu^* \langle X, A \rangle = \langle \mu^* X, \mu^* A \rangle$$

and

$$\mu_* \langle X, A \rangle = \langle \mu_* X, \mu_* A \rangle.$$

We also prove the general case in a modified form.

Theorem 12.4 *For a diffeomorphism μ on M, a tensor field S on M, vector fields $X_1, \ldots, X_q \in \mathcal{X}(M)$ and covector fields $A^1, \ldots, A^p \in \mathcal{K}(M)$, there holds*

$$\mu^*(S(A^1, \ldots, X_q)) = (\mu^* S)(\mu^* A^1, \ldots, \mu^* X_q)$$

and

$$\mu_*(S(A^1, \ldots, X_q)) = (\mu_* S)(\mu_* A^1, \ldots, \mu_* X_q).$$

Proof According to the Definitions 12.4 and 12.3, we have

$$(\mu^* S)(\mu^* A^1, \ldots, \mu^* X_q)$$
$$= ((T_\mu)^* \circ S \circ \mu)((T_\mu)^* \circ A^1 \circ \mu, \ldots, (T_\mu^{-1} \circ X_q \circ \mu)$$
$$= (S \circ \mu)(A^1 \circ \mu, \ldots, X_q \circ \mu) = \mu^*(S(A^1, \ldots, X_q))$$

and

$$(\mu_* S)(\mu_* A^1, \ldots, \mu_* X_q)$$
$$= ((T_\mu)_* \circ S \circ \mu^{-1})((T_\mu^*)^{-1} \circ A^1 \circ \mu^{-1}, \ldots, T_\mu \circ X_q \circ \mu^{-1})$$
$$= (S \circ \mu^{-1})(A^1 \circ \mu^{-1}, \ldots, X_q \circ \mu^{-1}) = \mu_*(S(A^1, \ldots, X_q)).$$

Theorem 12.5 *For a diffeomorphism μ, a vector field X and a scalar field f, there holds*

$$\mu^*(Xf) = (\mu^*X)(\mu^*f).$$

Proof We calculate the scalar fields on both sides. On the left we have

$$\mu^*(Xf)(P) = (Xf)(\mu(P)) = X(\mu(P))f$$

and on the right

$$(\mu^*X)(\mu^*f)(P) = (T_{\mu^{-1}} \circ X \circ \mu)(f \circ \mu)(P) = T_{\mu^{-1}}(X(\mu(P)))(f \circ \mu) = X(\mu(P))f.$$

12.3   Axiomatic Set Up

The Lie derivative of a field with respect to a vector field X at a point P is intended to quantify how the function value of the field to be derived changes for an observer who is currently at P and is moving along the flow X. For a scalar field f, this is the derivative of the function $g(t) = (f \circ \mu_t)(P)$ for $t = 0$, where $(\mu_t)_{t \in \mathbb{R}}$ is the flow of X. According to Definition 12.1, this is $g'(0) = X(P)f$. This motivates the following definition.

Definition 12.5 The **Lie derivative of the scalar field** f with respect to the vector field X is the scalar field

$$L_X f = Xf.$$

$\blacklozenge$

The Lie derivative for vector fields, covector fields and general tensor fields, the results from the similar calculation rules postulated as for the covariant derivative. The calculation rule

$$L_X(Yf) = (L_X Y)f + Y L_X f$$

is equivalent to

$$(L_X Y)f = XYf - YXf.$$

The Lie derivative of a vector field is thus given as follows.

Definition 12.6 The **Lie derivative of a vector field** Y with respect to the vector field X is the Lie bracket of X and Y,

$$L_X Y = [X, Y].$$

$\blacklozenge$

The calculational rules

$$L_X\langle Y, A\rangle = \langle L_X Y, A\rangle + \langle Y, \mathrm{\L}_X A\rangle$$

and

$$L_X(S(A^1, \ldots, X_q)) = (L_X S)(A^1, \ldots, X_q)$$

$$+ \sum_{i=1}^{p} S(A^1, \ldots, L_X A^i, \ldots, A^p, X_1, \ldots, X_q) + \sum_{j=1}^{q} S(A^1, \ldots, A^p, X_1, \ldots, L_X X_j, \ldots, X_q)$$

are enforced by the following two definitions.

Definition 12.7 The **Lie derivative of the covector field** A with respect to the vector field X is the covector field $L_X A$, characterised by

$$\langle Y, L_X A\rangle = X\langle Y, A\rangle - \langle [X, Y], A\rangle .$$

♦

Definition 12.8 The **Lie derivative of the tensor field** S with respect to the vector field X is

$$(L_X S)(A^1, \ldots, X_q) = X(S(A^1, \ldots, X_q))$$

$$- \sum_{i=1}^{p} S(A^1, \ldots, L_X A^i, \ldots, A^p, X_1, \ldots, X_q) - \sum_{j=1}^{q} S(A^1, \ldots, A^p, X_1, \ldots, [X, X_j], \ldots, X_q) .$$

♦

It still remains to be checked that the expressions for $L_X A$ and $L_X S$ actually define a covector field and a tensor field respectively, that is, the number

$$\langle Y, L_X A\rangle(P) = X(P)\langle Y, A\rangle - \langle [X, Y](P), A(P)\rangle$$

for the vector field Y only uses the vector field $Y(P)$ and that the number specified for $(L_X S)(A^1, \ldots, X_q)(P)$ except for X and S depends only on the linear forms $A^1(P), \ldots, A^p(P)$ and the vectors $X_1(P), \ldots, X_q(P)$. According to Theorem 4.1, the $\mathcal{F}$-homogeneity of the corresponding expressions must be shown. We have

$$X\langle fY, A\rangle - \langle [X, fY], A\rangle$$

$$= Xf\langle Y, A\rangle + fX\langle Y, A\rangle - f\langle [X, Y], A\rangle - \langle (Xf)Y, A\rangle = f(X\langle Y, A\rangle - \langle [X, Y], A\rangle)$$

and after an analogous calculation

$$(L_X S)(f^1 A^1, \ldots, g_q X_q) = f^1 \cdots f^p g_1 \cdots g_q (L_X S)(A^1, \ldots, X_q),$$

where in addition to

$$[X, g_j X_j] = g_j [X, X_j] + (X g_j) X_j$$

the following calculation rule, easily derivable from Definition 12.7,

$$L_X(fA) = f L_X A + (Xf)A$$

is used.

12.4   The Derivative Formula

At the beginning of the previous section, an interpretation of the Lie derivative was postulated, which in the case of a scalar field is expressed as

$$L_X f = \left. \frac{d}{dt} \right|_{t=0} (f \circ \mu_t) = \left. \frac{d}{dt} \right|_{t=0} (\mu_t^* f)$$

where μ_t is the flow of X. The subject of this section is the following theorem, which interprets the Lie derivative at $t = 0$ of a tensor field as

$$L_X S = \left. \frac{d}{dt} \right|_{t=0} (\mu_t^* S).$$

Theorem 12.6 *For a tensor field S and a vector field X with flow μ_t, the following* **derivation formula** *holds*

$$\frac{d}{dt} \mu_t^* S = \mu_t^* L_X S.$$

Proof Because of

$$\mu_{s+t}^* = (\mu_t \circ \mu_s)^* = \mu_s^* \circ \mu_t^*$$

the derivation formula to be proven obviously implies the exchange rule

$$\mu_t^* L_X S = L_X(\mu_t^* S).$$

We verify the derivation formula one after the other for scalar fields, vector fields, covector fields and general tensor fields. When proving the formula for the next object, we can use the derivation formula and the exchange rule for the objects processed up to that point.

For a scalar field f, we have, by the definitions of the pull-back and the flow, that

$$\frac{d}{dt}(\mu_t^* f) = \frac{d}{dt}(f \circ \mu_t) = (Xf) \circ \mu_t = \mu_t^*(Xf) = \mu_t^* L_X f .$$

For the proof of the formula for a vector field Y, we differentiate the equation from Theorem 12.5

$$(\mu_t^* Y)f = \mu_t^*(Y(\mu_{-t}^* f)) .$$

For the left side of the derivation formula, applied on f, we then get with an obvious product rule for differentiation

$$\frac{d}{dt}(\mu_t^* Y)f = \frac{d}{dt}(\mu_t^*(Y(\mu_{-t}^* f))) = \mu_t^* L_X(Y(\mu_{-t}^* f)) - \mu_t^*(Y(\mu_{-t}^* L_X f))$$

$$= \mu_t^*(X(Y(\mu_{-t}^* f))) - \mu_t^*(Y(X(\mu_{-t}^* f))) = \mu_t^*([X, Y](\mu_{-t}^* f)) = (\mu_t^*[X, Y])f .$$

The right hand side, applied to f, also gives

$$(\mu_t^* L_X Y)f = (\mu_t^*[X, Y])f .$$

In order to prove the derivation formula for a covector field A, the equation

$$\frac{d}{dt}\langle Y, \mu_t^* A \rangle = \langle Y, \mu_t^* L_X A \rangle$$

must be shown for vector field Y. The left hand side is, thanks to the derivation formula for scalar and vector fields,

$$\frac{d}{dt}\langle Y, \mu_t^* A \rangle = \frac{d}{dt}(\mu_t^*\langle \mu_{-t}^* Y, A \rangle) = \mu_t^* L_X \langle \mu_{-t}^* Y, A \rangle - \mu_t^*\langle \mu_{-t}^* L_X Y, A \rangle ,$$

and the right hand side is, according to the calculation rules for the Lie derivative,

$$\langle Y, \mu_t^* L_X A \rangle = \mu_t^*\langle \mu_{-t}^* Y, L_X A \rangle = \mu_t^*(L_X \langle \mu_{-t}^* Y, A \rangle) - \mu_t^*\langle L_X(\mu_{-t}^* Y), A \rangle ,$$

which is the same according to the exchange rule for vector fields.

Finally, we prove the derivation formula for a (p, q)-tensor field T by verifying the equation

$$\frac{d}{dt}(\mu_t^* T)(A^1, \ldots, X_q) = (\mu_t^* L_X T)(A^1, \ldots, X_q)$$

for covector fields $A^1, \ldots, A^p$ and vector fields $X_1, \ldots, X_q$. The left hand side is given by the following, because of the derivation formula for scalar fields

$$\frac{d}{dt}\left[\mu_t^*\{T(\mu_{-t}^*A^1, \ldots, \mu_{-t}^*X_q)\}\right]$$

$$= \mu_t^* L_X\{T(\mu_{-t}^*A^1, \ldots, \mu_{-t}^*X_q)\}$$

$$+ \sum_{i=1}^p \mu_t^* \left\{T(\mu_{-t}^*A^1, \ldots, \frac{d}{dt}\mu_{-t}^*A^i, \ldots, \mu_{-t}^*X_q)\right\}$$

$$+ \sum_{j=1}^q \mu_t^* \left\{T(\mu_{-t}^*A^1, \ldots, \frac{d}{dt}\mu_{-t}^*X_j, \ldots, X_q)\right\}.$$

Because of the derivation formula and exchange rule for covector and vector fields

$$\frac{d}{dt}\mu_{-t}^*A^i = -\mu_{-t}^* L_X A^i = -L_X(\mu_{-t}^*A^i)$$

and

$$\frac{d}{dt}\mu_{-t}^*X_j = -\mu_{-t}^* L_X X_j = -L_X(\mu_{-t}^*X_j).$$

Inserting these in the left hand side of the formula to be proven gives

$$\frac{d}{dt}(\mu_t^*T)(A^1, \ldots, X_q)$$

$$= \mu_t^* L_X\{T(\mu_{-t}^*A^1, \ldots, \mu_{-t}^*X_q)\}$$

$$- \sum_{i=1}^p \mu_t^*\{T(\mu_{-t}^*A^1, \ldots, L_X(\mu_{-t}^*A^i), \ldots, \mu_{-t}^*X_q)\}$$

$$- \sum_{j=1}^q \mu_t^*\{T(\mu_{-t}^*A^1, \ldots, L_X(\mu_{-t}^*X_j), \ldots, \mu_{-t}^*X_q)\}.$$

According to Definition 12.8, this is also the result on the right hand side

$$(\mu_t^* L_X T)(A^1, \ldots, X_q) = \mu_t^*[(L_X T)(\mu_{-t}^*A^1, \ldots, \mu_{-t}^*X_q)].$$

12.5   Component Representation

The Lie derivative of the scalar field f with respect to the vector field $X = X^i \partial_i$ is the scalar field $Xf = X^i \partial_i f$. The Lie derivative $L_X Y$ of the vector field $Y = Y^k \partial_k$ is the vector field $[X, Y]$ having the components $X^i \partial_i Y^k - Y^i \partial_i X^k$. According to

Definition 12.7, the components of the Lie derivative $L_X A$ of the covector field $A = A_j du^j$ are the scalars

$$\langle \partial_k, L_{X^i \partial_i}(A_j du^j)\rangle = X^i \partial_i \langle \partial_k, A_j du^j\rangle - \langle [X^i \partial_i, \partial_k], A_j du^j\rangle$$

$$= X^i \partial_i A_k + \langle \partial_k X^i \partial_i, A_j du^j\rangle = X^i \partial_i A_k + \partial_k X^i A_i.$$

Theorem 12.7 *The components of the Lie derivative $L_X S$ are computed from the components of the tensor field S via*

$$(L_X S)_{k_1 \cdots k_q}^{i_1 \cdots i_p} = X^l \partial_l S_{k_1 \cdots k_q}^{i_1 \cdots i_p}$$

$$- \partial_{l_1} X^{i_1} S_{k_1 \cdots k_q}^{l_1 i_2 \cdots i_p} - \cdots - \partial_{l_p} X^{i_p} S_{k_1 \cdots k_q}^{i_1 \cdots i_{p-1} l_p}$$

$$+ \partial_{k_1} X^{r_1} S_{r_1 k_2 \cdots k_q}^{i_1 \cdots i_p} + \cdots + \partial_{k_q} X^{r_q} S_{k_1 \cdots k_{q-1} r_q}^{i_1 \cdots i_p}.$$

Proof For vector fields and covector fields, this representation has already been proven, and can now be used. For the (p, q)-tensor field S, we have

$$(L_X S)_{k_1 \cdots k_q}^{i_1 \cdots i_p} = L_{X^i \partial_i} S(du^{i_1}, \ldots, du^{i_p}, \partial_{k_1}, \ldots, \partial_{k_q})$$

$$= X^i \partial_i S(du^{i_1}, \ldots, du^{i_p}, \partial_{k_1}, \ldots, \partial_{k_q})$$

$$- S(L_{X^i \partial_i} du^{i_1}, \ldots, \partial_{k_q}) - \cdots - S(\ldots, L_{X^i \partial_i} du^{i_p}, \ldots, \partial_{k_q})$$

$$- S(du^{i_1}, \ldots, L_{X^i \partial_i} \partial_{k_1}, \ldots, \partial_{k_q}) - \cdots - S(du^{i_1}, \ldots, L_{X^i \partial_i} \partial_{k_q}).$$

With

$$L_{X^i \partial_i} du^{i_p} = \partial_j X^{i_p} du^j$$

and

$$L_{X^i \partial_i} \partial_{k_p} = [X^i \partial_i, \partial_{k_q}] = -\partial_{k_q} X^i \partial_i$$

and this results in the desired representation.

12.6 Killing Vectors

If μ is an isometric mapping (or simply isometry), then we expect the arc length of the image curve $\mu \circ \gamma$ to be identical to the arc length of the original curve γ. The equality of the integrals

$$\int_{t_1}^{t_2} \sqrt{g(\mu(\gamma(t)))(T_\mu \gamma'(t), T_\mu \gamma'(t))}\, dt$$

and

$$\int_{t_1}^{t_2} \sqrt{g(\gamma(t))(\gamma'(t), \gamma'(t))} \, dt$$

can be enforced by the agreement of the integrands.

Definition 12.9 A C^∞-diffeomorphism μ on a semi-Riemannian manifold $[M, g]$ is an **isometry**, if for its tangent T_μ satisfies

$$g(\mu(P))(T_\mu x, T_\mu y) = g(P)(x, y)$$

for $P \in M$ and $x, y \in M_P$, that is $\mu^* g = g$.   ◆

Definition 12.10 A vector field X on a semi-Riemannian manifold $[M, g]$ with the property that $L_X g = 0$ is called **Killing vector field**.   ◆

Theorem 12.8 *A vector field is a Killing vector field if and only if its flow consists of isometries.*

Proof If all the diffeomorphisms μ_t of the flow of X are isometries, then it follows that $\mu_t^* g - g = 0$ and so the derivative formula (Theorem 12.6) implies that $L_X g = 0$. Conversely, suppose that $L_X g = 0$. Then for $x, y \in M_P$, by the derivative formula, the real-valued function $f(t) = \mu_t^* g(x, y)$ has derivative zero for all t, i.e., it is a constant, and so $\mu_t^* g = g$.

If a coordinate vector field is a Killing vector field, then this expresses a symmetry of the metric, and adding a constant to this coordinate is an isometry. In the Schwarzschild spacetime, ∂_t and ∂_φ are obviously Killing vector fields, but ∂_r and ∂_ϑ are not. To verify this systematically, we use the component representation established in Theorem 12.7,

$$(L_X g)_{ik} = X^j \partial_j g_{ik} + \partial_i X^j g_{jk} + \partial_k X^j g_{ij} \,.$$

For a coordinate vector field $X = \partial_l$, the condition $L_X g = 0$ is reduced to $\partial_l g_{ik} = 0$. While some metric components g_{ik} depend on r, respectively ϑ, they are all independent of t and φ.

We will now show that each Killing vector field produces a conservation theorem. The following characterisation is useful for this.

Theorem 12.9 *A vector field X is Killing if and only if for each pair of vector fields Y and Z, there holds $g(\nabla_Y X, Z) = -g(\nabla_Z X, Y)$.*

Proof According to Definition 12.8,

$$L_X g(Y, Z) = Xg(Y, Z) - g([X, Y], Z) - g(Y, [X, Z])$$

and according to the calculation rules for the covariant derivative,

$$Xg(Y, Z) = g(\nabla_Y X + [X, Y], Z) + g(Y, \nabla_Z X + [X, Z]),$$

giving finally, in light of the above that

$$L_X g(Y, Z) = g(\nabla_Y X, Z) + g(\nabla_Z X, Y).$$

Theorem 12.10 *For a Killing vector field X, the expression $g(X, \gamma')$ along a geodesic γ is constant.*

Proof The derivative with respect to the curve parameter s is

$$\frac{d}{ds} g(X(\gamma(s)), \gamma'(\gamma(s))) = \gamma' g(X, \gamma') = g(\nabla_{\gamma'} X, \gamma') + g(X, \nabla_{\gamma'} \gamma').$$

The last summand is zero because γ is a geodesic, while the penultimate one is zero because of Theorem 12.9.

In the Schwarzschild spacetime, the Killing vector fields ∂_t and ∂_φ lead to the conserved quantities

$$g(\partial_t, \gamma') = g(\partial_t, t' \partial_t) = t' h(r) = E$$

(see also Theorem 14.2) and

$$g(\partial_\varphi, \gamma') = g(\partial_\varphi, \varphi' \partial_\varphi) = \varphi' r^2 = -L$$

(see Sect. 10.5).

12.7   The Lie Derivative of Differential Forms

A differential form is a tensor field, and so a vector field can be used to find its Lie derivative. On the other hand, one also has the outer derivative for differential forms. It turns out that the Lie derivative can be traced back to the outer derivative. Incidentally, it follows from here that the Lie derivative of a differential form is again a differential form.

Definition 12.11 For a vector field X and a differential form F of order $(p + 1)$, the pth-order differential form

$$i_X F(X_1, \ldots, X_p) = F(X, X_1, \ldots, X_p)$$

is called the **inner product of X and F**.   ◆

Theorem 12.11 *For a vector field X and a differential form F, there holds*

$$L_X F = i_X dF + d i_X F$$

Proof Let the order of F be p. For vector fields $X_1, \ldots, X_p$ and with $X_0 = X$, we have

$$(i_X dF)(X_1, \ldots, X_p) = dF(X_0, X_1, \ldots, X_p)$$

$$= \sum_{i=0}^{p} (-1)^i X_i F(X_0, \ldots, X_{i-1}, X_{i+1}, \ldots, X_p)$$

$$+ \sum_{0 \le i < j \le p} (-1)^{i+j} F([X_i, X_j], X_0, \ldots, X_{i-1}, X_{i+1}, \ldots, X_{j-1}, X_{j+1}, \ldots, X_p)$$

and

$$d(i_X F)(X_1, \ldots, X_p)$$

$$= \sum_{i=1}^{p} (-1)^{i-1} X_i ((i_X F)(X_1, \ldots, X_{i-1}, X_{i+1}, \ldots, X_p))$$

$$+ \sum_{1 \le i < j \le p} (-1)^{i+j} (i_X F)([X_i, X_j], X_1, \ldots, X_{i-1}, X_{i+1}, \ldots, X_{j-1}, X_{j+1}, \ldots, X_p)$$

$$= \sum_{i=1}^{p} (-1)^{i-1} X_i F(X_0, X_1, \ldots, X_{i-1}, X_{i+1}, \ldots, X_p)$$

$$+ \sum_{1 \le i < j \le p} (-1)^{i+j} F(X_0, [X_i, X_j], X_1, \ldots, X_{i-1}, X_{i+1}, \ldots, X_{j-1}, X_{j+1}, \ldots, X_p).$$

The sum is

$$(i_X dF)(X_1, \ldots, X_p) + d(i_X F)(X_1, \ldots, X_p)$$

$$= X_0 F(X_1, \ldots, X_p) + \sum_{j=1}^{p} (-1)^j F([X_0, X_j], X_1, \ldots, X_{j-1}, X_{j+1}, \ldots, X_p)$$

$$= L_X(F(X_1, \ldots, X_p)) - \sum_{j=1}^{p} F(X_1, \ldots, X_{j-1}, L_X X_j, X_{j+1}, \ldots, X_p)$$

$$= (L_X F)(X_1, \ldots, X_p).$$

Integration on Manifolds

13

13.1 Introduction

The concept of a manifold encompasses curves and surfaces in three-dimensional Euclidean space. Thus the concept of an integral on manifolds must generalise the notion of curve and surface integrals.

As we know, an integral of the first kind of a real-valued integrand f along a curve $\mathcal{K}$ is calculated using a parametric representation $\gamma : [a, b] \to \mathcal{K}$ via

$$\int_{\mathcal{K}} f\, ds = \int_a^b f(\gamma(t)) \sqrt{\gamma'(t) \cdot \gamma'(t)}\, dt \,.$$

The surface integral of the first kind, with a parameter representation $(u, v) \to (x, y, z)$ on the parameter domain Γ for the surface $\mathcal{F}$ is given by

$$\iint_{\mathcal{F}} f\, do = \iint_{\Gamma} f(x(u, v), y(u, v), z(u, v)) \sqrt{(\partial_u \cdot \partial_u)(\partial_v \cdot \partial_v) - (\partial_u \cdot \partial_v)^2}\, du\, dv \,,$$

where the vectors ∂_u and ∂_v have the components $\frac{\partial x}{\partial u}$, $\frac{\partial y}{\partial u}$, $\frac{\partial z}{\partial u}$, respectively $\frac{\partial x}{\partial v}$, $\frac{\partial y}{\partial v}$, $\frac{\partial z}{\partial v}$ with respect to the Cartesian coordinates. These rules for the calculation of curve and surface integrals of the first kind can be described in a unified manner as follows: Choose a chart φ and calculate the n-fold integral

$$\int_{\varphi(M)} \cdots \int f(\varphi^{-1}(u^1, \ldots, u^n)) \sqrt{|g|}\, du^1 \ldots du^n$$

© The Author(s), under exclusive license to Springer Nature Switzerland AG 2023
R. Oloff, *The Geometry of Spacetime*, Graduate Texts in Physics,
https://doi.org/10.1007/978-3-031-16139-1_13

with

$$g = \det(\partial_i \cdot \partial_k) = \det(g(\partial_i, \partial_k)) \ .$$

For $n = 1$, this reduces to a curve integral and for $n = 2$, it is a surface integral. This general recipe requires a scalar product on the tangent spaces, and thus only makes sense in the case of a semi-Riemannian manifold if it can be described within a single chart, or at least on the support of f.

Curve and surface integrals of the second kind also use the dot product. However, this can be avoided if the integrands are interpreted appropriately. A curve integral of the second kind of a vector field F where the curve has a paramteric representation γ is given by

$$\int_{\mathcal{K}} F \, dr = \int_a^b F(\gamma(t)) \cdot \gamma'(t) \, dt \ .$$

If one uses the (location dependent) linear form $\alpha : v \to F \cdot v$ as the integrand instead of the vector function F, we obtain the formula

$$\int_{\mathcal{K}} \alpha = \int_a^b \alpha(\gamma'(t)) \, dt \ .$$

The situation is similar for the surface integrals of the second kind. In the formula

$$\iint_{\mathcal{F}} F \, d\vec{o} = \iint_{\Gamma} F(x(u, v), y(u, v), z(u, v)) \cdot (\partial_u \times \partial_v) \, du dv$$

the vector F is used as a factor in the triple product $F \cdot (\partial_u \times \partial_v)$. When the first factor is inserted, the triple product is a skew-symmetric, doubly covariant tensor. Thus the vector field F generates a 2-form ω, whose integral $\int \omega$ should be the integral $\iint F \, d\vec{o}$. Hence we have

$$\int_{\mathcal{F}} \omega = \iint_{\Gamma} \omega(x(u, v), y(u, v), z(u, v))(\partial_u, \partial_v) \, du dv \ .$$

Summarising, the curve and surface integrals of the second kind are encompassed in the following notion of integration: *To calculate the integral $\int \omega$ of an n-form ω on an oriented n-dimensional manifold M, choose a positively oriented chart φ that contains the support of ω (if possible; if not, then use a partition of unity in the manner described in the following section), and calculate the n-fold integral*

$$\int_M \omega = \int_{\varphi(M)} \omega(\varphi^{-1}(u^1, \dots, u^n))(\partial_1, \cdots, \partial_n) \, du^1 \cdots du^n \ .$$

The classical integral theorems will turn out to be special cases of the equation $\int_{\partial M} \omega = \int_M d\omega$ (Theorem 13.4). Here ω is an $(n-1)$-form on an n-dimensional oriented manifold with boundary ∂M (Sect. 13.4).

13.2   Partition of Unity

The previous section explained how to calculate integrals on a manifold covered by a single chart. Usually, however, no chart (U, φ) can be found such that the support

$$\operatorname{supp}\omega := \overline{\{P \in M : \ \omega(P) \neq 0\}}$$

is completely contained in U. The integrand ω must be broken down into several components, each of which is contained in a chart. This is done with the help of a so-called partition of unity.

Definition 13.1 Let $\mathcal{U} = (U_\alpha)_{\alpha \in \mathcal{A}}$ be an open cover of the manifold M, i.e., the set U_α are open and $M = \bigcup_{\alpha \in \mathcal{A}} U_\alpha$. A **partition of unity** with respect to $\mathcal{U}$ is a family $(\tau_\beta)_{\beta \in \mathcal{B}}$ of non-negative C^∞-functions τ_β with the following properties: ◆

(Z1) At every point $P \in M$ there exists an open set V with $P \in V$ and $\operatorname{supp}\tau_\beta \cap V = \emptyset$ for all but finitely many of the $\beta \in \mathcal{B}$.
(Z2) For every point $P \in M$,

$$\sum_{\beta \in \mathcal{B}} \tau_\beta(P) = 1 .$$

(Z3) For every $\beta \in \mathcal{B}$, there exists an $\alpha \in \mathcal{A}$ such that $\operatorname{supp}\tau_\beta \subseteq U_\alpha$.

We know that a subset of K of a topological space is called **compact** if every open cover of K has a finite subcover, that is if $(V_\gamma)_{\gamma \in \mathcal{C}}$ is such that the V_γ are open and $K \subseteq \cup V_\gamma$, then there exists $\gamma_1, \ldots, \gamma_m$ such that $K \subseteq V_{\gamma_1} \cup \ldots \cup V_{\gamma_m}$. The well-known Heine-Borel theorem states that compact subsets of $\mathbb{R}^n$ are precisely those that are both closed and bounded.

   To construct a partition of unity, one needs the axiom of second countability which was already mentioned in Sect. 1.2, and was included in the definition of a manifold. A collection of open sets $G_1, G_2, \ldots$ with the property that each open set in M is a union of sets from G_i is called a **basis of the topology**. The second countability axiom says that the topology has a countable basis. The Euclidean space $\mathbb{R}^n$ fulfills this axiom, since the open balls with a rational radius and centers all of whose coordinates are rational obviously form a basis. If the topology on a manifold can be generated using countably many charts, then the second countability axiom also applies to the manifold, since the open subsets of $\mathbb{R}^n$ that generate the topology of $\mathbb{R}^n$ are mapped to open subsets of M by Def. 1.6, which then also generate the topology of M. The closures of the above mentioned basis elements in $\mathbb{R}^n$ are compact, and this is also true for the corresponding basis elements of the manifold.

Theorem 13.1 *For every open cover* $\mathcal{U}$ *of a manifold* M, *there exists a partition of unity with respect to* $\mathcal{U}$.

Proof Let $G_1, G_2, \ldots$ be a basis for the topology of M, where the closures of these open sets are compact. We first construct successice compact subsets $K_1, K_2, \ldots$ of M, such that

$$\overset{\infty}{\underset{i=1}{\cup}} K_i = M$$

and the property that K_i is contained in the interior $\overset{\circ}{K}_{i+1}$ of K_{i+1}, comprising all the inner points of K_{i+1}. Since the compact set K_1 is covered by the open sets G_i, we can find a natural number n_1 such that

$$K_1 \subseteq G_1 \cup \ldots \cup G_{n_1} .$$

Also the compact set

$$K_2 := \overline{G}_1 \cup \ldots \cup \overline{G}_{n_1}$$

can be covered by open sets G_i, and thus there exists a natural number n_2 such that

$$K_2 \subseteq G_1 \cup \ldots \cup G_{n_1} \cup \ldots \cup G_{n_2} .$$

The set

$$K_3 := \overline{G}_1 \cup \ldots \cup \overline{G}_{n_2}$$

is compact and can be covered by

$$K_3 \subseteq G_1 \cup \ldots \cup G_{n_2} \cup \ldots \cup G_{n_3} ,$$

and so on. From

$$K_m := \overline{G}_1 \cup \ldots \cup \overline{G}_{n_{m-1}}$$

with the cover

$$K_m \subseteq G_1 \cup \ldots \cup G_{n_m}$$

we construct

$$K_{m+1} := \overline{G}_1 \cup \ldots \cup \overline{G}_{n_m} .$$

The sequence of compact sets K_i is constructed according to this principle. Clearly there holds that

$$\overset{\circ}{K}_{i+1} = G_1 \cup \ldots \cup G_{n_i} ,$$

and so this sequence has the desired properties.

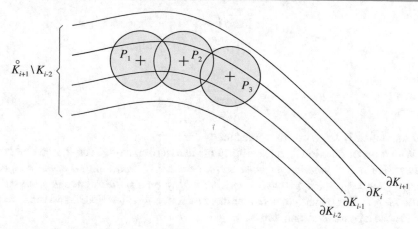

Fig. 13.1 Construction of a partition of unity

For a fixed index i, we choose for every point $P \in K_i \setminus \overset{\circ}{K}_{i-1}$ an open set U_P from the given open cover $\mathcal{U}$ with $P \in U_P$ and construct a non-negative C^{∞}-function λ_P with $\lambda_P(P) > 0$, whose support $\mathrm{supp}\lambda_P$ is contained in the open set $\left(\overset{\circ}{K}_{i+1} \setminus K_{i-2} \right) \cap U_P$. In the following we will proof that such a function always exists. The compact set $K_i \setminus \overset{\circ}{K}_{i-1}$ is covered by the open sets $\lambda_P^{-1}((0, \infty))$ for which a finite number of sets $\lambda_{P_k}^{-1}((0, \infty))$, $k = 1, \ldots, r_i$ is sufficient. This is illustrated in Fig. 13.1. With the notation $\lambda_k^i := \lambda_{P_k}$, $k = 1, \ldots, r_i$, we have

$$\sum_{k=1}^{r_i} \lambda_k^i(Q) > 0$$

for all $Q \in K_i \setminus \overset{\circ}{K}_{i-1}$. This construction is carried out for all indices i. This creates a countable system of non-negative C^{∞}-functions λ_k^i, which already have the properties (Z1) and (Z3). In particular the sum

$$\lambda(P) := \sum_{i=1}^{\infty} \sum_{k=1}^{r_i} \lambda_k^i(P)$$

can be formed. The property (Z2) is now also enforced by the normalisation

$$\tau_k^i(P) := \lambda_k^i(P)/\lambda(P).$$

Finally, it has to be shown that for $P \subset M$ and an open set U with $P \in U$, there actually exists a non-negative C^{∞}-function λ on M with $\lambda(P) > 0$ and $\mathrm{supp}\lambda \subset U$. With a chart, the problem can be converted from M over to $\mathbb{R}^n$, and solved there with the function

$$\lambda(u^1, \ldots, u^n) := \begin{cases} \exp\left(-\frac{1}{1-cr^2}\right) & \text{for } r < 1/\sqrt{c} \\ 0 & \text{otherwise} \end{cases}$$

with

$$r = \sqrt{\sum (u^k - u_0^k)^2}$$

and a sufficiently large positive constant c.

With a given function λ, a previously made assertion from Sect. 2.1 can be now proven: *For a C^∞-function f given on an open set U containing a point P, there exists a function $g \in \mathcal{F}(\mathcal{M})$ which matches with f on a suitably chosen open set V containing P.* It is enough to consider this problem only on $M = \mathbb{R}^n$ and for $P = 0$. We choose a positive ε such that

$$U_\varepsilon := \{u = (u^1, \ldots, u^n) : \ |u| < \varepsilon\} \subset \overline{U}_\varepsilon \subset U$$

and construct a C^∞-function h such that

$$h(u) = \begin{cases} 1 \text{ for } |u| \le \frac{\varepsilon}{2} \\ 0 \text{ for } |u| \ge \varepsilon \end{cases}.$$

To do this, we enforce the equation

$$\int \cdots \int_{|v| \le \frac{\varepsilon}{4}} \lambda_a(v^1, \ldots, v^n) \, dv^1 \ldots dv^n = 1$$

for the function

$$\lambda_a(v^1, \ldots, v^n) := \begin{cases} a \exp\left(\dfrac{-1}{1 - \frac{16}{\varepsilon^2}|v|^2}\right) & \text{for} |v| < \frac{\varepsilon}{4} \\ 0 & \text{otherwise} \end{cases}$$

by a suitable choice of the positive constant a and set

$$h(u^1, \ldots, u^n) := \int \cdots \int_{|v-u| \le \frac{3}{4}\varepsilon} \lambda_a(v^1, \ldots, v^n) \, dv^1 \ldots dv^n .$$

For $|u| \le \varepsilon/2$ the support of λ_a lies entirely in the integration region, and consequently there holds that $h(u) = 1$. For $|u| \ge \varepsilon$ the support and the integration region are disjoint, and so $h(u) = 0$. The function $g = fh$ is differentiable arbitrarily many times and matches with f on $U_{\varepsilon/2}$.

13.3   Integrals

In Sect. 13.1, we already saw that in the case of an integral over an oriented n-dimensional manifold, the integrand could be an n-form. The precise formulation of the concept of the integral depends on the smoothness properties of the permitted integrands. Here we will suppose that the integrand can be differentiated arbitrarily many times, and in particular that it is continuous, so that when we move the calculation over to a chart, we can use the notion of the Riemann integral in $\mathbb{R}^n$. In order to avoid infinite integrals, we also assume that the integrand support is compact. This can then be covered by finitely many positively oriented charts.

Definition 13.2 Let ω be an n-form which is infinitely often differentiable, with a compact support on the oriented n-dimensional manifold M. Let the positively oriented charts (U_k, φ_k), $k = 1, \ldots, m$ form a cover of the support, and $\tau_1, \ldots, \tau_m$ be a partition of unity with respect to this cover. The the integral $\int\limits_M \omega$ is defined by

$$\int\limits_M \omega = \sum_{k=1}^m \int \ldots \int\limits_{u \in \varphi_k(U_k)} \tau_k \omega \, du^1 \ldots du^n$$

where

$$\tau_k \omega(u) = \tau_k(\varphi_k^{-1}(u)) \, \omega(\varphi_k^{-1} u) \left(\frac{\partial}{\partial u^1}, \ldots, \frac{\partial}{\partial u^n} \right) .$$

$\blacklozenge$

It must of course now be shown that the expression formulataed in the definition is independent of the cover and the partition of unity. Let (V_l, ψ_l), $l = 1, \ldots, r$ be another cover with positively oriented charts, and let $\sigma_1, \ldots, \sigma_r$ be a partition of unity. Then we have

$$\sum_{k=1}^m \int \ldots \int\limits_{\varphi_k(U_k)} \tau_k \omega \, du^1 \ldots du^n$$

$$= \sum_{k=1}^m \int \ldots \int\limits_{\varphi_k(U_k)} (\tau_k \circ \varphi_k^{-1}) \left(\sum_{l=1}^r \sigma_l \circ \varphi_k^{-1} \right) (\omega \circ \varphi_k^{-1}) \left(\frac{\partial}{\partial u^1}, \ldots, \frac{\partial}{\partial u^n} \right) du^1 \ldots du^n$$

$$= \sum_{k,l} \int \ldots \int\limits_{\varphi_k(U_k \cap V_l)} \left((\sigma_l \tau_k) \circ \varphi_k^{-1} \right) (\omega \circ \varphi_k^{-1}) \left(\frac{\partial}{\partial u^1}, \ldots, \frac{\partial}{\partial u^n} \right) du^1 \ldots du^n$$

and analogously

$$\sum_{l=1}^{r} \int_{\psi_l(V_l)} \cdots \int \sigma_l \omega \, dv^1 \ldots dv^n$$

$$= \sum_{k,l} \int_{\psi_l(U_k \cap V_l)} \cdots \int ((\sigma_l \tau_k) \circ \psi_l^{-1})(\omega \circ \psi_l^{-1})\left(\frac{\partial}{\partial v^1}, \ldots, \frac{\partial}{\partial v^n}\right) dv^1 \ldots dv^n .$$

This reduces the problem to verifying the equation

$$\int_{\varphi(W)} \cdots \int \omega(\varphi^{-1}(u))\left(\frac{\partial}{\partial u^1}, \ldots, \frac{\partial}{\partial u^n}\right) du^1 \ldots du^n$$

$$= \int_{\psi(W)} \cdots \int \omega(\psi^{-1}(v))\left(\frac{\partial}{\partial v^1}, \ldots, \frac{\partial}{\partial v^n}\right) dv^1 \ldots dv^n .$$

The right hand side integrand can be converted to the new coordinate vector fields as follows:

$$\omega(\varphi^{-1}(u))\left(\frac{\partial}{\partial u^1}, \ldots, \frac{\partial}{\partial u^n}\right)$$

$$= \omega(\varphi^{-1}(u))\left(\frac{\partial v^{k_1}}{\partial u^1}\frac{\partial}{\partial v^{k_1}}, \ldots, \frac{\partial v^{k_n}}{\partial u^n}\frac{\partial}{\partial v^{k_n}}\right)$$

$$= \frac{\partial v^{k_1}}{\partial u^1} \cdots \frac{\partial v^{k_n}}{\partial u^n}\, \omega(\varphi^{-1}(u))\left(\frac{\partial}{\partial v^{k_1}}, \ldots, \frac{\partial}{\partial v^{k_n}}\right)$$

$$= \sum_{\mathcal{P}} \frac{\partial v^{\mathcal{P}(1)}}{\partial u^1} \cdots \frac{\partial v^{\mathcal{P}(n)}}{\partial u^n}\, \omega(\varphi^{-1}(u))\left(\frac{\partial}{\partial v^1}, \ldots, \frac{\partial}{\partial v^n}\right)\chi(\mathcal{P})$$

$$= \omega(\varphi^{-1}(u))\left(\frac{\partial}{\partial v^1}, \ldots, \frac{\partial}{\partial v^n}\right)\det\left(\frac{\partial v^k}{\partial u^i}\right) .$$

Since the determinant is positive, the equation to be proven follows from this with the usual transformation formula for Riemann integrals in $\mathbb{R}^n$.

The following properties of the integral can be seen directly from its definition.

Theorem 13.2(1) *For n-forms ω and ν on M and numbers a and b,*

$$\int_M (a\omega + b\nu) = a\int_M \omega + b\int_M \nu .$$

(2) *The oriented manifold which arises from the oriented manifold M by changing its orientation, is denoted by* $-M$. *Then*

$$\int_{-M} \omega = - \int_{M} \omega \ .$$

(3) *Let* μ *be an orientation-preserving diffeomorphism from* M_1 *to* M_2. *Then*

$$\int_{M_2} \mu^* \omega = \int_{M_1} \omega \ .$$

The notion of the integral introduced in Definition 13.2 generalises, as already explained in Sect. 13.1, curve and surface integrals of the second kind. The generalisation of the curve and surface integrals of the first kind requires a metric. Then the volume form V (Def.6.4) is also available with the fundamental tensor g. For a smooth, real-valued function f with a compact support, the product fV is suitable as an integrand.

Theorem 13.3 *Let* f *be a real-valued, arbitrarily many times differentiable function, having a compact support on an oriented n-dimensional semi-Riemannian manifold* $[M, g]$ *with a volume form* V. *Let the positively oriented charts* (U_k, φ_k), $k = 1 \ldots, m$ *cover the support, and let* $\tau_1, \ldots, \tau_m$ *be a partition of unity with respect to this covering. Then*

$$\int_{M} fV = \sum_{k=1}^{m} \int \ldots \int_{u \in \varphi_k(U_k)} \tau_k f w \, du^1 \ldots du^n$$

with

$$\tau_k f w(u) = \tau_k(\varphi_k^{-1}(u)) f(\varphi_k^{-1}(u)) \sqrt{\left| \det \left(g \left(\frac{\partial}{\partial u^i}, \frac{\partial}{\partial u^j} \right) \right) \right|} \ .$$

Proof. We want to show

$$V(\partial_1, \ldots, \partial_n) = \sqrt{\left| \det \left(g \left(\partial_i, \partial_j \right) \right) \right|} \ .$$

Let $e_1, \ldots, e_n$ be a positively oriented orthogonal basis for M_P with $g(e_i, e_i) = \varepsilon_i = \pm 1$. We have for the left hand side that

$$V(\partial_1, \ldots, \partial_n) = V\left(\sum_{k_1} g(\partial_1, \varepsilon_{k_1} e_{k_1}) e_{k_1}, \ldots, \sum_{k_n} g(\partial_n, \varepsilon_{k_n} e_{k_n}) e_{k_n}\right)$$

$$= \sum_{k_1, \ldots, k_n} g(\partial_1, \varepsilon_{k_1} e_{k_1}) \cdots g(\partial_n, \varepsilon_{k_n} e_{k_n}) V(e_{k_1}, \ldots, e_{k_n})$$

$$= \sum_{\mathcal{Q}} \chi(\mathcal{Q}) g(\partial_1, \varepsilon_{\mathcal{Q}(1)} e_{\mathcal{Q}(1)}) \cdots g(\partial_n, \varepsilon_{\mathcal{Q}(n)} e_{\mathcal{Q}(n)}) = \det G$$

where G is the matrix $G = (g(\partial_i, \varepsilon_j e_j))$. From

$$g(\partial_i, \partial_j) = g\left(\sum_l g(\partial_i, \varepsilon_l e_l) e_l, \sum_k g(\partial_j, \varepsilon_k e_k) e_k\right) = \sum_{l,k} g(\partial_i, \varepsilon_l e_l) g(e_l, e_k) g(\varepsilon_k e_k, \partial_j)$$

we see that the matrix $(g(\partial_i, \partial_j))$ is the product of the matrices G, $\mathrm{diag}(\varepsilon_1, \ldots, \varepsilon_n)$ and G^T. Hence it follows that

$$|\det(g(\partial_i, \partial_j))| = (\det G)^2 .$$

Since the bases $e_1, \ldots, e_n$ and $\partial_1, \ldots, \partial_n$ have the same orientation, $\det G$ is positive. This shows that the right hand side of the equality we had to prove is

$$\sqrt{|\det(g(\partial_i, \partial_j))|} = \det G$$

as well.

13.4   Manifolds with Boundary

For us, the standard example of a manifold has been a curved surface. It could be a closed surface like the surface of a sphere, an unbounded surface like a cylinder, or a subset of such a surface. In the latter case, we did not allow the subset to have, what is colloquially referred to as a 'border'. In this section, we rectify this, and include the boundary in our considerations. It will be shown that this border is itself a manifold with a dimension reduced by one. In particular, we can form integrals on the boundary as well, and in the next section we will learn about the abstract version of the well-known classical integral theorems of differential and integral calculus.

If we want to cover a surface including its (smooth) boundary with charts (U, φ), we will have to allow image sets $\varphi(U)$ which may not always be open in $\mathbb{R}^2$, but can be sets G which are the intersection of an open set G^* with a closed half-plane (Fig. 13.2). To be precise, and also to include higher dimensions, let

$$\mathbb{R}^n_- = \{(u^1, \ldots, u^n) \in \mathbb{R}^n : u^1 \le 0\} ,$$

Fig. 13.2 The boundary ∂G
of set G open in $\mathbb{R}^n_-$

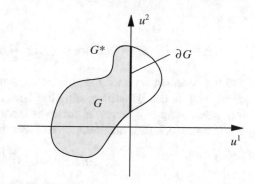

and call a set G of the form $G = G^* \cap \mathbb{R}^n_-$ where G^* is an open subset of $\mathbb{R}^n$, **open in $\mathbb{R}^n_-$**. Although this is not so common in topology, in the context of manifolds with boundary, given an open set G in $\mathbb{R}^n_-$, we call the set

$$\partial G := \{(u^1, \dots, u^n) \in G : u^1 = 0\}$$

as the **boundary** of G. At such boundary points, we introduce a concept of differentiability with the agreement that a function given on an open set G in $\mathbb{R}^n_-$ is differentiable if it can be extended to a differentiable function on an open set containing G.

From the atlas of a **manifold with boundary** M, one demands that the transition maps $\varphi \circ \psi^{-1}$ arising from a change of charts are C^∞-diffeomorphisms between open sets in $\mathbb{R}^n_-$. The points $P \in M$ for which, in a chart (U, φ) with $P \in U$, there holds that $\varphi(P) \in \partial(\varphi(U))$, form the **boundary** ∂M of the manifold M with boundary. This property of a point is clearly independent of the choice of a chart.

Every chart (U, φ) of M with $U \cap \partial M \neq \emptyset$ creates a chart of ∂M by restricting φ to $U \cap \partial M$. Altogether, we obtain an atlas for the boundary ∂M of a manifold M with boundary arises from an atlas for M, and we see that the boundary ∂M is itself an $(n-1)$-dimensional manifold.

An orientation for a manifold M with boundary can also be transferred to its boundary ∂M as follows: Let (U, φ) be an admissible chart in the given orientation of M and suppose that $P \in U \cap \partial M$. The the basis $\partial_2, \dots, \partial_n$ is in the orientation of $(\partial M)_P$. Clearly, this means that the tangent vectors $x_1, \dots, x_{n-1} \in (\partial M)_P$ form a positively oriented basis if and only if for every outwardly directed tangent vector $y \in M_P$, the basis $y, x_1, \dots, x_{n-1}$ is positively oriented with respect to M.

13.5 Integral Theorems

The classical integral theorem by Stokes states that for a vector field $X = X^i \partial_i$ on a surface $\mathcal{F}$ with a boundary curve $\mathcal{K}$, there holds

$$\int_{\mathcal{K}} X \, dr = \iint_{\mathcal{F}} \mathrm{rot} X \, d\vec{o}$$

where

$$\operatorname{rot}X = (\partial_2 X^3 - \partial_3 X^2)\partial_1 + (\partial_3 X^1 - \partial_1 X^3)\partial_2 + (\partial_1 X^2 - \partial_2 X^1)\partial_3 \,,$$

and the coordinate vector fields ∂_i refer to the Cartesian coordinates in $\mathbb{R}^3$. The curve integral can be written using the 1-form ω with the components $\omega_i = X^i$ in the terminology of differential forms as the integral $\int \omega$ over the boundary curve. By Theorem 6.9, the outer derivative of ω is the 2-form $d\omega$ with the components

$$(d\omega)_{23} = \partial_2 X^3 - \partial_3 X^2$$

$$(d\omega)_{13} = \partial_1 X^3 - \partial_3 X^1$$

$$(d\omega)_{12} = \partial_1 X^2 - \partial_2 X^1 \,.$$

By inserting two vector fields Y and Z, we obtain the scalar field

$$
\begin{aligned}
d\omega(Y, Z) &= d\omega(Y^i \partial_i, Z^k \partial_k) \\
&= (\partial_2 X^3 - \partial_3 X^2)Y^2 Z^3 + (\partial_1 X^3 - \partial_3 X^1)Y^1 Z^3 + (\partial_1 X^2 - \partial_2 X^1)Y^1 Z^2 \\
&= \operatorname{rot}X \cdot (Y \times Z) \,.
\end{aligned}
$$

Consequently, the flux integral $\iint \operatorname{rot}X \, d\vec{o}$ can be understood as the integral $\int d\omega$ over the surface. Hence, the classical integral theorem of Stokes, in the language of differential forms, is

$$\int_{\partial \mathcal{F}} \omega = \int_{\mathcal{F}} d\omega \,.$$

This equation also holds in general on manifolds with boundaries. This is stated in the following theorem, which is also called the **Stokes Integral Theorem**.

Theorem 13.4 *For a differential form ω of order $n-1$ on an n-dimensional oriented manifold M with boundary, there holds*

$$\int_{\partial M} \omega = \int_{M} d\omega \,.$$

Proof Without loss of generality, we may assume that the support of the integrand ω can be covered by a single chart, since otherwise the compact support can be covered with a finite number of charts and using a corresponding partition of unity $\tau_1 + \ldots + \tau_m = 1$, the differential form can then be decomposed into the sum $\omega = \tau_1 \omega + \ldots + \tau_m \omega$, and we can then apply the following considerations to $\tau_k \omega$. Let (U, φ) be a chart with $\operatorname{supp}\omega \subseteq U$, and $\varphi(U)$ open in $\mathbb{R}^n_-$. According to the definition of the integral and Theorem 6.10, we have

$$\int_M d\omega = \int \cdots \int_{(-\infty,0]\times\mathbb{R}^{n-1}} (d\omega \circ \varphi^{-1})(\partial_1,\ldots,\partial_n)\, du^1 \ldots du^n$$

$$= \sum_{l=1}^{n} (-1)^{l-1} \int \cdots \int_{(-\infty,0]\times\mathbb{R}^{n-1}} \frac{\partial}{\partial u^l}(\omega \circ \varphi^{-1})(\ldots,\partial_{l-1},\partial_{l+1},\ldots)\, du^1 \ldots du^n .$$

Since $\mathrm{supp}(\omega \circ \varphi^{-1})$ is contained in an open set in $\mathbb{R}^n_-$, we have

$$\int_{-\infty}^{0} \frac{\partial}{\partial u^1}(\omega \circ \varphi^{-1})(\partial_2,\ldots,\partial_n)\, du^1 = (\omega \circ \varphi^{-1})(\partial_2,\ldots,\partial_n)\Big|_{u^1=0}$$

and

$$\int_{-\infty}^{+\infty} \frac{\partial}{\partial u^l}(\omega \circ \varphi^{-1})(\ldots,\partial_{l-1},\partial_{l+1},\ldots)\, du^l = 0$$

for $l \neq 1$ and thus altogether

$$\int_M d\omega = \int \cdots \int_{\mathbb{R}^{n-1}} (\omega \circ \varphi^{-1})(\partial_2,\ldots,\partial_n)\Big|_{u^1=0} du^2 \ldots du^n = \int_{\partial M} \omega .$$

Theorem 13.4 also contains the classical integral theorem of Gauss. Given a bounded domain $\mathcal{G}$ with a smooth surface $\partial\mathcal{G}$ and a vector field

$$f = f^1\partial_1 + f^2\partial_2 + f^3\partial_3$$

therein, there holds

$$\iint_{\partial\mathcal{G}} f\, d\vec{o} = \int_{\partial\mathcal{G}} *Jf = \int_{\mathcal{G}} d*Jf = \iiint_{\mathcal{G}} (*d*Jf)\, dx^1 dx^2 dx^3$$

with

$$*d*Jf = *d*(f^1 dx^1 + f^2 dx^2 + f^3 dx^3) = *d(f^1 dx^2 \wedge dx^3 + f^2 dx^3 \wedge dx^1$$
$$+ f^3 dx^1 \wedge dx^2)$$

$$= *((\partial_1 f^1 + \partial_2 f^2 + \partial_3 f^3)dx^1 \wedge dx^2 \wedge dx^3) = \partial_1 f^1 + \partial_2 f^2 + \partial_3 f^3 = \mathrm{div} f .$$

13.6   Extremal Principles

It is known that important equations in Physics can be derived from extremal principles. This also applies to the theory of relativity. We will show in the following that the Euler-Lagrange euqation for the integral of the curvature scalar contain the Einstein field equation for empty space.

According to the definitions in Sect. 8.3 and Theorems 8.2 and 7.4, the curvature scalar S is a function of the metric coefficients and their first and second partial derivatives. However, if we want to apply the classical variational calculation to these variables, then we need to formulate partial derivatives of S with respect to g_{ab}, $g_{ab,c}$ and $g_{ab,cd}$, which is extremely tedious. The partial derivatives with respect to g^{ab}, Γ^c_{ab} and $\Gamma^c_{ab,d}$ are much easier to determine. Therefore, we consider S as a function

$$S = g^{ik}\mathrm{Ric}_{ik} = g^{ik}(\Gamma^l_{ki,l} - \Gamma^l_{li,k} + \Gamma^r_{ki}\Gamma^l_{lr} - \Gamma^r_{li}\Gamma^l_{kr})$$

of these variables and adopt the viewpoint that the integral

$$\int SV = \int_{\mathcal{G}} S(g^{ab}(u),\, \Gamma^c_{ab}(u),\, \Gamma^c_{ab,d}(u))\sqrt{-g}\, du$$

where $g = \det(g_{ik})$ is stationary. This is an integral of the form

$$\int_{\mathcal{G}} L\left(f_i(u),\, \frac{\partial f_i}{\partial u^k}(u)\right) du\ ,\qquad i = 1, ..., m\ ,\qquad k = 1, ..., n\ ,$$

where $\mathcal{G}$ is an integration domain in $\mathbb{R}^n$ (in our case $\mathbb{R}^4$) and L is a formula with $m + nm$ inputs.

We now summarise the arguments that lead to the Euler-Lagrange equations in the context of the classical calculus of variations. If the integral is extremal for the functions $f_1, ..., f_m$, then for every system of functions $\hat{f}_1, ..., \hat{f}_m$ that vanish on the boundary $\partial\mathcal{G}$ of the domain, the function

$$F(\varepsilon) = \int_{\mathcal{G}} L\left((f_i + \varepsilon\hat{f}_i)(u),\, \frac{\partial}{\partial u^k}(f_i + \varepsilon\hat{f}_i)(u)\right) du$$

is extremal at $\varepsilon = 0$. Therefore we have

$$0 = F'(0) = \int_{\mathcal{G}} \left(\sum_{i=1}^{m} \frac{\partial L}{\partial f_i}\cdot\hat{f}_i + \sum_{k=1}^{n}\sum_{i=1}^{m} \frac{\partial L}{\partial f_{i,k}}\cdot\hat{f}_{i,k}\right) du$$

and since

$$\frac{\partial L}{\partial f_{i,k}}\cdot\frac{\partial\hat{f}_i}{\partial u^k} = \frac{\partial}{\partial u^k}\left(\frac{\partial L}{\partial f_{i,k}}\cdot\hat{f}_i\right) - \left(\frac{\partial}{\partial u^k}\frac{\partial L}{\partial f_{i,k}}\right)\cdot\hat{f}_i$$

finally

$$0 = \sum_{i=1}^{m} \int_{\mathcal{G}} \left(\frac{\partial L}{\partial f_i} - \sum_{k=1}^{n} \frac{\partial}{\partial u^k} \left(\frac{\partial L}{\partial f_{i,k}} \right) \right) \cdot \hat{f}_i \, du + \sum_{i=1}^{m} \int_{\mathcal{G}} \sum_{k=1}^{n} \frac{\partial}{\partial u^k} \left(\frac{\partial L}{\partial f_{i,k}} \cdot \hat{f}_i \right) du .$$

By the classical Gauss integral theorem, the integrals in the second summand are surface integrals of vector integrands which vanish on the boundary $\partial \mathcal{G}$. Therefore only the first summand remains, and as the functions $\hat{f}_i$ were chosen arbitrarily, we obtain the **Euler-Lagrange euqations**

$$\frac{\partial L}{\partial f_i} = \sum_{k=1}^{n} \frac{\partial}{\partial u^k} \frac{\partial L}{\partial f_{i,k}}$$

In our case, they are

(EL1) $\dfrac{\partial}{\partial g^{ab}} (\sqrt{-g} \, g^{ik} \mathrm{Ric}_{ik}) = 0$

and

(EL2) $\sqrt{-g} \, g^{ik} \dfrac{\partial \mathrm{Ric}_{ik}}{\partial \Gamma^{c}_{ab}} = \displaystyle\sum_{j=0}^{3} \partial_j \left(\sqrt{-g} \, g^{ik} \dfrac{\partial \mathrm{Ric}_{ik}}{\partial \Gamma^{c}_{ab,j}} \right).$

To solve the equation (EL1) we have to determine the partial derivatives of $g = \det(g_{ik})$ with respect to g^{ab}. By expanding along a column we have

$$\frac{1}{g} = \det(g^{ik}) = \sum_{i=0}^{3} g^{ik} G^{ik}$$

where the G^{ik} are the minors, and so it follows that

$$\frac{\partial(1/g)}{\partial g^{ab}} = G^{ab} .$$

Since (g_{ik}) is the inverse of (g^{ik}), we have

$$g_{ba} = \frac{1}{1/g} G^{ab} ,$$

and so altogether

$$\frac{\partial}{\partial g^{ab}} \left(\frac{1}{g} \right) = \frac{1}{g} g_{ba} .$$

As

$$0 = \frac{\partial}{\partial g^{ab}} \left(g \cdot \frac{1}{g} \right) = \frac{1}{g} \frac{\partial g}{\partial g^{ab}} + g \frac{\partial(1/g)}{\partial g^{ab}}$$

it follows that

$$\frac{\partial g}{\partial g^{ab}} = -g\,g_{ba}$$

and

$$\frac{\partial \sqrt{-g}}{\partial g^{ab}} = \frac{-1}{2\sqrt{-g}} \frac{\partial g}{\partial g^{ab}} = -\frac{1}{2}\sqrt{-g}\,g_{ba} \ .$$

Therefore the Euler-Lagrange equations (EL1) become

$$0 = \frac{\partial}{\partial g^{ab}}(\sqrt{-g}\,g^{ik}\mathrm{Ric}_{ik}) = -\tfrac{1}{2}\sqrt{-g}\,g_{ba}g^{ik}\mathrm{Ric}_{ik} + \sqrt{-g}\,\mathrm{Ric}_{ba} = \sqrt{-g}\left(\mathrm{Ric}_{ba} - \tfrac{1}{2}Sg_{ba}\right)$$

and hence

$$\mathrm{Ric}_{ba} - \tfrac{1}{2}Sg_{ba} = 0 \ .$$

These are the Einstein field equation in components.

It remains to be shown that the variables Γ^r_{ij} assume the values of the Christoffel symbols

$$\Gamma^r_{ij} = \tfrac{1}{2}g^{kr}(\partial_i g_{jk} + \partial_j g_{ik} - \partial_k g_{ij})$$

if the Euler-Lagrange equations are fulfilled. For a fixed point P we can choose a chart around P, such that at the point P all the Christoffel symbols are zero and the matrix of the metric components is diagonal, i.e.,

$$(g_{ik}) = \mathrm{diag}(g_0, g_1, g_2, g_3)$$

and

$$(g^{ik}) = \mathrm{diag}(g^0, g^1, g^2, g^3) \quad \text{with} \quad g = 1/g_i \ .$$

We will show that such a coordinate system exists only at the end of this section. Now we must verify the equations

$$\tfrac{1}{2}g^{kr}(\partial_i g_{jk} + \partial_j g^{ik} - \partial_k g_{ij}) = 0$$

at the point P. We will show that $\partial_i g_{jk} = 0$.

The Euler-Lagrange equations (EL2) reduce to

$$\sum_{j=1}^{3} \partial_j \left(\sqrt{-g}\,g^{ik}\frac{\partial \mathrm{Ric}_{ik}}{\partial \Gamma^c_{ab,j}} \right) = 0$$

at the point P. The formula for Ric_{ik} shows that

$$0 = \sum_{j=0}^{3} \partial_j \left(\sqrt{-g}\,g^{ik}\frac{\partial \mathrm{Ric}_{ik}}{\partial \Gamma^a_{ab,j}} \right) = \partial_a(\sqrt{-g}\,g^{ba}) - \sum_{j=0}^{3} \partial_j(\sqrt{-g}\,g^{bj}) = -\sum_{j \neq a} \partial_j(\sqrt{-g}\,g^{bj})$$

and for $a \neq c$

$$0 = \sum_{j=0}^{3} \partial_j \left(\sqrt{-g}\, g^{ik} \frac{\partial \mathrm{Ric}_{ik}}{\partial \Gamma^c_{ab,j}} \right) = \partial_c(\sqrt{-g}\, g^{ba}) \quad .$$

The formula

$$\frac{\partial g^{ij}}{\partial g_{rs}} = -g^{ir} g^{sj}$$

is useful in future calculations, and can be justified as follows: From the constancy of the expressions $g^{ik} g_{kl}$, it follows that

$$\frac{\partial}{\partial g_{rs}}(g^{ik} g_{kl}) = 0$$

and thus

$$\frac{\partial g^{ik}}{\partial g_{rs}} g_{kl} = \begin{cases} -g^{ir} & \text{for} \quad l = s \\ 0 & \text{otherwise} \end{cases} .$$

This can be resolved to yield

$$\frac{\partial g^{ij}}{\partial g_{rs}} = \frac{\partial g^{ik}}{\partial g_{rs}} g_{kl}\, g^{lj} = -g^{ir} g^{sj} \quad .$$

With the formula

$$\frac{\partial \sqrt{-g}}{\partial g_{ab}} = -\frac{1}{2} g_{ba}$$

already used in the evaluation of (EL1), we continue to work on the Euler-Lagrange equations (EL2). For $a \neq c$ we obtain

$$0 = \partial_c(\sqrt{-g}\, g^{ba}) = g^{ba} \frac{\partial \sqrt{-g}}{\partial g^{ij}} \frac{\partial g^{ij}}{\partial g_{rs}} \partial_c g_{rs} + \sqrt{-g} \frac{\partial g^{ba}}{\partial g_{rs}} \partial_c g_{rs}$$

$$= \sqrt{-g}\, (\tfrac{1}{2} g^{ba} g_{ji}\, g^{ir} g^{sj} - g^{br} g^{sa}) \partial_c g_{rs} = \sqrt{-g} \left(\tfrac{1}{2} g^{ba} \sum_k g^k\, \partial_c g_{kk} - g^b\, g^a\, \partial_c g_{ba} \right) .$$

For $a \neq b$, this is

$$0 = -\sqrt{-g}\, g^b\, g^a\, \partial_c g_{aa} \quad .$$

This shows that for $j \neq k$ and arbitrarily i $\partial_i g_{jk} = 0$. For $a = b$, we obtain

$$\frac{1}{2} \sum_k g^k\, \partial_c g_{kk} = g^a\, \partial_c g_{aa} \quad .$$

For $c = 0$ these are three equations that can be written in the form

$$\begin{pmatrix} -1 & 1 & 1 \\ & 1 & -1 & 1 \\ & 1 & 1 & -1 \end{pmatrix} \begin{pmatrix} g^1\,\partial_0 g_{11} \\ g^2\,\partial_0 g_{22} \\ g^3\,\partial_0 g_{33} \end{pmatrix} = \begin{pmatrix} 0 \\ 0 \\ 0 \end{pmatrix}.$$

From this is follows that $g^a\,\partial_0 g_{aa} = 0$ for $a = 1, 2, 3$. Analogous considerations also apply for $c = 1, 2, 3$. Altogether, this results in $\partial_i g_{jj} = 0$ for $i \neq j$.

We now have the equations (EL2) with $a = c$ available. For $a \neq b$ they deliver

$$0 = -\sum_{j \neq a} \partial_j(\sqrt{-g}\,g^{bj}) = -\sqrt{-g}\sum_{j \neq a} \left(\frac{1}{2}\,g^{bj}\,g_{vu}\,g^{ur}\,g^{sv} - g^{br}\,g^{sj} \right) \partial_j g_{rs}$$

$$-\sqrt{-g}\left(\frac{1}{2}g^b \sum_r g^r\,\partial_b g_{rr} - g^b\,g^b\,\partial_b g_{bb} \right) = \frac{1}{2}\sqrt{-g}\,g^b\,g^b\,\partial_b g_{bb}$$

and thus also $\partial_i g_{ii} = 0$.

Finally, as already announced earlier, and already used, we have to construct a chart around a given point P, at which the Christoffel symbolds are zero and at which the component matrix (g_{ik}) is diagonal. We do this in the context of an n-dimensional semi-Riemannian manifold. Let a chart be given with the coordinates $u^1, \ldots, u^n$. The point P has coordinates $u_o^1, \ldots, u_o^n$. We try to introduce new coordinates $\bar{u}^1, \ldots, \bar{u}^n$ by requiring

$$u^k = u_o^k + \bar{u}^k - \tfrac{1}{2}\Gamma_{ij}^k(P)\bar{u}^i\bar{u}^j.$$

The C^∞-map, which maps the independent variables $\bar{u}^k$ to the dependent u^k, maps the null vector to $(u_0^1, \ldots, u_0^n)$. At this point, the functional matrix $(\partial u^k/\partial \bar{u}^l)$ is the identity matrix, and so this map has a local inverse, so that one can (locally) solve for the independent variables. In the domain of definition for the new chart, we have

$$\frac{\partial u^k}{\partial \bar{u}^l} = \delta_l^k - \Gamma_{lj}^k(P)\bar{u}^j$$

and

$$\frac{\partial^2 u^k}{\partial \bar{u}^r \partial \bar{u}^l} = -\Gamma_{lr}^k(P).$$

The coordinate vector fields $\bar{\partial}_l$ of the new coordinates are represented in terms of the old ones as $\bar{\partial}_l = (\partial u^k/\partial \bar{u}^l)\partial_k$. By the chain rule, we have

$$\bar{\partial}_l\frac{\partial u^t}{\partial \bar{u}^s} = \frac{\partial u^k}{\partial \bar{u}^l}\,\partial_k\frac{\partial u^t}{\partial \bar{u}^s} = \frac{\partial u^k}{\partial \bar{u}^l}\frac{\partial^2 u^t}{\partial \bar{u}^r \partial \bar{u}^s}\frac{\partial \bar{u}^r}{\partial u^k} = -\frac{\partial u^k}{\partial \bar{u}^l}\frac{\partial \bar{u}^r}{\partial u^k}\Gamma_{sr}^t(P).$$

At point P this means

$$\bar{\partial}_l(P) \frac{\partial u^t}{\partial \bar{u}^s} = -\frac{\partial u^k}{\partial \bar{u}^l}(0,\ldots,0) \frac{\partial \bar{u}^r}{\partial u^k}(u_0^1,\ldots,u_0^n) \, \Gamma_{sr}^t(P) = -\Gamma_{sl}^t(P) = -\Gamma_{ls}^t(P)$$

$$= -\frac{\partial u^i}{\partial \bar{u}^l}(0,\ldots,0) \frac{\partial u^k}{\partial \bar{u}^s}(0,\ldots,0)\Gamma_{ik}^t(P) \, ,$$

and from the transformation formula given in Theorem 7.5 it can be seen that the Christoffel symbols for the coordinates $\bar{u}^1,\ldots,\bar{u}^n$ at P are zero.

To diagonalise the matrix $(g_{ik}(P))$, a second transformation from the coordinates $\bar{u}^k$ to the final coordinates $\hat{u}^l = \beta_k^l \bar{u}^k$ is now necessary. For the constant coefficients β_k^l, we have $\bar{\partial}_k = \beta_k^l \hat{\partial}_l$ and $\hat{\partial}_l = \alpha_l^k \bar{\partial}_k$, where the matrix (α_l^k) is the inverse to (β_k^l). These coefficients should be chosen so that the matrix

$$(g(P)(\hat{\partial}_i, \hat{\partial}_k)) = (g(P)(\alpha_i^r \bar{\partial}_r, \alpha_k^s \bar{\partial}_s)) = (\alpha_i^r \, g(P)(\bar{\partial}_r, \bar{\partial}_s) \, \alpha_k^s)$$

is diagonal. This is obviously possible. Since the coefficients α_l^k are constants, it can be seen from the formula in Theorem 7.5 that the Christoffel symbols with respect to the coordinates $\hat{u}^k$ at the point P are also again zero. So this chart has all the required properties.

Nonrotating Black Holes

14

14.1 The Schwarzschild Half Plane

Already in Chap. 10 we examined geodesics in the Schwarzschild spacetime. In this chapter we adopt the viewpoint that the total mass of the non-rotating fixed star that generates the spacetime is concentrated at $r = 0$. Thus the region when the Schwarzschild radius r is near the critical value 2μ and below becomes the focus of interest. This number 2μ is often called the Schwarzschild radius (of the fixed star). Since we have already reserved this name for the radial coordinate, we call this number the **Schwarzschild horizon**.

For $r > 2\mu$, we have the Schwarzschild metric available on the manifold $\mathbb{R} \times (2\mu, \infty) \times S_2$. Its validity is based on the plausibility considerations given in Sect. 9.7 within the framework of the Einstein field equation and also experimental confirmation for large values of r. The formula for the Schwarzschild metric are also formally meaningful on the manifold $\mathbb{R} \times (0, 2\mu) \times S_2$. We adopt the viewpoint that they are also valid there. At first this seems speculative, especially since the spaces $\mathbb{R} \times (2\mu, \infty) \times S_2$ and $\mathbb{R} \times (0, 2\mu) \times S_2$ are not connected. However, at the end of this section, it will become clear that the difficulties at the points $r = 2\mu$ are largely due to the coordinate system used.

To describe a radially moving particle or photon, you only need the functions $t(\tau)$ and $r(\tau)$ in addition to specifying a point on S_2. Since knowledge of the point from S_2 is unnecessary on the grounds of symmetry, in a fundamental investigation of such motion, the essential information about a world line of a radial particle is contained in a curve in the $r-t$-plane. The Schwarzschild spacetime $\mathbb{R} \times ((0, 2\mu) \cup (2\mu, \infty)) \times S_2$ in this case is thus reduced to the so-called **Schwarzschild half plane** $\mathbb{R} \times ((0, 2\mu) \cup (2\mu, \infty))$, equipped with the indefinite metric g, which is defined by

Fig. 14.1 Future cones in
the Schwarzschild half plane

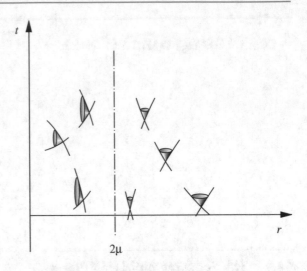

$$g(\partial_t, \partial_t) = 1 - 2\mu/r$$
$$g(\partial_r, \partial_r) = -(1 - 2\mu/r)^{-1}.$$
$$g(\partial_t, \partial_r) = 0$$

Tangent vectors $a\partial_t + b\partial_r$ are timelike if and only if

$$(1 - 2\mu/r)a^2 > (1 - 2\mu/r)^{-1}b^2.$$

For $r > 2\mu$, this means

$$|b/a| < 1 - 2\mu/r$$

and for $0 < r < 2\mu$,

$$|b/a| > 2\mu/r - 1.$$

The time-orientation of the outer Schwarzschild spacetime implies that for $r > 2\mu$
the timelike vectors $a\partial_t + b\partial_r$ with $a > 0$ are future-directed. For $0 < r < 2\mu$, we
declare that the timelike vectors with $b < 0$ are future-directed. This decision has the
consequence that particles and photons can no longer leave the region $0 < r < 2\mu$.
The future cones are marked in Fig. 14.1.

The geodesic equations can be read from Theorem 10.5 with the specialisation
$\vartheta' = \varphi' = 0$. We have

$$t'' = \frac{2\mu}{r(2\mu - r)}t'r'$$

and

$$r'' = \frac{(2\mu - r)\mu}{r^3}(t')^2 + \frac{\mu}{r(r - 2\mu)}(r')^2$$

under the constraint

$$(r - 2\mu)(t')^2 + \frac{r^2}{2\mu - r}(r')^2 = r$$

for particles and

$$(r - 2\mu)^2(t')^2 = r^2(r')^2$$

for light rays.

Theorem 14.1 *The light rays in the Schwarzschild half plane are the graphs of the functions*

$$t(r) = \pm(r + 2\mu \log |r - 2\mu| + c).$$

Proof For the derivative dt/dr we must have

$$\frac{dt}{dr} = \frac{t'}{r'} = \pm\frac{r}{r - 2\mu} = \pm\left(1 + \frac{2\mu}{r - 2\mu}\right),$$

and the functions $t(r)$ result upon integration.

These functions are sketched in Fig. 14.2. For $r > 2\mu$, the function

$$t = r + 2\mu \log(r - 2\mu) + c$$

describes an outward light ray. Near 2μ the differential quotient dr/dt is only slightly above 0, and for $r \to \infty$, it converges to 1. The other sign variant

$$t = -r - 2\mu \log(r - 2\mu) + c$$

describes an incoming light ray. Both from the formula and from Fig. 14.2, it can be seen that for $r \to 2\mu$ the Schwarzschild time t tends to ∞, and the light beam does not reach the Schwarzschild horizon in finite time, at least not in the sense of a resting observer. The effect is even clearer for an incoming particle. As its speed γ' must lie inside the light cone, the t-coordinate grows even faster with decreasing r along the incident trajectory. We imagine that the particle sends light signals to the observer $1/\sqrt{1 - 2\mu/r_0}\partial_t$ located at $r = r_0$, called the **Schwarzschild observer**. The t-coordinate of the intersection of the outgoing light ray with the perpendicular $r = r_0$, is larger than the t-coordinate of the position of the particle and converges even faster to ∞ (Fig. 14.2). This is the observers proper time except for the factor $\sqrt{1 - 2\mu/r_0}$. So he has the impression that the particle does not reach the Schwarzschild horizon, but slows down the closer it gets to this horizon. Because of the gravitational red shift, the signals get weaker and the picture gets darker and darker.

An observer accompanying the particle experiences a completely different scenario. We will show that the particle reaches not only the horizon $r = 2\mu$ but even the central singularity $r = 0$ in finite proper time.

Fig. 14.2 Light rays in the
Schwarzschild half plane

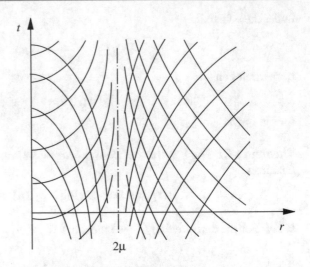

Theorem 14.2

(1) *For particles and photons, $h(r)t' = (1 - 2\mu/r)t'$ is constant.*
(2) *With the constant $E = h(r)t'$, we have for radial particles*

$$E^2 - 1 = (r')^2 - 2\mu/r$$

and for radial photons

$$E^2 = (r')^2 .$$

Proof The derivative

$$(h(r)t')' = \frac{2\mu}{r^2}r't' + \left(1 - \frac{2\mu}{r}\right)t''$$

is zero according to the geodesic equation $t'' = -2\mu/(r^2 h(r))t'r'$. From the constraint

$$1 = g(\gamma', \gamma') = h(r)(t')^2 - \frac{1}{h(r)}(r')^2$$

for particles, it follows that

$$(h(r)t')^2 - (r')^2 = h(r) ,$$

and thus

$$E^2 - (r')^2 = 1 - 2\mu/r .$$

For photons, this results in $E^2 - (r')^2 = 0$.

The number $E = h(r)t'$ is positive for each particle γ, because both $h(r)$ and t' are positive in the case $r > 2\mu$. For $0 < r < 2\mu$, $h(r)$ is negative. As E is positive, t' must be negative there. In Fig. 14.2, the curves satisfying the equation

$$t(r) = -r - 2\mu \log(2\mu - r) + c$$

represent incoming photons in the black hole. For a particle that has entered the black hole, the Schwarzschild time runs backwards. Apparently, this concept of time is not very suitable for describing the processes in the black hole.

The number E can be interpreted as energy per rest mass, measured by a Schwarzschild observer with a very large radial coordinate r, because for $(1/\sqrt{1 - 2\mu/r})\, \partial_t$, the particle $\gamma' = t'\partial_t + r'\partial_r$ has the rest mass m and the energy

$$mg\left(\gamma', \frac{1}{\sqrt{1 - 2\mu/r}}\, \partial_t\right) = mt'\sqrt{1 - 2\mu/r} = \frac{mE}{\sqrt{1 - 2\mu/r}},$$

and this expression converges to mE as $r \to \infty$.

Another interpretation of the number E of a particle follows from Theorem 14.2(2). In case $E \geq 1$, the particle can assume any Schwarzschild radius r. In the limit case $E = 1$, r' must become 0 for $r \to \infty$, and one could say that the incident particle started at ∞ with $r' = 0$. For $E < 1$, the inequality $E^2 - 1 \geq -2\mu/r$ is exactly fulfilled for r such that $r \leq 2\mu/(1 - E^2)$, so that the Schwarzschild radius r is bounded above. This is the case when the particle started at $r = 2\mu/(1 - E^2)$ with $r' = 0$. Since the relative velocity of the radial particle $t'\partial_t + r'\partial_r$ is $v = (r'/(t'\sqrt{1 - 2\mu/r}))\, \partial_r$ for the observer $(1/\sqrt{1 - 2\mu/r})\, \partial_t$ according to Theorem 5.1, this means that the particle started at the specified value of r with the relative velocity zero.

We now show that a radially incident particle reaches the singularity $r = 0$ in finite proper time from any point. For a particle with $E = 1$ it is very easy to represent the proper time τ as a function of the Schwarzschild radius r. It follows from Theorem 14.2(2) that $(d\tau/dr)^2 = r/(2\mu)$. Since τ increases with decreasing r for the incoming particle, the negative root of $r/(2\mu)$ must be chosen. The primitive functions are

$$\tau(r) = -\frac{2/3}{\sqrt{2\mu}} r^{3/2} + c\,.$$

Thus the particle that started at $r = \infty$ with relative velocity zero needs the proper time

$$\tau(0) - \tau(r_0) = \frac{1}{3}\sqrt{2/\mu}\, r_0^{3/2}$$

for the path from the position $r = r_0$ to the singularity $r = 0$. For a radial particle that starts at r_0 with the relative velocity zero, the proper time cannot be expressed so elementarily in terms of the Schwarzschild radius. However, the graph of the function $\tau(r)$ can be interpreted surprisingly simply geometrically. According to the following theorem, this graph is a cycloid except for a distortion factor (Fig. 14.3).

Fig. 14.3 Schwarzschild radius r and proper time τ for a radially falling particle

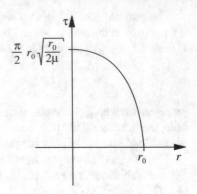

Theorem 14.3 *For $0 < r < r_0$, let $\tau(r)$ denote the proper time needed for a particle starting at r_0 with the relative velocity zero along the path up to the Schwarzschild radius r. Then*

$$r(\eta) = \frac{1}{2}r_0(1 + \cos \eta)$$

and

$$\tau(\eta) = \frac{1}{2}r_0\sqrt{r_0/(2\mu)}\,(\eta + \sin \eta)$$

for $0 < \eta < \pi$ give a parametric representation of the graph of the function $\tau(r)$.

Proof We show that the expression given for the parameter representation gives $dr/d\tau$ and this actually satisfies the differential equation arising from Theorem 14.2(2). The left-hand side there is

$$E^2 - 1 = -2\mu/r_0\,.$$

With

$$\frac{dr}{d\tau} = \frac{dr/d\eta}{d\tau/d\eta} = \frac{-\sin \eta}{\sqrt{r_0/(2\mu)}\,(1 + \cos \eta)}$$

we obtain that the right-hand side matches the left-hand side value, since

$$\left(\frac{dr}{d\tau}\right)^2 - \frac{2\mu}{r} = \frac{2\mu}{r_0}\frac{\sin^2\eta}{(1+\cos \eta)^2} - \frac{4\mu}{r_0(1+\cos \eta)} = \frac{2\mu}{r_0}\cdot\frac{\sin^2\eta - 2(1+\cos \eta)}{1 + 2\cos \eta + \cos^2\eta} = -\frac{2\mu}{r_0}\,.$$

14.2 Optics of Black Holes

A black hole appears as a black disc to an observer outside. The angular radius will now be shown to depend on the mass μ and the distance of the observer. Simultaneously, we also determine the angular radius of a fixed star, which of course differs only slightly from the Euclidean viewpoint.

The starting point are the geodesic equations of the Schwarzschild spacetime with $\vartheta = \pi/2$. The equation

$$t'' = -\frac{2\mu}{r^2 h(r)} t' r'$$

is equivalent to $h(r)t' = E$, by Theorem 14.2, and $\varphi'' = -(2/r)r'\varphi'$ is equivalent to $r^2\varphi' = L$ (see the proof of Theorem 10.9). The constraint

$$h^2(r)(t')^2 = (r')^2 + h(r)r^2(\varphi')^2$$

for light rays is rewritten with the constants E and L as

$$(r')^2 = E^2 - L^2 h(r)/r^2 \ .$$

In contrast to a particle, the numbers E and L have no physical meaning for a photon, because they are not uniquely determined by the light ray, and with γ, also $\bar{\gamma}(\sigma) = \gamma(c\sigma)$ fulfills the geodesic equations, together with the constraint

$$g(\bar{\gamma}', \bar{\gamma}') = g(c\gamma', c\gamma') = c^2 g(\gamma', \gamma') = 0 \ .$$

However, the quotient L/E is uniquely determined, and this also plays a role in the following theorems.

Theorem 14.4 *The function $r(\varphi)$, which may be determined locally, which describes the path of a photon in the plane $\vartheta = \pi/2$, satisfies the differential equation*

$$\left(\frac{dr}{d\varphi}\right)^2 = \frac{E^2}{L^2} r^4 - h(r)r^2 \ .$$

Proof The above representation of $(r')^2$ and $\varphi' = L/r^2$ should be used in $(dr/d\varphi)^2 = (r')^2/(\varphi')^2$.

Since the right-hand side of the differential equation must not become negative, only the region where $\sqrt{h(r)}/r \leq E/|L|$ is accessible to the photon. The function $\sqrt{h(r)}/r$, which is sketched in Fig. 14.4, also plays a role in the next theorem. It expresses the sine of the angle α indicated in Fig. 14.5 in terms of L/E and r.

Theorem 14.5 *Let $\gamma' = t'\partial_t + r'\partial_r + \varphi'\partial_\varphi$ be a photon in the equatorial plane $\vartheta = \pi/2$. Then in the region $r > 2\mu$, the angle α between $r'\partial_r + \varphi'\partial_\varphi$ and $-\partial_r$ in the sense of the Euclidean space $\partial_t^\perp$ satisfies*

$$\sin \alpha = \frac{|L|}{E} \cdot \frac{\sqrt{h(r)}}{r} \ .$$

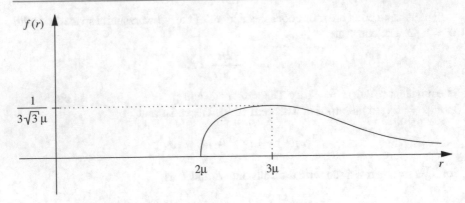

Fig. 14.4 The function $f(r) = \sqrt{1 - 2\mu/r}/r$

Fig. 14.5 The angle α
between $r'\partial_r + \varphi'\partial_\varphi$ and ∂_r

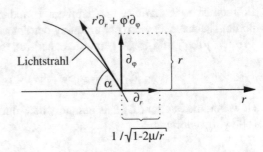

Proof We have

$$\sin\alpha = \frac{|\varphi'\partial_\varphi|}{|r'\partial_r + \varphi'\partial_\varphi|} = \frac{|\varphi'|r}{\sqrt{(r')^2/h(r) + (\varphi')^2 r^2}} = \frac{|\varphi'|r}{\sqrt{h(r)}\,t'} = \frac{|L|/r}{E/\sqrt{h(r)}} = \frac{|L|}{E}\frac{\sqrt{h(r)}}{r}.$$

In the sense of Euclidean geometry, a fixed star with radius R whose center is at a distance r_0 from an observer, has an angular radius β, with $\sin\beta = R/r_0$. In the theory of relativity, in the normal case when $R \gg 2\mu$, this has a slightly larger value.

Theorem 14.6 *Let $r_0 > R > 3\mu$. The fixed star with the mass μ and the radius R has for an observer at $r = r_0$ the angular radius β with*

$$\sin\beta = \frac{R}{r_0} \cdot \sqrt{\frac{h(r_0)}{h(R)}}.$$

Proof We determine the quotient $|L|/E$ for a photon tangential to the fixed star, and then calculate $\sin\beta$ by Theorem 14.5. It makes sense to set the derivative $dr/d\varphi$ to zero in the equation in Theorem 14.4 and to solve for $|L|/E$ from the resulting equation with $r = R$, giving $|L|/E = R/\sqrt{h(R)}$. In fact a photon with $|L|/E > R/\sqrt{h(R)}$ does not reach the region with $r \le R$, and so it misses the fixed star.

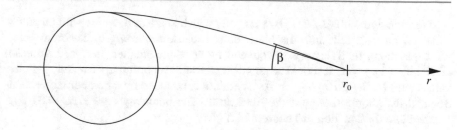

Fig. 14.6 The angular radius β of a fixed star of a black hole

In the other case $|L|/E < R/\sqrt{h(R)}$, a positive number ε can be found such that $E^2/L^2 > h(R)/R^2 + \varepsilon^2$. By Theorem 14.4, for $r \geq R$, we have

$$\left(\frac{dr}{d\varphi}\right)^2 = r^4\left(\frac{E^2}{L^2} - \frac{h(r)}{r^2}\right) > R^4\left(\frac{h(R)}{R^2} + \varepsilon^2 - \frac{h(r)}{r^2}\right) \geq R^4\varepsilon^2.$$

Thus $|dr/d\varphi|$ can be estimated from below up to a positive constant. Therefore, the photon from $r = r_0$ must reach the region $r \leq R$. This proves $|L|/E = R/\sqrt{h(R)}$ for the boundary case of a tangential photon, and the formula for $\sin\beta$ follows, as already mentioned.

Finally, we now deal with the appearance of a black hole. It will be seen that for an observer with $r_0 > 3\mu$, it looks like a black sphere with radius 3μ. Around this sphere, the starry sky behind is of course altered more or less due to the deflection of the light rays.

We first examine if there are circular rays of light. The assumption $r = r_0$, $\vartheta = \pi/2$, $\varphi = \tau$, using the constraint, leads to $t' = r_0/\sqrt{h(r_0)}$. The r'' geodesic equation then gives $\mu = r_0 h(r_0)$, and so $r_0 = 3\mu$. A photon, that starts at $r_0 = 3\mu$ with the angle $\pi/2$, remains on the circular path $r = 3\mu$. An observer at $r_0 = 3\mu$, who looks in the direction with the angle $\pi/2$, sees the back of his own head (of course dramatically distorted). For this circular path, we have

$$\frac{L}{E} = \frac{(3\mu)^2}{3\mu\sqrt{h(3\mu)}} = 3\sqrt{3}\,\mu\,.$$

According to Theorem 14.5, for a photon starting at $r_0 = 3\mu$ with a slightly smaller angle than $\pi/2$, L/E is also somewhat smaller than $3\sqrt{3}\,\mu$. Again from Theorem 14.5 and the graph sketched in Fig. 14.4, it can now be seen that the angle α continues to fall along the trajectory of the photon, at least not grow again. So the photon falls into the black hole. An observer, who at $r_0 = 3\mu$ looks in a direction with an angle smaller than $\pi/2$, sees darkness (more precisely, one would have to say that the observer does not encounter photons from these directions). For analogous reasons, an initial angle larger than $\pi/2$ at 3μ will continue to increase along the path, so that the photon flies past the black hole, and an observer who looks in this direction sees the starry sky. Summarising, the observer at 3μ sees the black hole as if he was standing directly in front of a huge black wall.

Every photon with $(L/E)/(3\sqrt{3}\,\mu) < 1$, which starts at $r_0 > 3\mu$ with an angle smaller than $\pi/2$, falls into the black hole, because according to Theorem 14.5, the angle along its trajectory is bounded by an angle smaller than $\pi/2$. In case $(L/E)/(3\sqrt{3}\,\mu) > 1$, it should be noted that the radial coordinate r can only assume values with $(L/E)\sqrt{h(r)}/r \leq 1$. As a result, r is bounded by a number larger than 3μ, and the photon flies past the black hole. The boundary case $L/E = 3\sqrt{3}\,\mu$ inserted into the equation in Theorem 14.5 gives

$$\sin\beta = 3\sqrt{3}\,\mu\sqrt{h(r_0)}/r_0$$

for the angular radius β of the black hole. Analogously, this formula can also be shown for r_0 between 2μ and 3μ. Summarising, we have the following theorem.

Theorem 14.7 *For the angular radius β of the black hole with mass μ, which an observer sees at $r > 2\mu$, satisfies*

$$\sin\beta = 3\mu\sqrt{3h(r)}/r\ ,$$

and for $r > 3\mu$, β is smaller than $\pi/2$, and for $r < 3\mu$, β is larger than $\pi/2$.

In the vicinity of the singularity $r = 0$, there are enormous tidal forces that finally tear apart every extensive body, including its atoms. In the example after Theorem 10.8, we had calculated the tidal force operator for the Schwarzschild metric. With respect to the basis e_1, e_2, e_3 obtained from $\partial_r, \partial_\vartheta, \partial_\varphi$ by normalisation, it had the matrix

$$(F_i^k) = \text{diag}\left(\frac{2\mu}{r^3}, -\frac{\mu}{r^3}, -\frac{\mu}{r^3}\right)\ .$$

The signs of the diagonal elements are to be interpreted to mean that an expansive stress occurs in the direction of the singularity and a compressive stress transversally to it. For $r \to 0$, the tidal forces tend to ∞. They depend continuously on r, and for falling r, they grow monotonously. Nothing unusual happens at $\mu = r$, and so the tidal force actually has nothing to do with the black hole phenomenon. The larger the mass μ, the lower are the tidal forces when entering the black hole.

14.3 The Kruskal Plane

The components of the fundamental tensor of the Schwarzschild half plane are singular at $r = 2\mu$. In Sect. 14.1, we found that a particle does not reach this singularity in finite Schwarzschild time. However, we also showed there that the particle crosses this mark in finite proper time and even reaches the central singularity $r = 0$. This discrepancy now suggests that a singularity occurs at $r = 2\mu$ in the Schwarzschild half plane only because the Schwarzschild coordinate system is not suitable there.

We will now stepwise construct another chart that covers both the external region $r > 2\mu$ and the internal region $0 < r < 2\mu$. The results of the transformations in the individual steps are shown in Fig. 14.7. We first describe the transformation in the outer region. In the first step, we keep the Schwarzschild time t, but replace the Schwarzschild radius r with the new coordinate

$$r^* := r + 2\mu \log(r - 2\mu).$$

The half plane $r > 2\mu$ becomes the entire r^*-t-plane. In the r^*-t-plane, the outgoing null geodesics are the straight lines $t = r^* + c$, and the straight lines $t + r^* = c$ perpendicular to these represent the incoming null geodesics. In the second step, we introduce the new coordinates $U = t - r^*$ and $V = t + r^*$. Except for the factor $\sqrt{2}$, this is a rotation composed with a reflection. In the third step, the coordinates U and V are transformed separately into $\tilde{u} = -e^{-U/4\mu}$ and $\tilde{v} = e^{V/4\mu}$. This is a bijective map of the entire $U-V$-plane onto the second quadrant of the $\tilde{u}-\tilde{v}$-plane. The outgoing and the incoming null geodesics are then still parallel to the second and to the first coordinate axes, respectively, but now they are only half-lines. The fourth and last transformation delivers the coordinates $u = \frac{1}{2}(\tilde{v} - \tilde{u})$, $v = \frac{1}{2}(\tilde{v} + \tilde{u})$. This is again essentially a rotation composed with a reflection. From the second quadrant of the $\tilde{u}-\tilde{v}$-plane, the region we get in the $u-v$-plane is $|u| > |v|$, $u > 0$. When used together, we obtain

$$u = \frac{1}{2}\left(e^{\frac{t+r^*}{4\mu}} + e^{\frac{-t+r^*}{4\mu}}\right) = \cosh\frac{t}{4\mu}\; e^{\frac{r}{4\mu}}\; e^{\frac{1}{2}\log(r-2\mu)} = \sqrt{r - 2\mu}\; e^{\frac{r}{4\mu}}\cosh\frac{t}{4\mu}$$

and analogously

$$v = \sqrt{r - 2\mu}\; e^{\frac{r}{4\mu}}\sinh\frac{t}{4\mu}\;.$$

The outgoing light rays are half-lines $v = u - c$, and the incoming ones are given by $u + v = c$, each with a positive constant c. The straight line $r = c$ with $c > 2\mu$ becomes the right branch of the hyperbola

$$u^2 - v^2 = (c - 2\mu)e^{c/2\mu}\;.$$

The half-line $t = c$ in the $u-v$-plane corresponds to a beam starting from the origin with the slope $\tanh(c/4\mu)$.

From the partial derivatives

$$\frac{\partial u}{\partial t} = \frac{v}{4\mu}\;,\quad \frac{\partial v}{\partial t} = \frac{u}{4\mu}\;,\quad \frac{\partial u}{\partial r} = \frac{u}{4\mu} + \frac{u}{2r - 4\mu} = \frac{u}{4\mu h(r)}\;,\quad \frac{\partial v}{\partial r} = \frac{v}{4\mu h(r)}$$

it follows, by using the chain rule for functions of two variables, that

$$\partial_t = \frac{v}{4\mu}\partial_u + \frac{u}{4\mu}\partial_v \quad\text{and}\quad \partial_r = \frac{u}{4\mu h(r)}\partial_u + \frac{v}{4\mu h(r)}\partial_v\;.$$

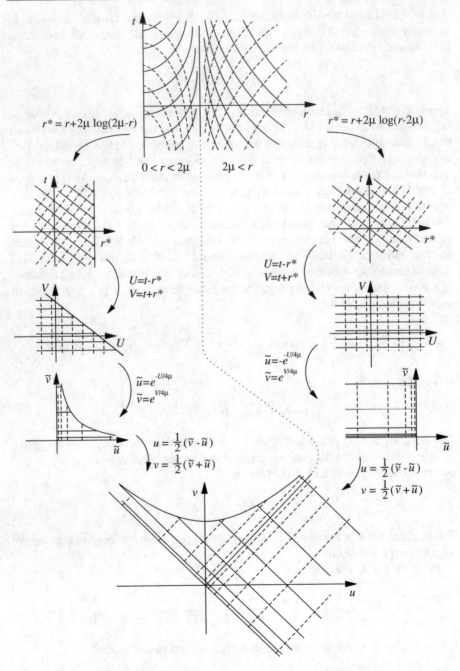

Fig. 14.7 From Schwarzschild to Kruskal

For ∂_u and ∂_v, we obtain

$$\partial_u = \frac{4\mu}{r e^{r/2\mu}} \left(u \partial_r - \frac{v}{h(r)} \partial_t \right)$$

and

$$\partial_v = \frac{4\mu}{r e^{r/2\mu}} \left(\frac{u}{h(r)} \partial_t - v \partial_r \right).$$

The components of the metric with respect to this chart are given by

$$g(\partial_u, \partial_u) = \frac{16\mu^2}{r^2 e^{r/\mu}} g \left(u\partial_r - \frac{v}{h(r)} \partial_t , \; u\partial_r - \frac{v}{h(r)} \partial_t \right) = \frac{16\mu^2}{r^2 e^{r/\mu}} \cdot \frac{v^2 - u^2}{h(r)} = -\frac{16\mu^2}{r e^{r/2\mu}},$$

and analogously,

$$g(\partial_v, \partial_v) = \frac{16\mu^2}{r e^{r/2\mu}}$$

and also $g(\partial_u, \partial_v) = 0$. The tangent vectors $a\partial_u + b\partial_v$ are timelike if $|b| > |a|$. If one follows the transformation of the future cone from the Schwarzschild spacetime through all the sub-steps to the Kruskal plane, then one can see that the timelike vector ∂_v is future-pointing and thus also all timelike vectors $a\partial_u + b\partial_v$ with $b > 0$.

The transformation of the interior of the Schwarzschild spacetime again happens in four steps. First r is replaced by $r^* := r + 2\mu \log(2\mu - r)$, and then the coordinates $U = t - r^*$ and $V = t + r^*$ are introduced. In the third step there is a difference compared to the corresponding transformation used to handle the outer region, since for the new variables $\tilde{u}$ and $\tilde{v}$, we now have $\tilde{u} = e^{-U/4\mu}$ and $\tilde{v} = e^{V/4\mu}$. Finally, the final variables u and v are introduced by $u = \frac{1}{2}(\tilde{v} - \tilde{u})$ and $v = \frac{1}{2}(\tilde{v} + \tilde{u})$. The entire transformation is given by

$$u = \sqrt{2\mu - r} \; e^{\frac{r}{4\mu}} \sinh \frac{t}{4\mu}$$

and

$$v = \sqrt{2\mu - r} \; e^{\frac{r}{4\mu}} \cosh \frac{t}{4\mu}.$$

The region $0 < r < 2\mu$ corresponds to the region with $|u| < v < \sqrt{u^2 + 2\mu}$ in the $u-v$-plane. The formulae for the metric components are the same as in the previous case. In Sect. 14.1, we interpreted the equations $t = -r^* + c$ as incoming photos in the interior region $0 < r < 2\mu$. In the $u-v$-plane, these are the lines $u + v = c$ with a positive constant c. This fits the point of view in the previous case. Overall, one can say that an incident radial photon moves in the $u-v$-plane along the straight line $u + v = c > 0$ in the direction with increasing coordinate v. The photon easily crosses the line $u = v$, which corresponds to the mark $r = 2\mu$, and finally reaches the singularity $r = 0$. In the interior region $0 < r < 2\mu$, there are also the outer null

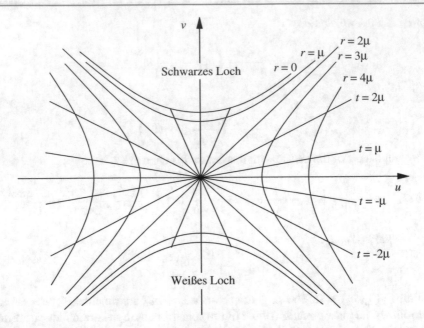

Fig. 14.8 Black and white hole in the Kruskal plane

geodesics $t = r^* + c$, which are given by $v = u + c$ with $c > 0$ in the $u-v$-plane. There is initially no interpretation for this.

The facts described here lead to a radical but speculative generalisation. In the $u - v$-plane, the region $v^2 < u^2 + 2\mu$ (see Fig. 14.8) is equipped with the metric

$$g(\partial_v, \partial_v) = \frac{16\mu^2}{re^{r/2\mu}} = -g(\partial_u, \partial_u)$$

and $g(\partial_u, \partial_v) = 0$, where the positive number r is characterised by

$$(r - 2\mu)e^{r/2\mu} = u^2 - v^2 \text{ for } |u| \geq |v|$$

and

$$(2\mu - r)e^{r/2\mu} = v^2 - u^2 \text{ for } |u| \leq |v| .$$

Null geodesics are the parallel line families $v = u + c$ and $u + v = c$. The future cones consist of the tangent vectors $a\partial_u + b\partial_v$ with $b > |a|$. Particles and photons must leave the region $v < -|u|$, and no particle or photon can ever enter this region. This region thus represents a so-called **white hole**. In addition to the outer region $u > |v|$, there is another outer region $u < -|v|$. This can also be used to interpret the null geodesic $v = u + c$ in the black hole, they are photons that have fallen into the black hole from the second outer region. No photon or particle can get into the outer region from one outer region.

Cosmology

15.1   Spaces of Constant Curvature

The Gaussian curvature of a surface can be generalised to the so-called sectional curvatures for higher-dimensional manifolds. If these sectional curvatures are constant, then the curvature tensor has a fairly simple form. We restrict ourselves here to Riemannian manifolds, as we need the concepts for the manifold $\mathbb{R}^3$ equipped with a suitable Riemannian metric.

Theorem 15.1 *Let R be the covariant curvature tensor of a Riemannian manifold M, and let E be a two-dimensional subspace of a tangent space M_P. Then for every basis x_1, x_2 in E, the expression*

$$K_E = \frac{R(x_1, x_2, x_1, x_2)}{g(x_1, x_1)g(x_2, x_2) - (g(x_1, x_2))^2}$$

has the same value.

Proof If we use a different basis $y_i = \lambda_i^1 x_1 + \lambda_i^2 x_2$, $i = 1, 2$, and use the skew-symmetry of R with respect to the last two and first two arguments of R (Theorem 8.5), then we get for the numerator

$$R(y_1, y_2, y_1, y_2) = (\lambda_1^1 \lambda_2^2 - \lambda_1^2 \lambda_2^1) R(x_1, x_2, y_1, y_2) = (\lambda_1^1 \lambda_2^2 - \lambda_1^2 \lambda_2^1)^2 R(x_1, x_2, x_1, x_2)$$

and for the denominator

$$g(y_1, y_1)g(y_2, y_2) - (g(y_1, y_2))^2 = \begin{vmatrix} g(y_1, y_1) & g(y_1, y_2) \\ g(y_2, y_1) & g(y_2, y_2) \end{vmatrix}$$

© The Author(s), under exclusive license to Springer Nature Switzerland AG 2023
R. Oloff, *The Geometry of Spacetime*, Graduate Texts in Physics,
https://doi.org/10.1007/978-3-031-16139-1_15

$$= \begin{vmatrix} \lambda_1^1 & \lambda_1^2 \\ \lambda_2^1 & \lambda_2^2 \end{vmatrix} \cdot \begin{vmatrix} g(x_1,x_1) & g(x_1,x_2) \\ g(x_2,x_1) & g(x_2,x_2) \end{vmatrix} \cdot \begin{vmatrix} \lambda_1^1 & \lambda_1^2 \\ \lambda_1^2 & \lambda_2^2 \end{vmatrix}$$

$$= (\lambda_1^1 \lambda_2^2 - \lambda_1^2 \lambda_2^1)^2 (g(x_1,x_1)g(x_2,x_2) - (g(x_1,x_2))^2) \, .$$

Thus the quotient is the same number as for the basis x_1, x_2.

The number K_E, depending only on the subspace E, is called the **sectional curvature for** E. In the case of a curved surface, using coordinate vector fields ∂_1 and ∂_2, which are orthonormal at point P, in light of the considerations in the proof of Theorem 8.16,

$$K_{M_P} = R_{1212} = R_{212}^1 = \det W = K \, .$$

If all sectional curvatures K_E coincide at a point P (and with K denoting the common value), then for all tangent vectors $x, y \in M_P$

$$R(x, y, x, y) = K(g(x,x)g(y,y) - (g(x,y))^2) \, .$$

It turns out that the covariant curvature tensor at this point can then be reconstructed from the number K.

Theorem 15.2 *If at a point P of a Riemannian manifold M, all the sectional curvatures K_E are equal to K, then for all tangent vectors $x, y, z, v \in M_P$*

$$R(x, y, z, v) = K(g(x,z)g(y,v) - g(x,v)g(y,z)) \, .$$

Proof The fourfold covariant tensor

$$S(x, y, z, v) = g(x,z)g(y,v) - g(x,v)g(y,z)$$

has the properties

$$S(y, x, z, v) = -S(x, y, z, v)$$
$$S(x, y, v, z) = -S(x, y, z, v)$$
$$S(z, v, x, y) = S(x, y, z, v)$$
$$S(x, y, z, v) + S(x, v, y, z) + S(x, z, v, y) = 0 \, ,$$

which are also possessed by R (Theorem 8.5). Hence these properties are also transferred to the combination $T = R - KS$. By the hypothesis of the theorem, we know that $T(x, y, x, y) = 0$ for all $x, y \in M_P$, and we wish to show $T = 0$. From

$$0 = T(x+z, y, x+z, y) = T(x, y, x, y) + T(z, y, z, y) + T(x, y, z, y) + T(z, y, x, y) \quad = 2T(x, y, z, y)$$

it follows that $T(x, y, z, y) = 0$. Since

$$0 = T(x, y+v, z, y+v) = T(x, y, z, v) + T(x, v, z, y)$$

we obtain

$$T(x, y, z, v) = T(x, v, y, z) = T(x, z, v, y)$$

and $3T(x, y, z, v) = 0$. Thus $T = 0$.

If the sectional curvatures match in each of the tangent spaces M_P, then one could still expect the common value of these sectional curvatures of the two-dimensional subspaces of M_P to depend on P. Surprisingly, this is not the case.

Theorem 15.3 *Let M be a connected N-dimensional $(n > 2)$ Riemannian manifold. For $P \in M$, let $K(P)$ be the sectional curvature for all two-dimensional subspaces of M_P. Then the scalar field $K(.)$ is constant on M.*

Proof We will show that $\partial_i K = 0$ for any chart with ∂_i denoting the coordinate vector fields. By Theorem 15.2,

$$R_{ijkl} = K(g_{ik}g_{jl} - g_{il}g_{jk}) .$$

Using $\nabla g = 0$, we get by covariant differentiation that

$$R_{ijkl;m} = (\partial_m K)(g_{ik}g_{jl} - g_{il}g_{jk}) .$$

If the analogous expressions for $R_{ijlm;k}$ and $R_{ijmk;l}$ are added, then one obtains, using the second Bianchi identity (Theorem 11.9) that

$$0 = (\partial_m K)(g_{ik}g_{jl} - g_{il}g_{jk}) + (\partial_k K)(g_{il}g_{jm} - g_{im}g_{jl}) + (\partial_l K)(g_{im}g_{jk} - g_{ik}g_{jm}) .$$

We now multiply by $g^{ik}g^{jl}$ (pull up two indices and twice contract). Using the identities

$$(g_{ik}g^{ik})(g_{jl}g^{jl}) = n^2$$

$$(g_{il}g^{ik}g_{jk})g^{jl} = g_{jl}g^{jl} = n$$

$$g_{il}g^{ik}g_{jm}g^{jl} = g_{jm}g^{jk}$$

$$g_{im}g^{ik}(g_{jl}g^{jl}) = g_{im}g^{ik}n$$

$$g_{im}g^{ik}g_{jk}g^{jl} = g_{im}g^{il}$$

$$(g_{ik}g^{ik})g_{jm}g^{jl} = ng_{jm}g^{jl}$$

we obtain

$$0 = \partial_m K(n^2 - n) + \partial_k K(g_{jm}g^{jk} - g_{im}g^{ik}n) + \partial_l K(g_{im}g^{il} - ng_{jm}g^{jl})$$

$$= n(n-1)\partial_m K + (1-n)\partial_m K + (1-n)\partial_m K = (n-2)(n-1)\partial_m K .$$

Definition 15.1 A connected Riemannian manifold M is said to have a **constant curvature** K, if for the covariant tensor R at every point $P \in M$ and for arbitrary tangent vectors $x, y, z, v \in M_P$, we have

$$R(x, y, z, v) = K(g(x, z)g(y, v) - g(x, v)g(y, z)) \, .$$

◆

Theorem 15.4 *For an n-dimensional Riemannian manifold with a constant curvature K, the Ricci tensor is given by*

$$\mathrm{Ric} = (n - 1)Kg \, .$$

Proof We have

$$\mathrm{Ric}_{ik} = R^j_{ijk} = g^{jl} R_{lijk} = K(g_{lj}g_{ik} - g_{lk}g_{ij})g^{jl} = K(ng_{ik} - g_{ik}) = K(n - 1)g_{ik} \, .$$

Example 1 We determine a metric g for $\mathbb{R}^3$, such that it creates a Riemannian manifold of constant curvature, where we start with the assumption that

$$(g_{ik}) = \mathrm{diag}(a(r), r^2, r^2 \sin^2\vartheta)$$

with respect to the spherical coordinate chart r, ϑ, φ. To determine the covariant derivative, we refer to the formulae from Sect. 9.7 and obtain

$$\nabla_{\partial_r}\partial_r = \frac{a'(r)}{2a(r)}\, \partial_r \, , \quad \nabla_{\partial_\vartheta}\partial_\vartheta = -\frac{r}{a(r)}\, \partial_r \, ,$$

$$\nabla_{\partial_\varphi}\partial_\varphi = -\frac{r}{a(r)}\sin^2\vartheta \, \partial_r - \sin\vartheta\cos\vartheta \, \partial_\vartheta \, ,$$

$$\nabla_{\partial_r}\partial_\vartheta = \frac{1}{r}\, \partial_\vartheta \, , \quad \nabla_{\partial_r}\partial_\varphi = \frac{1}{r}\, \partial_\varphi \, , \quad \nabla_{\partial_\vartheta}\partial_\varphi = \cot\vartheta \, \partial_\varphi \, .$$

The matrices of the three relevant curvature operators $R(\partial_r, \partial_\vartheta)$, $R(\partial_r, \partial_\varphi)$ and $R(\partial_\vartheta, \partial_\varphi)$ are then given by

$$(R^k_{i12}) = \begin{pmatrix} 0 & \dfrac{ra'(r)}{2a^2(r)} & 0 \\[2mm] -\dfrac{a'(r)}{2ra(r)} & 0 & 0 \\[2mm] 0 & 0 & 0 \end{pmatrix}$$

$$(R^k_{i13}) = \begin{pmatrix} 0 & 0 & \dfrac{ra'(r)}{2a^2(r)}\sin^2\vartheta \\[2mm] 0 & 0 & 0 \\[2mm] -\dfrac{a'(r)}{2ra(r)} & 0 & 0 \end{pmatrix}$$

$$
(R_{i23}^k) = \begin{pmatrix} 0 & 0 & 0 \\ 0 & 0 & \left(1 - \frac{1}{a(r)}\right)\sin^2\vartheta \\ 0 & \frac{1}{a(r)} - 1 & 0 \end{pmatrix} .
$$

Thus the Ricci tensor is given by

$$
(\mathrm{Ric}_{ik}) = \mathrm{diag}\left(\frac{a'(r)}{ra(r)}, \; \frac{ra'(r)}{2a^2(r)} + 1 - \frac{1}{a(r)}, \; \left(\frac{ra'(r)}{2a^2(r)} + 1 - \frac{1}{a(r)} \right)\sin^2\vartheta \right) .
$$

The following two equations follow from Theorem 15.4:

$$
\frac{a'(r)}{ra(r)} = 2Ka(r)
$$

and

$$
\frac{ra'(r)}{2a^2(r)} + 1 - \frac{1}{a(r)} = 2Kr^2 .
$$

The first inserted into the second gives

$$
a(r) = \frac{1}{1 - Kr^2} .
$$

Conversely, suppose $a(r)$ is chosen to be this value. The components $R_{ijkl} = g_{is}R_{jkl}^s$ can essentially be read off from the given matrices of the curvature operators, and the constant curvature can therefore be actually confirmed. Summarising, we can conclude that $\mathbb{R}^3$ with the metric given by

$$
(g_{ik}) = \mathrm{diag}\left(\frac{1}{1 - Kr^2}, \; r^2, \; r^2\sin^2\vartheta \right)
$$

in this chart is a manifold of constant curvature K. The radial coordinate r can of course no longer be interpreted as the distance to the origin, since the curve $r = s$, $0 \leq s \leq r_0$, ϑ and φ being held constant, has the arc length

$$
\int_0^{r_0} \sqrt{g(\partial_r, \partial_r)}\, ds = \int_0^{r_0} 1/\sqrt{1 - Ks^2}\, ds = \begin{cases} \frac{1}{\sqrt{K}}\arcsin(\sqrt{K}\, r_0) & \text{f"ur } K > 0 \\ \frac{1}{\sqrt{-K}}\mathrm{arcsinh}(\sqrt{-K}\, r_0) & \text{f"ur } K < 0 \end{cases} .
$$

Example 2 We anticipate that the surface of a sphere in $\mathbb{R}^4$, is equipped with the Euclidean metric, has a constant curvature. The generalised spherical coordinates

$$\xi^1 = R_0 \sin\chi \sin\vartheta \sin\varphi$$

$$\xi^2 = R_0 \sin\chi \sin\vartheta \cos\varphi$$

$$\xi^3 = R_0 \sin\chi \cos\vartheta$$

$$\xi^4 = R_0 \cos\chi$$

provide a chart with the coordinates $r := R_0 \sin\chi$, ϑ and φ. We have

$$\partial_r = \sin\vartheta \sin\varphi\, \partial_{\xi^1} + \sin\vartheta \cos\varphi\, \partial_{\xi^2} + \cos\vartheta\, \partial_{\xi^3} - \frac{r}{\sqrt{R_0^2 - r^2}}\, \partial_{\xi^4}$$

$$\partial_\vartheta = r\cos\vartheta \sin\varphi\, \partial_{\xi^1} + r\cos\vartheta \cos\varphi\, \partial_{\xi^2} - r\sin\vartheta\, \partial_{\xi^3}$$

$$\partial_\varphi = r\sin\vartheta \cos\varphi\, \partial_{\xi^1} - r\sin\vartheta \sin\varphi\, \partial_{\xi^2}$$

and obtain $\partial_r \cdot \partial_\vartheta = \partial_r \cdot \partial_\varphi = \partial_\vartheta \cdot \partial_\varphi = 0$

$$\partial_r \cdot \partial_r = 1 + \frac{r^2}{R_0^2 - r^2} = \frac{R_0^2}{R_0^2 - r^2} = \frac{1}{1 - (1/R_0^2)r^2}$$

$$\partial_\vartheta \cdot \partial_\vartheta = r^2$$

$$\partial_\varphi \cdot \partial_\varphi = r^2 \sin^2\vartheta \; .$$

Hence this manifold has the constant curvature $K = 1/R_0^2$. This is the same result as for the surface of a sphere in $\mathbb{R}^3$.

15.2   The Robertson-Walker Metric

In space, matter is concentrated in star systems with the sun, planets and moons, in galaxies, in galaxy clusters, etc. However, astronomical observations show that when averaged over orders of magnitudes of 10^8 to 10^9 light years, matter is evenly distributed again to a certain degree of accuracy. This is the starting point for the so-called **cosmological principle**. It says that density and pressure are spatially constant (homogeneous) and that all directions are therefore equally important from every point (isotropic). Relative speeds, such as the rotation of planets around a fixed star or the movement of a fixed star in a galaxy, are neglected. Matter is viewed as an ideal flow whose particles fall freely along geodesics that do not cross. With the so-called 'comoving coordinates' $\xi^0, \xi^1, \xi^2, \xi^3$, the flow vector has the components $(1, 0, 0, 0)$ and the subspace $\partial_0^\perp$ orthogonal to the observer ∂_0 is spanned by $\partial_1, \partial_2, \partial_3$. Owing to the isotropy, $\partial_0^\perp$, with the metric $-g$, must be a space with a constant curvature. The manifold given in the following definition has all these properties.

Definition 15.2 Let $M = \mathbb{R} \times \mathbb{R}^3$. With respect to the chart $(t, Q) \to (t, r, \vartheta, \varphi)$, where r, ϑ, φ denote the spherical coordinates, the **Robertson-Walker metric** g is defined by

$$(g_{ik}) = \text{diag}\left(1, \frac{-S^2(t)}{1 - Kr^2}, \; -S^2(t)\, r^2, \; -S^2(t)r^2 \sin^2\vartheta\right).$$

The positive number $S(t)$, dependent on t, is called the **scale factor**.   ◆

By calculating the Christoffel symbols with the formulae mentioned in Sect. 7.5 for the Robertson-Walker metric, we obtain

$$\nabla_{\partial_t}\partial_t = 0$$

$$\nabla_{\partial_r}\partial_r = \frac{S(t)S'(t)}{1 - Kr^2}\partial_t + \frac{Kr}{1 - Kr^2}\partial_r$$

$$\nabla_{\partial_\vartheta}\partial_\vartheta = S(t)S'(t)r^2\partial_t - (1 - Kr^2)r\,\partial_r$$

$$\nabla_{\partial_\varphi}\partial_\varphi = S(t)S'(t)r^2 \sin^2\vartheta\,\partial_t - (1 - Kr^2)r \sin^2\vartheta\,\partial_r - \sin\vartheta \cos\vartheta\,\partial_\vartheta$$

$$\nabla_{\partial_t}\partial_r = \frac{S'(t)}{S(t)}\partial_r$$

$$\nabla_{\partial_t}\partial_\vartheta = \frac{S'(t)}{S(t)}\partial_\vartheta$$

$$\nabla_{\partial_t}\partial_\varphi = \frac{S'(t)}{S(t)}\partial_\varphi$$

$$\nabla_{\partial_r}\partial_\vartheta = \frac{1}{r}\partial_\vartheta$$

$$\nabla_{\partial_r}\partial_\varphi = \frac{1}{r}\partial_\varphi$$

$$\nabla_{\partial_\vartheta}\partial_\varphi = \cot\vartheta\,\partial_\varphi.$$

To determine the curvature tensor, the matrices of six curvature operators must be calculated in accordance with Definition 8.1. The following components are non-zero

$$R^1_{001} = R^2_{002} = R^3_{003} = S''(t)/S(t)$$

$$R^0_{101} = S''(t)S(t)/(1 - Kr^2)$$

$$R^0_{202} = S''(t)S(t)r^2$$

$$R^0_{303} = S''(t)S(t)r^2 \sin^2\vartheta$$

$$R^1_{212} = ((S'(t))^2 + K)r^2 = -R^3_{223}$$

$$R^2_{112} = R^3_{113} = -((S'(t))^2 + K)/(1 - Kr^2)$$

$$R^1_{313} = R^2_{323} = ((S'(t))^2 + K)r^2 \sin^2\vartheta.$$

Another twelve components arise from the skew-symmetry $R^l_{ijk} = -R^l_{ikj}$, and all others are zero. The Ricci tensor has the matrix representation

$$(\mathrm{Ric}_{ik}) = \mathrm{diag}\left(-\frac{3S''(t)}{S(t)}\ ,\ \frac{a(t)}{1-Kr^2}\ ,\ a(t)r^2\ ,\ a(t)r^2\sin^2\vartheta\right)$$

where

$$a(t) = S''(t)S(t) + 2(S'(t))^2 + 2K\ .$$

For fixed coordinates r, ϑ, φ, the curve γ, which assigns to the number t the point with the coordinates t, r, ϑ, φ, is a geodesic, since

$$\nabla_{\gamma'(t)}\gamma'(t) = \nabla_{\partial_t}\partial_t = 0\ .$$

The tangent vector ∂_t represents a particle or an observer, which is at rest relative to the fixed stars surrounding it. The time t which passes for him is called **cosmic time**.

For the observer ∂_t, the universe at time t is the Riemannian manifold $\mathbb{R}^3$ (at least locally, as long as the expression $1 - Kr^2$ stays positive), the metric being determined by the Robertson-Walker metric g as follows: Each tangent space to $\mathbb{R}^3$ is a three-dimensional subspace of the corresponding tangent space of the Robertson-Walker spacetime, and $-g$ with $t = t_0$ is restricted to this subspace. As a result, $\mathbb{R}^3$ becomes a space with the constant curvature $K/S^2(t_0)$, which can be justified as follows: Without the prefactor $S^2(t_0)$, the constant curvature would be K. The prefactor does not influence the Christoffel symbols, and therefore not the components of the Riemannian curvature tensor (see Theorem 8.2). By pulling the index, the factor $S^2(t_0)$ appears in the covariant curvature tensor. Clearly the equation in Definition 15.1 then holds with $K/S^2(t_0)$ instead of K.

The curve $t = t_0$, $r = s$ for $0 \le s \le r_0$, $\vartheta = \vartheta_0$, $\varphi = \varphi_0$, has, with respect to $-g$, the arc length $S(t_0)I$, where

$$I = \int_0^{r_0} \frac{ds}{\sqrt{1-Ks^2}}\ .$$

In $\mathbb{R}^3$, with the metric agreed upon here, we have

$$\nabla_{\partial_r}\partial_r = \frac{Kr}{1-Kr^2}\partial_r$$

and so for the unit tangent vector $(\sqrt{1-Kr^2}/S(t))\,\partial_r$ $(\sqrt{1-Kr^2}/S(t))\,\partial_r$

$$\nabla_{(\sqrt{1-Kr^2}/S(t))\,\partial_r}(\sqrt{1-Kr^2}/S(t))\,\partial_r$$

$$= \frac{\sqrt{1-Kr^2}}{S(t)}\left(\partial_r\frac{\sqrt{1-Kr^2}}{S(t)}\,\partial_r + \frac{\sqrt{1-Kr^2}}{S(t)}\nabla_{\partial_r}\partial_r\right) = 0\ .$$

The above mentioned curve is, as expected a geodesic, and the arc length $S(t_0)I$ is the distance that a fixed star located at the point with the coordinates r_0, ϑ_0, φ_0 has from the observer at the origin at time t_0. The time derivative of this distance is $S'(t_0)I$, based on the size of the distance, this is $S'(t_0)/S(t_0)$. This is the so-called **Hubble constant** $H(t_0)$. The **Hubble law** is a consequence of the above: *The 'recessional velocity' of a fixed star at a time instant t is* $H(t)$ *times the position vector of the location of the fixed star.* Of course, this only applies if certain individual motion, e.g. within a galaxy, can be ignored. So it applies to very distant fixed stars, or better galaxies. The term 'recessional velocity' suggests that the currently observed Hubble parameter is positive. Astronomical measurements show that its reciprocal $1/H(t)$ in the epoch we observe is $(18 \pm 2) \times 10^9$ years. Roughly speaking, this means that an object located at a distance of $1, 8 \cdot 10^{10}$ km recedes one kilometer per year.

In the context of nonrelativistic physics, the recessional velocity of distant galaxies would result in a reduced speed of light emitted by such galaxies and thus a red shift of the spectrum at the receiver by the Doppler effect. This phenomenon, the **cosmological red shift**, actually occurs. The remainder of this section is devoted to explaining this red shift within the realm of general relativity.

According to the cosmological principle, we may assume that the receiver has the position $r=0$ and that the transmitter emits the light signal from a position $r=r_1$ radially towards $r=0$. The null geodesic is characterised by $\varphi'=0$, $\vartheta'=0$ and

$$(t')^2 - \frac{S^2(t)}{1 - Kr^2} (r')^2 = 0 .$$

For the function $t(r)$ describing the photons, this gives the differential equation

$$\frac{dt}{dr} = -\frac{S(t)}{\sqrt{1 - Kr^2}} .$$

This can be solved by separating the variables. The function $t(r)$ can thus be interpreted in the manner described below.

A photon that has been emitted at r_1 at time t_1, and received at $r=0$ at time t_3, is characterised by

$$\int_{t_1}^{t_3} \frac{dt}{S(t)} = \int_0^{r_1} \frac{dr}{\sqrt{1 - Kr^2}} .$$

A photon emitted there at a slightly later time $t_2 > t_1$ reaches the origin $r=0$ at time $t_4 > t_3$, and again

$$\int_{t_2}^{t_4} \frac{dt}{S(t)} = \int_0^{r_1} \frac{dr}{\sqrt{1 - Kr^2}} .$$

The integrals on the left-hand side are thus the same, giving

$$\int\limits_{t_1}^{t_2} \frac{dt}{S(t)} = \int\limits_{t_3}^{t_4} \frac{dt}{S(t)} .$$

So there exists $t_{12} \in (t_1, t_2)$ and $t_{34} \in (t_3, t_4)$ such that

$$\frac{t_2 - t_1}{S(t_{12})} = \frac{t_4 - t_3}{S(t_{34})} .$$

It should be that t_2 is close to t_1 and therefore t_4 is close to t_3. Then the last equation yields for the frequency ν of light that

$$\frac{\nu(t_1)}{\nu(t_3)} = \frac{S(t_3)}{S(t_1)} .$$

As $S(t_1) < S(t_3)$, the light received at $r=0$ thus has a lower frequency than the frequency at its emission.

This effect can also be quantified in terms of the distance D between the transmitter and the receiver. We assume that r_1 and $t_3 - t_1$ are not too large and approximate

$$\frac{S(t_3)}{S(t_1)} \approx \frac{S(t_3)}{S(t_3) + (t_1 - t_3)S'(t_3)} \approx 1 + (t_3 - t_1)S'(t_3)/S(t_3) \approx 1 + S'(t_3) \int\limits_{t_1}^{t_3} \frac{dt}{S(t)}$$

$$= 1 + S'(t_3) \int\limits_{0}^{r_1} \frac{dr}{\sqrt{1 - Kr^2}} \approx 1 + S'(t_3)r_1 = 1 + \frac{S'(t_3)}{S(t_3)} S(t_3)r_1 \approx 1 + H(t_3)D .$$

Summarising: *An observer receives light, which is emitted from a galaxy at a distance D with the frequency at emission ν, with the frequency ν_0 at reception. Then if H_0 is the Hubble parameter, we have the approximation*

$$\nu/\nu_0 \approx 1 + H_0 D .$$

15.3   Universe Models

In the Robertson-Walker metric given in the previous section, the scale factor $S(t)$ and the number K are still undetermined. The Einstein field equation now gives rise to requirements to be satisfied. For the sake of simplicity, here we assume that the cosmological constant is zero, and so we will use the Einstein equation having the form $G = 8\pi T$. The starting point is the already calculated diagonal matrix of the Ricci tensor with respect to the comoving coordinates t, r, ϑ, φ. From the components

$$\text{Ric}_0^0 = -3S''(t)/S(t)$$

and

$$\text{Ric}_1^1 = \text{Ric}_2^2 = \text{Ric}_3^3 = -a(t)/S^2(t)$$

of the mixed Ricci tensor, the curvature scalar $-3(S''/S + a/S^2)$ can be read off. The Einstein tensor is given by

$$(G_{ik}) = \text{diag}\left(3\,\frac{(S'(t))^2 + K}{S^2(t)}\,,\ \frac{b(t)}{1 - Kr^2}\,,\ b(t)r^2\,,\ b(t)r^2\sin^2\vartheta\right)$$

where

$$b(t) = -2S''(t)S(t) - (S'(t))^2 - K\,.$$

As already mentioned, matter is regarded as an ideal flow (see Sect. 9.1) with the density ϱ and the pressure p, both depending only on the cosmic time t by the cosmological principle, and having the vector field Z, which, with respect to the comoving coordinates, has the 4-tuple of components $(1, 0, 0, 0)$. The energy-momentum tensor T with respect to these coordinates is the diagonal matrix

$$(T_{ik}) = \text{diag}\left(\varrho(t)\,,\ \frac{p(t)S^2(t)}{1 - Kr^2}\,,\ p(t)S^2(t)r^2\,,\ p(t)S^2(t)r^2\sin^2\vartheta\right)\,.$$

The equation $G_{ik} = 8\pi T_{ik}$ for components implies for $(i, k) = (0, 0)$ that

$$\frac{3}{S^2(t)}((S'(t))^2 + K) = 8\pi\varrho(t)$$

and for $(1, 1), (2, 2)$ and $(3, 3)$ that

$$2S''(t)S(t) + (S'(t))^2 + K = -8\pi S^2(t)p(t)\,.$$

Owing to the presence of $(S')^2$, we call the first equation the **energy equation,** and as the second equation contains S'', we call it the **equation of motion.**

Henceforth we restrict ourselves to the special case when $p = 0$. Physically, this means that the matter generating spacetime is incoherent, there is no interaction between the particles, and the particles behave like dust. This point of view largely corresponds to the real situation, and the pressure is actually negligible compared to the density.

Theorem 15.5 *In the special case* $p = 0$, *the product* $\varrho(t)S^3(t)$ *is constant for the Robertson-Walker spacetime, and with the constant* $C = \varrho S^3$, *we have the* **Friedmann differential equation**

$$(S'(t))^2 + K = \frac{8}{3}\pi C/S(t)\,.$$

Proof The energy equation can be written in the form

$$(S'(t))^2 + K = \frac{8}{3}\pi S^3(t)\varrho(t)/S(t),$$

and so only the constancy of the product $S^3(t)\varrho(t)$ is to be shown. By differentiating the energy equation in the form

$$(S')^2 + K = \frac{8}{3}\pi S^2 \varrho$$

and upon multiplying the result by S, we obtain

$$2SS'S'' = \frac{8}{3}\pi(2S^2 S'\varrho + S^3 \varrho').$$

By using the equation of motion in the left-hand side, and also the energy equation, we obtain

$$2SS'S'' = -S'((S')^2 + K) = -\frac{8}{3}\pi S' S^2 \varrho,$$

and so this gives

$$0 = \frac{8}{3}\pi(3S^2 S'\varrho + S^3 \varrho').$$

Thus

$$0 = 3S^2 S'\varrho + S^3 \varrho' = (S^3 \varrho)'.$$

In the solution of Friedmann's differential equation, we must distinguish the three cases $K = 0$, $K \gg 0$ or $K \ll 0$. In each case, the so-called **big bang** occurs: The scalar factor S is zero for a certain value, which is the origin of the time measurement.

In the case $K = 0$, the differential equation can be separated, and for the initial condition $S(0) = 0$, it gives the solution

$$S(t) = t^{\frac{2}{3}}\sqrt[3]{6\pi C}$$

(See Fig. 15.1). The energy equation gives the density

$$\varrho(t) = \frac{3}{8\pi}(S'(t))^2/S^2(t) = \frac{1}{6\pi t^2}.$$

The Robertson-Walker metric with this scale factor has the following component matrix with respect to the previously used coordinates

$$(g_{ik}) = \text{diag}(1, -ct^{\frac{4}{3}}, -ct^{\frac{4}{3}}r^2, -ct^{\frac{4}{3}}r^2 \sin^2\vartheta)$$

Fig. 15.1 Solutions S(t) of the Friedmann differential equation

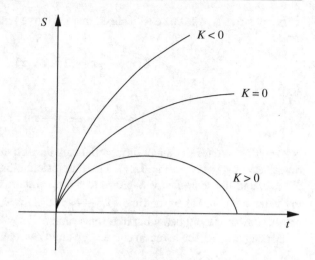

where $c = (6\pi C)^{\frac{2}{3}}$. If we use Cartesian coordinates in the subspace $\partial_t^\perp$ instead of spherical coordinates, then we obtain the component matrix

$$(g_{ik}) = \mathrm{diag}(1, -ct^{\frac{4}{3}}, -ct^{\frac{4}{3}}, -ct^{\frac{4}{3}}) .$$

This is the Einstein-de Sitter spacetime, which was already mentioned as an example in Sect. 4.5. The matrix of the associated Einstein tensor was then calculated in Sect. 8.5:

$$(G_{ik}) = \mathrm{diag}\left(\frac{4}{3}\frac{1}{t^2}, 0, 0, 0\right) .$$

The representation for G given here gives the same result, because we have for the scale factor S that

$$3(S'(t))^2 / S^2(t) = \frac{4}{3} \cdot \frac{1}{t^2}$$

and

$$-2S''(t)S(t) - (S'(t))^2 = 0 .$$

The Hubble function is

$$H(t) = S'(t) / S(t) = \frac{2}{3} \cdot \frac{1}{t} .$$

With the value of H given in the previous section, this model results in the age of the current cosmos of approximately 12 billion years (One currently assumes the age of current cosmos of 15 18 billion years. Thus the model with $\Lambda = 0$ and $K = 0$ seems to be an incomplete description of reality.).

For positive K, one can easily check that the curve with the parametric representation

$$S(\tau) = \frac{4}{3}\pi\frac{C}{K}(1 - \cos\tau)$$

and

$$t(\tau) = \frac{4}{3}\pi\frac{C}{K\sqrt{K}}(\tau - \sin\tau)\,,$$

is a solution to the Friedmann differential equation. Apart from the factor $\sqrt{K}$, this describes a cycloid (see Fig. 15.1). The time calculation begins with the big bang at $t=0$, and the scale factor S grows to the maximum value $\frac{8}{3}\pi\frac{C}{K}$ and then starts to decrease again. When the time $\frac{8}{3}\pi^2\frac{C}{K\sqrt{K}}$ has elapsed, the cosmos has collapsed again, similar to a big bang, but in reverse order.

For negative K, the solution curve is given in parametric form as

$$S(\tau) = \frac{4}{3}\pi\frac{C}{-K}(\cosh\tau - 1)$$

and

$$t(\tau) = \frac{4}{3}\pi\frac{C}{(-K)^{\frac{3}{2}}}(\sinh\tau - \tau)\,.$$

This behave qualitatively in the same manner as the $K=0$ case, but the expansion happens faster.

Rotating Black Holes

16

16.1 The Kerr Metric

The gravitational field generated by a non-rotating fixed star with the mass μ and the radius R in the exterior region $r > R$ is described by the (outer) Schwarzschild metric with the parameter μ. If this fixed star collapses into a black hole, nothing changes in the formula for this metric, but the restriction $r > R$ is relaxed to $r > 0$, apart from the difficulty with $r = 2\mu$. A (relativistic) characterisation of a rotating fixed star is not yet known. However, the metric that describes a black hole that is the result of the collapse of a rotating fixed star with mass μ and angular momentum J is known. This is the Kerr metric (R. Kerr 1963) with the parameters μ and $a = J/\mu$, which is the focus of this chapter. The derivation of this metric is lengthy and goes beyond the scope of this book, but we refer the interested reader to [Ch] §§52–55.

Definition 16.1 The **Kerr metric** in the Boyer-Lindquist coordinates t, r, ϑ, φ has the component matrix

$$
\begin{pmatrix}
g_{tt} & 0 & 0 & g_{t\varphi} \\
0 & g_{rr} & 0 & 0 \\
0 & 0 & g_{\vartheta\vartheta} & 0 \\
g_{t\varphi} & 0 & 0 & g_{\varphi\varphi}
\end{pmatrix}
=
\begin{pmatrix}
1 - \frac{2\mu r}{\rho^2} & 0 & 0 & \frac{2\mu r a s^2}{\rho^2} \\
0 & -\frac{\rho^2}{\Delta} & 0 & 0 \\
0 & 0 & -\rho^2 & 0 \\
\frac{2\mu r a s^2}{\rho^2} & 0 & 0 & -(r^2 + a^2 + \frac{2\mu r a^2 s^2}{\rho^2})s^2
\end{pmatrix}
$$

where

$$
s = \sin\vartheta, \qquad c = \cos\vartheta, \qquad \rho^2 = r^2 + a^2 c^2
$$

© The Author(s), under exclusive license to Springer Nature Switzerland AG 2023
R. Oloff, *The Geometry of Spacetime*, Graduate Texts in Physics,
https://doi.org/10.1007/978-3-031-16139-1_16

and

$$\Delta = r^2 - 2\mu r + a^2 = (r - \mu)^2 + a^2 - \mu^2 .$$

◆

We have

$$\begin{vmatrix} g_{tt} & g_{t\varphi} \\ g_{t\varphi} & g_{\varphi\varphi} \end{vmatrix} = -(r^2 + a^2 + \frac{2\mu r a^2 s^2}{\rho^2})s^2 + \frac{2\mu r (r^2 + a^2) s^2}{\rho^2}$$

$$= -(r^2 + a^2)s^2 + \frac{2\mu r}{\rho^2}(r^2 + a^2 c^2)s^2 = -\Delta s^2 .$$

Thus these coordinates fail on the axis of rotation $s = 0$ and when $\Delta = 0$. There are three cases for the zeros of Δ. In general, we can limit ourselves to $a \neq 0$, for otherwise it is a non-rotating black hole (Chap. 14). When $|a| < \mu$ (slowly rotating) there are two zeros

$$r_+ = \mu + \sqrt{\mu^2 - a^2} \quad \text{and} \quad r_- = \mu - \sqrt{\mu^2 - a^2}$$

with $0 < r_- < \mu < r_+ < 2\mu$. In the case $|a| > \mu$ (rapidly rotating), Δ has no zeros, and in the borderline case $|a| = \mu$, Δ has only one zero, at μ. We will focus on the most interesting case, when $|a| < \mu$, here. Then there are no metric difficulties with $r = 0$ except when $\vartheta = \pi/2$, and negative values of r are also allowed for the Kerr spacetime. With this in mind, the Kerr spacetime, insofar as it can be described with the Boyer-Lindquist coordinates, consist of three **Boyer-Lindquist blocks**, $\mathbb{R} \times (r_+, \infty) \times S_2$ (Region I), $\mathbb{R} \times (r_-, r_+) \times S_2$ (Region II) and $\mathbb{R} \times (-\infty, r_-) \times S_2$ (Region III). For the geometric illustration for a fixed t, we transform the radial coordinate r into e^r and use e^r, ϑ, φ as the spherical coordinates. The two event horizons at $r = r_\pm$ are spherical surfaces with the radius e^{r_+}, respectively e^{r_-}, and the **ring singularity** $r = 0$ and $\vartheta = \pi/2$ is the equator of the spherical surface with the radius $e^0 = 1$ (Fig. 16.1).

The Kerr metric has the required signature, because in regions I and III, the submatrix formed by the matrix elements g_{tt}, $g_{\varphi\varphi}$, $g_{t\varphi}$ is indefinite and the other two diagonal elements g_{rr} and $g_{\vartheta\vartheta}$ are negative. In the region II, this submatrix is negative definite and g_{rr} is positive while $g_{\vartheta\vartheta}$ is negative.

The matrix element g_{tt} is not only negative in the region II. Clearly $g_{tt} > 0$ if and only if $|r - \mu| > \sqrt{\mu^2 - a^2 c^2}$ (Fig. 16.2). This means that the coordinate vector field ∂_t loses the property of being timelike if r decreases before the event horizon $r = r_+$ is reached. The part of the region I where

$$r_+ = \mu + \sqrt{\mu^2 - a^2} < r < \mu + \sqrt{\mu^2 - a^2 c^2}$$

is called the **ergosphere**.

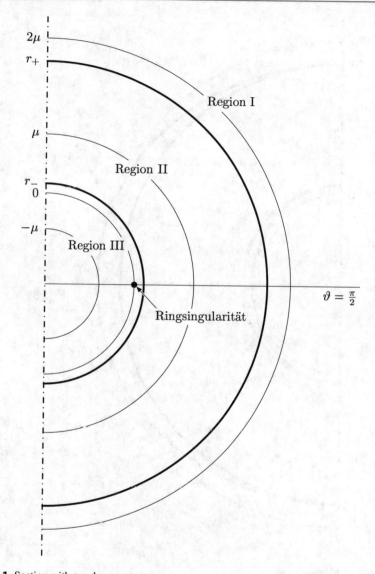

Fig. 16.1 Section with t and φ constant

The matrix element $g_{\varphi\varphi}$ is negative for $r \geq 0$ and outside the axis of rotation. Also for $r \leq -\mu$, we have $g_{\varphi\varphi} < 0$, because then we have

$$(r^2 + a^2c^2)(r^2 + a^2) > 2\mu(-r)a^2s^2$$

thanks to

$$2\mu(-r)a^2s^2 \leq 2(-r)^2a^2 = 2a^2r^2$$

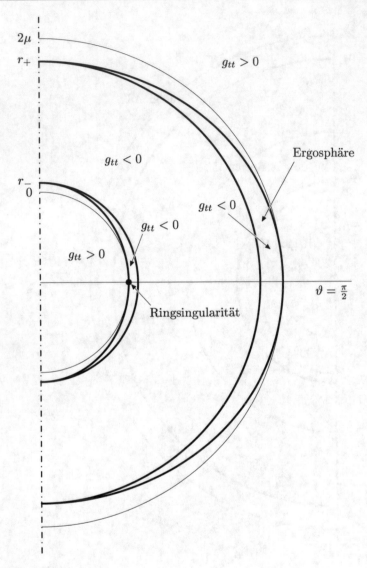

Fig. 16.2 Sign behavior of g_{tt}

and

$$(r^2 + a^2c^2)(r^2 + a^2) = r^4 + r^2a^2(1 + c^2) + a^4c^2 > a^2r^2 + r^2a^2 = 2a^2r^2 \ .$$

Between $-\mu$ and 0, $g_{\varphi\varphi}$ can assume positive values (Fig. 16.3), i.e., we have

$$(r^2 + a^2)(r^2 + a^2c^2) < 2\mu(-r)a^2s^2 \ .$$

The left side $f(r)$ is a symmetric and convex parabola with apex at height a^4c^2 and the right side $g(r)$ is a decreasing linear function passing through the origin. Whether

Fig. 16.3 Sign behavior of $g_{\varphi\varphi}$

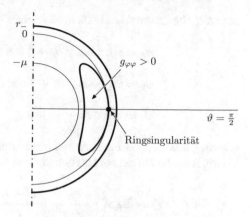

the straight line intersects, touches or misses the parabola depends on the angular coordinate ϑ (Fig. 16.3). In any case, it is indisputable, and only that will be used in the next section, that there are negative values of r, and for some such r and $\vartheta = \pi/2$ $g_{\varphi\varphi}$ is positive, because there holds

$$\lim_{r \to 0-} g_{\varphi\varphi}(r, \pi/2) = \lim_{r \to 0-} [-(r^2 + a^2 + 2\mu a^2/r)] = +\infty .$$

16.2   Other Representations of the Kerr Metric

The failure of the Boyer-Lindquist coordinates on the axis of rotation when $\sin \vartheta = 0$ is only due to the usage of spherical coordinates r, ϑ, φ. To describe the metric on the axis of rotation, we can use Cartesian coordinates

$$x = r \sin \vartheta \cos \varphi$$
$$y = r \sin \vartheta \sin \varphi$$
$$z = r \cos \vartheta$$

and obtain

$$\partial_r = \sin \vartheta \cos \varphi \, \partial_x + \sin \vartheta \sin \varphi \, \partial_y + \cos \vartheta \, \partial_z$$

$$\partial_\vartheta = r \cos \vartheta \cos \varphi \, \partial_x + r \cos \vartheta \sin \varphi \, \partial_y - r \sin \vartheta \, \partial_z$$

$$\partial_\varphi = -r \sin \vartheta \sin \varphi \, \partial_x + r \sin \vartheta \cos \varphi \, \partial_y \quad .$$

The coefficient matrix

$$\begin{pmatrix} 1 & 0 & 0 \\ 0 & r & 0 \\ 0 & 0 & r \sin \vartheta \end{pmatrix} \begin{pmatrix} \sin \vartheta \cos \varphi & \sin \vartheta \sin \varphi & \cos \vartheta \\ \cos \vartheta \cos \varphi & \cos \vartheta \sin \varphi & - \sin \vartheta \\ - \sin \varphi & \cos \varphi \end{pmatrix}$$

can easily be inverted, and this inverse gives the following representation

$$\partial_x = \sin\vartheta\cos\varphi\,\partial_r + \tfrac{1}{r}\cos\vartheta\cos\varphi\,\partial_\vartheta - \tfrac{\sin\varphi}{r\sin\vartheta}\,\partial_\varphi$$

$$\partial_y = \sin\vartheta\sin\varphi\,\partial_r + \tfrac{1}{r}\cos\vartheta\sin\varphi\,\partial_\vartheta + \tfrac{\cos\varphi}{r\sin\vartheta}\,\partial_\varphi$$

$$\partial_z = \cos\vartheta\,\partial_r - \tfrac{1}{r}\sin\vartheta\,\partial_\vartheta \qquad .$$

Inserting this into the metric gives the component matrix (g_{ik}), which has the following form on the axis of rotation $\sin\vartheta = 0$:

$$(g_{ik}) = \text{diag}\left(\frac{\Delta}{r^2+a^2}, -\frac{r^2+a^2}{r^2}, -\frac{r^2+a^2}{r^2}, -\frac{r^2+a^2}{\Delta}\right)$$

This coordinate system only fails when $r = 0$ and $\Delta = 0$.

When we last discussed the Kruskal plane, it was apparent that the singularities of the metric components of the Schwarzschild spacetime are only caused by the Schwarzschild coordinates. So it should not come as a surprise that the difficulties at the event horizons r_+ and r_- can be avoided by using other coordinates. These new coordinates we will use are called the **Kerr-star coordinates** $t^*, r, \vartheta, \varphi^*$. We will now introduce them, and they will be used in the next section.

We choose functions T and F of r with

$$\frac{dT}{dr} = \frac{r^2+a^2}{\Delta} \quad \text{and} \quad \frac{dF}{dr} = \frac{a}{\Delta}$$

and set

$$t^*(t,r) = t + T(r) \quad \text{and} \quad \varphi^*(\varphi,r) = \varphi + F(r) .$$

By the chain rule, the Kerr-star coordinate vector fields $\partial_{t^*}^*, \partial_r^*, \partial_\vartheta^*, \partial_{\varphi^*}^*$ are related to the Boyer-Lindquist coordinate vector fields $\partial_t, \partial_r, \partial_\vartheta, \partial_\varphi$ by

$$\begin{pmatrix} \partial_t & \partial_r & \partial_\vartheta & \partial_\varphi \end{pmatrix} = \begin{pmatrix} \partial_{t^*}^* & \partial_r^* & \partial_\vartheta^* & \partial_{\varphi^*}^* \end{pmatrix} \begin{pmatrix} 1 & \frac{r^2+a^2}{\Delta} & 0 & 0 \\ 0 & 1 & 0 & 0 \\ 0 & 0 & 1 & 0 \\ 0 & \frac{a}{\Delta} & 0 & 1 \end{pmatrix} .$$

The coordinate vector fields

$$\partial_{t^*}^* = \partial_t, \quad \partial_\vartheta^* = \partial_\vartheta, \quad \partial_{\varphi^*}^* = \partial_\varphi, \quad \partial_r^* = \partial_r - \frac{r^2+a^2}{\Delta}\partial_t - \frac{a}{\Delta}\partial_\varphi = \partial_r - \frac{1}{\Delta}V$$

(where V is as in Definition 16.2) are to be used in the metric. Taking into account the calculation rules in Theorem 16.1, the component matrix

$$(g^*_{ik}) = \begin{pmatrix} 1 - \frac{2\mu r}{\rho^2} & -1 & 0 & \frac{2\mu r a s^2}{\rho^2} \\ -1 & 0 & 0 & as^2 \\ 0 & 0 & -\rho^2 & 0 \\ \frac{2\mu r a s^2}{\rho^2} & as^2 & 0 & -\left(r^2 + a^2 + \frac{2\mu r a^2 s^2}{\rho^2}\right)s^2 \end{pmatrix}$$

is obtained, and this does not have singularities at $r = r_\pm$ any longer.

In Sect. 16.4 we will calculate the curvature of the Kerr spacetime with the connection forms and curvature forms. It is then advantageous to use the following to vector fields instead of the Boyer-Lindquist coordinate vector fields ∂_t and ∂_φ.

Definition 16.2 The **canonical vector fields** V and W are defined by

$$V = (r^2 + a^2)\partial_t + a\partial_\varphi \quad \text{and} \quad W = \partial_\varphi + as^2\partial_t .$$

◆

The following rules of calculation can be read from the representation of the metric.

Theorem 16.1 *The following hold for the metric components in the Boyer-Lindquist coordinates:*

(1) $g(W, \partial_\varphi) = g_{\varphi\varphi} + as^2 g_{t\varphi} = -(r^2 + a^2)s^2$

(2) $g(W, \partial_t) = g_{t\varphi} + as^2 g_{tt} = as^2$

(3) $g(V, \partial_\varphi) = ag_{\varphi\varphi} + (r^2 + a^2)g_{t\varphi} = -\Delta as^2$

(4) $g(V, \partial_t) = ag_{t\varphi} + (r^2 + a^2)g_{tt} = \Delta$.

From these formulae, we read off $g(V, W) = 0$, $g(V, V) = \Delta\rho^2$, $g(W, W) = -s^2\rho^2$. Using an approach with unknown coefficients, one also obtains

$$\partial_t = \frac{1}{\rho^2}V - \frac{a}{\rho^2}W \quad \text{and} \quad \partial_\varphi = -\frac{as^2}{\rho^2}V + \frac{r^2 + a^2}{\rho^2}W .$$

Theorem 16.2 *The vector fields* $E_0 = V/(\rho\sqrt{|\Delta|})$, $E_1 = (\sqrt{|\Delta|}/\rho)\partial_r$, $E_2 = (1/\rho)\partial_\vartheta$, $E_3 = W/(\rho s)$ *form a semi-orthonormal basis with respect to*

$$(g_{ik}) = \text{diag}\,(\text{sgn}\Delta, -\text{sgn}\Delta, -1, -1) .$$

16.3   Causal Structure

Future-pointing vectors are to be determined among the timelike vectors. Far outside the rotating black hole, the vectors of the form

$$\alpha \partial_t + \beta \partial_r + \frac{\gamma}{r} \partial_\vartheta + \frac{\delta}{rs} \partial_\varphi$$

with

$$0 < \left(1 - \frac{2\mu r}{\rho^2}\right) \alpha^2 - \frac{\rho^2}{\Delta} \beta^2 - \frac{\rho^2}{r^2} \gamma^2 - \left(1 + \frac{a^2}{r^2} + \frac{2\mu a^2 s^2}{\rho^2 r}\right) \delta^2 + \frac{4\mu a s}{\rho^2} \alpha \delta$$

$$\approx \alpha^2 - \beta^2 - \gamma^2 - \delta^2$$

are timelike, and they are future-pointing if and only if α is positive. In the ergosphere, ∂_t is no longer timelike, but there too, each future-pointing vector has a positive ∂_t-coefficient α, because otherwise, there would then exist a future-pointing vector

$$v = \beta \partial_r + \frac{\gamma}{r} \partial_\vartheta + \frac{\delta}{rs} \partial_\varphi$$

somewhere in the region I ($r > r_+$), which would contradict the inequality $g(v, v) \leq 0$. The fact that for all future-pointing vectors in region I, the coefficient α is positive, has the interpretation that the t coordinate is growing monotonicaly for each particle (observer), which is not surprising.

It is however surprising that the angular coordinate φ is monotonic for a particle in the ergosphere. To avoid repeated case distinction, we now assume $a > 0$. By the above inequality, $\alpha \delta$ must be positive in the ergosphere. Together with $\alpha > 0$ this means $\delta > 0$, and so φ must grow monotonously.

In region II ($r_- < r < r_+$), ∂_r is timelike. Analogous to the Schwarzschild space-time $-\partial_r$ is to be seen as future-pointing. This also results from the way regions I and II are joined together by introducing the Kerr-star coordinates. If one transfers the adjective 'future-pointing' to light-like vectors $v \neq 0$ that can be approximated by future-pointing timelike vectors, then $-\partial_{r*}^*$ is region I is future-pointing, because

$$g(-\partial_{r*}^*, \partial_t) = -g(\partial_r - \frac{V}{\Delta}, \partial_t) = \frac{1}{\Delta} g(V, \partial_t) = 1$$

is positive. On grounds of constant transition, $-\partial_{r*}^*$ should also be future-pointing in regions II and III. In region II, $-\partial_r$ is also future-pointing, because

$$g(-\partial_{r*}^*, -\partial_r) = g(\partial_r, \partial_r) = -\frac{\rho^2}{\Delta}$$

is positive there.

Since $-\partial_r$ is future-pointing in region II, every particle there must have a negative ∂_r-component, so the r-coordinate is monotonically decreasing. In particular, this means that a transition from region II to region I or from region III to region II is impossible.

The light-like vector $-\partial_{r*}^{*}$ should also be future-pointing in region III. Because of

$$g(-\partial_{r*}^{*}, V) = g\left(-\partial_r + \frac{V}{\Delta}, V\right) = \frac{1}{\Delta} g(V, V) = \rho^2 > 0$$

the timelike vector V is also future-pointing. It now turns out that the time-orientation thus determined in region III is so pathological that it contains an effect that is called a **time machine** in the science fiction literature.

Theorem 16.3 *In region III one can get from any point to any other point along a curve γ with future-pointing timelike tangent vectors γ'.*

Proof We choose a negative number $\bar{r}$, so that $g_{\varphi\varphi}$ is positive on the circular path $r = \bar{r}$ and $\vartheta = \pi/2$. The time machine is based on the fact that we can turn back the 'time' t by revolving along the circular path. The sought-after curve γ from $(t_0, r_0, \vartheta_0, \varphi_0)$ to $(t_1, r_1, \vartheta_1, \varphi_1)$ consists of three sections $\gamma_1, \gamma_2, \gamma_3$. γ_1 runs from $(t_0, r_0, \vartheta_0, \varphi_0)$ to $(\hat{t}, \bar{r}, \pi/2, \hat{\varphi})$, γ_2 follows the circular path from $(\hat{t}, \bar{r}, \pi/2, \hat{\varphi})$ to $(\check{t}, \bar{r}, \pi/2, \check{\varphi})$ and γ_3 finally leads to the endpoint $(t_1, r_1, \vartheta_1, \varphi_1)$. In detail, the curves γ_i are constructed as follows:

For γ_1, on $[0,1]$, we choose smooth functions, $r_1(.)$ with $r_1(0) = r_0$ and $r_1(1) = \bar{r}$, $\vartheta_1(.)$ with $\vartheta_1(0) = \vartheta_0$ and $\vartheta_1(1) = \pi/2$, and $\tau_1(.)$ with $\tau_1(0) = 0$ and $\tau_1'(s) = r_1^2(s) + a^2$. Then, with a positive constant c_1, let

$$\gamma_1(s) = (t_0 + c_1\tau_1(s), r_1(s), \vartheta_1(s), \varphi_0 + c_1 as).$$

From

$$\gamma_1' = c_1\tau_1'\partial_t + r_1'\partial_r + \vartheta_1'\partial_\vartheta + c_1 a\partial_\varphi = c_1 V + r_1'\partial_r + \vartheta_1'\partial_\vartheta$$

it follows that

$$g(\gamma_1', \gamma_1') = c_1^2 \Delta\rho^2 + (r_1')^2(-\rho^2/\Delta) + (\vartheta_1')^2(-\rho^2).$$

Since the negative summands are bounded below on $[0,1]$, $g(\gamma_1', \gamma_1') > 0$ can be guaranteed by choosing c_1 large enough. Then γ_1' is timelike, and because of

$$g(\gamma_1', V) = c_1 \Delta\rho^2 > 0$$

also future-pointing. The curve γ_1 ends at the point $\gamma_1(1) = (\hat{t}, \bar{r}, \pi/2, \hat{\varphi})$ with

$$\hat{t} = t_0 + c_1\tau_1(1) > t_0 \quad \text{and} \quad \hat{\varphi} = \varphi_0 + c_1 a > \varphi_0.$$

Analogously, γ_3 is defined with functions $r_3(.)$, $\vartheta_3(.)$ and $\tau_3(.)$, for which there holds $r_3(0) = \bar{r}$, $r_3(1) = r_1$, $\vartheta_3(0) = \pi/2$, $\vartheta_3(1) = \vartheta_1$, $\tau_3(0) = 0$ and $\tau_3'(s) = r_3{}^2(s) + a^2$. The constant c_3 is again chosen to be large enough to ensure time-likeness, and γ_3 is then

$$\gamma_3(s) = (\check{t} + c_3\tau_3(s), \ r_3(s), \ \vartheta_3(s), \ \check{\varphi} + c_3 a),$$

where $\check{t}$ and $\check{\varphi}$ are chosen such that

$$\check{t} + c_3\tau_3(1) = t_1 \quad \text{and} \quad \check{\varphi} + c_3 a = \varphi_1$$

hold.

The middle section γ_2 is taken as

$$\gamma_2(s) = (\hat{t} + s(\check{t} - \hat{t}), \ \bar{r}, \ \pi/2, \ \hat{\varphi} - c_2 s).$$

Then we have

$$\gamma_2' = (\check{t} - \hat{t})\partial_t - c_2\partial_\varphi$$

and thus

$$g(\gamma_2', \gamma_2') = (\check{t} - \hat{t})^2 g_{tt} - 2(\check{t} - \hat{t})c_2 g_{t\varphi} + c_2{}^2 g_{\varphi\varphi}$$

and

$$g(\gamma_2', V) = (\check{t} - \hat{t})g(\partial_t, V) - c_2 g(\partial_\varphi, V) = (\check{t} - \hat{t})\Delta + c_2\Delta a.$$

So if we choose the constant c_2 large enough, γ_2' is actually future-pointing timelike. Finally, we have to make sure that the φ-coordinate $\hat{\varphi} - c_2$ of the endpoint of γ_2 differs from the φ-coordinate $\check{\varphi}$ of the starting point of γ_3 only by an integer multiple of 2π.

Since there is no way from region III back to region I, the practical importance of this time machine is limited. However, it is interesting that the laws of general relativity do not rule out such a mixture of future and past in a global sense. Locally, i.e., in the individual tangent space, however, the future and the past are strictly separated, a point of view which is already anchored in the special theory of relativity.

16.4   Covariant Derivative and Curvature

With some effort the Christoffel symbols can be calculated from the metric components with respect to the Boyer-Lindquist coordinates. This then describes the covariant differentiation of vector fields.

Theorem 16.4 *The following hold for the coordinate vector fields of the Boyer-Lindquist coordinates:*

$$\nabla_{\partial_t}\partial_t = \frac{\Delta\mu(2r^2 - \rho^2)}{\rho^6}\,\partial_r - \frac{2\mu r a^2 sc}{\rho^6}\,\partial_\vartheta$$

$$\nabla_{\partial_r}\partial_r = \left(\frac{r}{\rho^2} - \frac{r-\mu}{\Delta}\right)\partial_r + \frac{a^2 sc}{\Delta\rho^2}\,\partial_\vartheta$$

$$\nabla_{\partial_\vartheta}\partial_\vartheta = -\frac{\Delta r}{\rho^2}\,\partial_r - \frac{a^2 sc}{\rho^2}\,\partial_\vartheta$$

$$\nabla_{\partial_\varphi}\partial_\varphi = \frac{\Delta s^2}{\rho^2}\left(\frac{\mu a^2 s^2(2r^2 - \rho^2)}{\rho^4} - r\right)\partial_r - \frac{sc}{\rho^2}\left(r^2 + a^2 + \frac{2\mu r a^2 s^2(2\rho^2 + a^2 s^2)}{\rho^4}\right)\partial_\vartheta$$

$$\nabla_{\partial_t}\partial_r = \frac{\mu(2r^2 - \rho^2)(r^2 + a^2)}{\Delta\rho^4}\,\partial_t + \frac{\mu a(2r^2 - \rho^2)}{\Delta\rho^4}\,\partial_\varphi$$

$$\nabla_{\partial_t}\partial_\vartheta = -\frac{2\mu r a^2 sc}{\rho^4}\,\partial_t + \frac{2\mu rac}{\rho^4 s}\,\partial_\varphi$$

$$\nabla_{\partial_t}\partial_\varphi = \frac{\Delta\mu a s^2(\rho^2 - 2r^2)}{\rho^6}\,\partial_r + \frac{2\mu rasc(r^2 + a^2)}{\rho^6}\,\partial_\vartheta$$

$$\nabla_{\partial_r}\partial_\vartheta = -\frac{a^2 sc}{\rho^2}\,\partial_r + \frac{r}{\rho^2}\,\partial_\vartheta$$

$$\nabla_{\partial_r}\partial_\varphi = \frac{\mu a s^2}{\Delta\rho^4}[(r^2 + a^2)(\rho^2 - 2r^2) - 2r^2\rho^2]\partial_t + \frac{r\rho^4 + \mu a^2 s^2\rho^2 - 2\mu r^2(r^2 + a^2)}{\Delta\rho^4}\,\partial_\varphi$$

$$\nabla_{\partial_\vartheta}\partial_\varphi = \frac{2\mu r a^3 s^3 c}{\rho^4}\,\partial_t +$$

$$+ \frac{c}{\Delta\rho^4 s}[(r^2 + a^2)\rho^4 + 2\mu r(a^4 s^2 + (2a^2 s^2 - r^2 - a^2)\rho^2) - 4\mu^2 r^2 a^2 s^2]\partial_\varphi\ .$$

The Christoffel symbols are captured in the given equations for the covariant derivatives. With their knowledge, the four geodesic equations can also be written down. However, these are far too complicated for the determination of explicit solutions. In the next section we will describe other tools that are useful for determining geodesics. The Christoffel symbols are also not recommended for the calculation of the curvature tensor. Instead, we will determine the connection forms and curvature forms for the semi-orthonormal basis E_0, E_1, E_2, E_3 given in Theorem 16.2.

By Theorem 8.21, the connection forms are calculated from the exterior derivatives of the dual basis, and their calculation require knowledge of the Lie brackets of the basis elements.

Theorem 16.5 *For the vector fields E_0, E_1, E_2, E_3, there holds*

$$[E_0, E_1] = \left(-\frac{2r\sqrt{|\Delta|}}{\rho^3} - |\Delta| \frac{\partial}{\partial r} \frac{1}{\rho\sqrt{|\Delta|}} \right) E_0 + \frac{2ras}{\rho^3} E_3$$

$$[E_0, E_2] = -\frac{a^2 sc}{\rho^3} E_0$$

$$[E_0, E_3] = 0$$

$$[E_1, E_2] = -\frac{a^2 sc}{\rho^3} E_1 - \frac{r\sqrt{|\Delta|}}{\rho^3} E_2$$

$$[E_1, E_3] = -\frac{r\sqrt{|\Delta|}}{\rho^3} E_3$$

$$[E_2, E_3] = \frac{2ac\sqrt{|\Delta|}}{\rho^3} E_0 - \frac{(r^2 + a^2)c}{\rho^3 s} E_3 \,.$$

Proof We verify the first equation as an example. We have

$$[E_0, E_1] = \left[\frac{r^2 + a^2}{\rho\sqrt{|\Delta|}} \partial_t + \frac{a}{\rho\sqrt{|\Delta|}} \partial_\varphi, \frac{\sqrt{|\Delta|}}{\rho} \partial_r \right] = -\frac{\sqrt{|\Delta|}}{\rho} \partial_r \left(\frac{r^2 + a^2}{\rho\sqrt{|\Delta|}} \partial_t + \frac{a}{\rho\sqrt{|\Delta|}} \partial_\varphi \right)$$

$$= -\frac{2r}{\rho^2} \partial_t - \frac{\sqrt{|\Delta|}}{\rho} \frac{\partial}{\partial r} \frac{1}{\rho\sqrt{|\Delta|}} V = -\frac{2r}{\rho^4} (V - aW) - \frac{\sqrt{|\Delta|}}{\rho} \frac{\partial}{\partial r} \frac{1}{\rho\sqrt{|\Delta|}} V$$

$$= -\frac{2r\sqrt{|\Delta|}}{\rho^3} E_0 + \frac{2ras}{\rho^3} E_3 - |\Delta| \frac{\partial}{\partial r} \frac{1}{\rho\sqrt{|\Delta|}} E_0 \,.$$

By Theorem 8.21, the Lie brackets $[E_i, E_j]$ are used to determine the values of the connection forms on the basis elements and thus these forms themselves. Finally, the covariant derivatives $\nabla_{E_i} E_j$ can thereby be written down. We demonstrate this for $i = j = 0$. To calculate $\nabla_{E_0} E_0$ we need the values of the forms $\omega_0{}^k$ on E_0. Note that

$$(\varepsilon_0, \varepsilon_1, \varepsilon_2, \varepsilon_3) = (\varepsilon, -\varepsilon, -1, -1)$$

where

$$\varepsilon = \text{sgn}\Delta \,.$$

By Theorem 8.21(1), $\omega_0{}^0$ disappears. We have

$$\langle E_0, \omega_0{}^1 \rangle = -\frac{1}{2} \left(\varepsilon_1 \varepsilon_0 \langle [E_0, E_1], A^0 \rangle + \varepsilon_1 \varepsilon_0 \langle [E_0, E_1], A^0 \rangle \right) = \langle [E_0, E_1], A^0 \rangle$$

and analogously

$$\langle E_0, \omega_0{}^2 \rangle = \varepsilon \langle [E_0, E_2], A^0 \rangle \quad \text{and} \quad \langle E_0, \omega_0{}^3 \rangle = \varepsilon \langle [E_0, E_3], A^0 \rangle \,,$$

and altogether

$$\nabla_{E_0} E_0 = \left(-\frac{2r\sqrt{|\Delta|}}{\rho^3} - |\Delta|\frac{\partial}{\partial r}\frac{1}{\rho\sqrt{|\Delta|}}\right) E_1 - \frac{\varepsilon a^2 sc}{\rho^3} E_2 \,.$$

Theorem 16.6 *The covariant differentiation in the Kerr spacetime is characterised by*

$$\nabla_{E_0} E_0 = \left(-\frac{2r\sqrt{|\Delta|}}{\rho^3} - |\Delta|\frac{\partial}{\partial r}\frac{1}{\rho\sqrt{|\Delta|}}\right) E_1 - \frac{\varepsilon a^2 sc}{\rho^3} E_2 \,, \qquad \nabla_{E_1} E_1 = \frac{\varepsilon a^2 sc}{\rho^3} E_2 \,,$$

$$\nabla_{E_2} E_2 = -\frac{\varepsilon r\sqrt{|\Delta|}}{\rho^3} E_1 \,, \qquad \nabla_{E_3} E_3 = -\frac{\varepsilon r\sqrt{|\Delta|}}{\rho^3} E_1 - \frac{(r^2 + a^2)c}{s\rho^3} E_2 \,,$$

$$\nabla_{E_0} E_1 = \left(-\frac{2r\sqrt{|\Delta|}}{\rho^3} - |\Delta|\frac{\partial}{\partial r}\frac{1}{\rho\sqrt{|\Delta|}}\right) E_0 + \frac{ars}{\rho^3} E_3 \,, \qquad \nabla_{E_1} E_0 = -\frac{ars}{\rho^3} E_3 \,,$$

$$\nabla_{E_0} E_2 = -\frac{a^2 sc}{\rho^3} E_0 + \frac{\varepsilon ac\sqrt{|\Delta|}}{\rho^3} E_3 \,, \qquad \nabla_{E_2} E_0 = \frac{\varepsilon ac\sqrt{|\Delta|}}{\rho^3} E_3 \,,$$

$$\nabla_{E_0} E_3 = -\frac{\varepsilon ars}{\rho^3} E_1 - \frac{\varepsilon ac\sqrt{|\Delta|}}{\rho^3} E_2 = \nabla_{E_3} E_0 \,,$$

$$\nabla_{E_1} E_2 = -\frac{a^2 sc}{\rho^3} E_1 \,, \qquad \nabla_{E_2} E_1 = \frac{r\sqrt{|\Delta|}}{\rho^3} E_2 \,,$$

$$\nabla_{E_1} E_3 = -\frac{\varepsilon ars}{\rho^3} E_0 \,, \qquad \nabla_{E_3} E_1 = -\frac{\varepsilon ars}{\rho^3} E_0 + \frac{r\sqrt{|\Delta|}}{\rho^3} E_3 \,,$$

$$\nabla_{E_2} E_3 = \frac{ac\sqrt{|\Delta|}}{\rho^3} E_0 \,, \qquad \nabla_{E_3} E_2 = -\frac{ac\sqrt{|\Delta|}}{\rho^3} E_0 + \frac{(r^2 + a^2)c}{s\rho^3} E_3 \,.$$

That the Kerr spacetime describes a rotating black hole was stipulated without proof at the beginning of this chapter. Also, for this to be true, a vacuum must be present, making this spacetime Ricci flat.

Theorem 16.7 *In the Kerr spacetime,* $\mathrm{Ric} = 0$.

Proof The connection forms can be read from the formulae given in Theorem 16.6. We have

$$\omega_0{}^1 = \left(-\frac{2r\sqrt{|\Delta|}}{\rho^3} - |\Delta| \frac{\partial}{\partial r} \frac{1}{\rho\sqrt{|\Delta|}} \right) A^0 + \frac{\varepsilon r a s}{\rho^3} A^3 = \omega_1{}^0$$

$$\omega_0{}^2 = \qquad -\frac{\varepsilon a^2 s c}{\rho^3} A^0 - \frac{\varepsilon a c \sqrt{|\Delta|}}{\rho^3} A^3 \qquad = \varepsilon \omega_2{}^0$$

$$\omega_0{}^3 = \qquad -\frac{a r s}{\rho^3} A^1 + \frac{\varepsilon a c \sqrt{|\Delta|}}{\rho^3} A^2 \qquad = \varepsilon \omega_3{}^0$$

$$\omega_1{}^2 = \qquad \frac{\varepsilon a^2 s c}{\rho^3} A^1 + \frac{r \sqrt{|\Delta|}}{\rho^3} A^2 \qquad = -\varepsilon \omega_2{}^1$$

$$\omega_1{}^3 = \qquad \frac{a r s}{\rho^3} A^0 + \frac{r \sqrt{|\Delta|}}{\rho^3} A^3 \qquad = -\varepsilon \omega_3{}^1$$

$$\omega_2{}^3 = \qquad \frac{\varepsilon a c \sqrt{|\Delta|}}{\rho^3} A^0 + \frac{(r^2 + a^2)c}{s\rho^3} A^3 \qquad = -\omega_3{}^2 .$$

A lengthy calculation according to Definition 8.9 gives the curvature forms as

$$\Omega_1^0 = 2\varepsilon g A^0 \wedge A^1 - 2\varepsilon h A^2 \wedge A^3$$

$$\Omega_2^0 = -g A^0 \wedge A^2 + \varepsilon h A^3 \wedge A^1$$

$$\Omega_3^0 = -g A^0 \wedge A^3 - \varepsilon h A^1 \wedge A^2$$

$$\Omega_3^2 = 2h A^0 \wedge A^1 + 2g A^2 \wedge A^3$$

$$\Omega_1^3 = -h A^0 \wedge A^2 - \varepsilon g A^3 \wedge A^1$$

$$\Omega_2^1 = -\varepsilon h A^0 \wedge A^3 - g A^1 \wedge A^2$$

where

$$g(r, \vartheta) = \frac{\mu r}{\rho^6}(r^2 - 3a^2 c^2) \quad \text{and} \quad h(r, \vartheta) = \frac{\mu a c}{\rho^6}(3r^2 - a^2 c^2) .$$

From this, for every index pair i and k,

$$\mathrm{Ric}_{ik} = \Omega_i^j(E_j, E_k) = 0$$

can be read off.

16.5   Conservation Theorems

There are quantities that do not change for a photon or a freely falling particle γ in Kerr spacetime. A first such conserved quantity is obviously $Q = g(\gamma', \gamma')$, and we use this notation throughout. Two further conserved quantities are generated by the Killing vector fields ∂_t and ∂_φ. As a generalisation of the situation in the Schwarzschild spacetime, we also have here

$$E = g(\partial_t, \gamma') = g_{tt} t' + g_{t\varphi} \varphi'$$

and

$$L = -g(\partial_\varphi, \gamma') = -g_{t\varphi} t' - g_{\varphi\varphi} \varphi' \,.$$

The aim of the following study is to formulate and justify a fourth conserved quantity.

For a geodesic γ in Kerr spacetime, we have the functions

$$R(r) = g(V, \gamma') = g((r^2 + a^2)\partial_t + a\partial_\varphi, \gamma') = (r^2 + a^2)E - aL$$

and

$$D(\vartheta) = -g(W, \gamma') = -g(\partial_\varphi + as^2\partial_t, \gamma') = L - as^2 E \,.$$

Obviously, we have

$$R + aD = \rho^2 E \,.$$

The functions R and D can be expressed in terms of t' and φ'.

Theorem 16.8 *For a geodesic γ, we have*

$$-at' + (r^2 + a^2)\varphi' = D/s^2 \quad \text{and} \quad t' - as^2\varphi' = R/\Delta \,.$$

Proof By Theorem 16.1, we have

$$R = g(V, t'\partial_t + \varphi'\partial_\varphi) = \Delta t' - \Delta as^2\varphi'$$

and

$$D = -g(W, t'\partial_t + \varphi'\partial_\varphi) = -as^2 t' + (r^2 + a^2)s^2\varphi' \,.$$

Since the coefficient matrix determinant is $\rho^2 \neq 0$, the two equations can be solved for t' and φ'.

Theorem 16.9 *For a geodesic γ we have*

$$\rho^2\varphi' = D/s^2 + aR/\Delta \quad \text{and} \quad \rho^2 t' = aD + (r^2 + a^2)R/\Delta \,.$$

Theorem 16.10 *For a geodesic* γ, *we have*

$$Q = -\frac{\rho^2}{\Delta}(r')^2 - \rho^2(\vartheta')^2 + \frac{R^2}{\Delta\rho^2} - \frac{D^2}{s^2\rho^2} .$$

Proof The vector

$$\gamma' = t'\partial_t + r'\partial_r + \vartheta'\partial_\vartheta + \varphi'\partial_\varphi$$

can be written as

$$\gamma' = r'\partial_r + \vartheta'\partial_\vartheta + \frac{g(\gamma', V)}{g(V, V)}V + \frac{g(\gamma', W)}{g(W, W)}W$$

with the orthonormal system $\partial_r, \partial_\vartheta, V, W$. Thus

$$Q = g(\gamma', \gamma') = (r')^2 g(\partial_r, \partial_r) + (\vartheta')^2 g(\partial_\vartheta, \partial_\vartheta) + \frac{(g(\gamma', V))^2}{g(V, V)} + \frac{(g(\gamma', W))^2}{g(W, W)}$$

$$= -\frac{\rho^2}{\Delta}(r')^2 - \rho^2(\vartheta')^2 + \frac{R^2}{\Delta\rho^2} - \frac{D^2}{s^2\rho^2} .$$

The next theorem has the most cumbersome proof.

Theorem 16.11 *For the geodesic* γ, *we have*

$$\rho^2(\rho^2\vartheta')' = -\frac{1}{2\vartheta'}\left(\frac{D^2}{s^2}\right)' + a^2 sc Q .$$

Proof In the proof of Theorem 10.3, we had determined the Euler-Lagrange equations for the geodesic problem. In the notation used at the time, the Euler equations were

$$\frac{1}{2}\left(\frac{\partial}{\partial x^i}g_{jk}\right)(\xi^j)'(\xi^k)' = \left(\frac{\partial}{\partial x^j}g_{ik}\right)(\xi^j)'(\xi^k)' + g_{ij}(\xi^j)'' .$$

The ϑ-Euler equation is

$$\frac{1}{2}\partial_\vartheta g_{tt}(t')^2 + \frac{1}{2}\partial_\vartheta g_{rr}(r')^2 + \frac{1}{2}\partial_\vartheta g_{\vartheta\vartheta}(\vartheta')^2 + \frac{1}{2}\partial_\vartheta g_{\varphi\varphi}(\varphi')^2 + \partial_\vartheta g_{t\varphi}t'\varphi'$$

$$= (\partial_t g_{\vartheta\vartheta}t' + \partial_r g_{\vartheta\vartheta}r' + \partial_\vartheta g_{\vartheta\vartheta}\vartheta' + \partial_\varphi g_{\vartheta\vartheta}\varphi')\vartheta' + g_{\vartheta\vartheta}\vartheta'' = (g_{\vartheta\vartheta}\vartheta')' ,$$

and so

$$-(\rho^2\vartheta')' = A + B$$

where

$$A = \frac{1}{2}(\partial_\vartheta g_{rr}(r')^2 + \partial_\vartheta g_{\vartheta\vartheta}(\vartheta')^2) = \frac{1}{2}\left(\frac{2a^2 sc}{\Delta}(r')^2 + 2a^2 sc(\vartheta')^2\right)$$

$$= a^2 sc\left(\frac{(r')^2}{\Delta} + (\vartheta')^2\right) = \frac{a^2 sc}{\rho^4}\left(\frac{R^2}{\Delta} - \frac{D^2}{s^2} - \rho^2 Q\right)$$

(Theorem 16.10) and

$$B = \frac{1}{2}(\partial_\vartheta g_{tt}t' + \partial_\vartheta g_{t\varphi}\varphi')t' + \frac{1}{2}(\partial_\vartheta g_{t\varphi}t' + \partial_\vartheta g_{\varphi\varphi}\varphi')\varphi'$$

$$= \frac{1}{2}\left(\partial_\vartheta g_{tt}t' + \partial_\vartheta\left(-\frac{a^2+r^2}{a}g_{tt}\right)\varphi'\right)t' + \frac{1}{2}\left(\partial_\vartheta g_{t\varphi}t' + \partial_\vartheta\left(-\frac{a^2+r^2}{a}g_{t\varphi} - \Delta s^2\right)\varphi'\right)\varphi'$$

$$= \frac{1}{2a}\partial_\vartheta g_{tt}(at' - (a^2+r^2)\varphi')t' + \frac{1}{2a}\partial_\vartheta g_{t\varphi}(at' - (a^2+r^2)\varphi')\varphi' - \Delta sc(\varphi')^2$$

$$= \frac{1}{2a}(at' - (a^2+r^2)\varphi')(\partial_\vartheta g_{tt}t' + \partial_\vartheta g_{t\varphi}\varphi') - \Delta sc(\varphi')^2$$

$$= -\frac{D}{2as^2}C - \frac{\Delta sc}{\rho^4}\left(\frac{D}{s^2} + \frac{aR}{\Delta}\right)^2$$

(Theorem 16.1(4) and (3), Theorems 16.8 and 16.9) with

$$C = \partial_\vartheta g_{tt}t' + \partial_\vartheta g_{t\varphi}\varphi'$$

$$= \frac{1}{\rho^2}\partial_\vartheta g_{tt}(aD + (a^2+r^2)R/\Delta) + \frac{1}{\rho^2}\partial_\vartheta g_{t\varphi}(D/s^2 + aR/\Delta)$$

$$= \frac{D}{s^2\rho^2}(as^2\partial_\vartheta g_{tt} + \partial_\vartheta g_{t\varphi}) + \frac{R}{\Delta\rho^2}\partial_\vartheta(ag_{t\varphi} + (a^2+r^2)g_{tt})$$

$$= \frac{2acD}{s\rho^2}(1 - g_{tt}) + \frac{R}{\Delta\rho^2}\partial_\vartheta\Delta$$

$$= \frac{4ac\mu r D}{s\rho^4}$$

(Theorem 16.9 and Theorem 16.1(2) and (4)). As an interim result, we have

$$-(\rho^2\vartheta')' = \frac{a^2 sc}{\rho^4}\left(\frac{R^2}{\Delta} - \frac{D^2}{s^2} - \rho^2 Q\right) - \frac{2\mu rcD^2}{s^3\rho^4} - \frac{\Delta sc}{\rho^4}\left(\frac{D}{s^2} + \frac{aR}{\Delta}\right)^2 .$$

By expanding the last square and sorting terms by powers of D, we get

$$(\rho^2\vartheta')' = \frac{cD^2}{s^3\rho^4}(2\mu r + \Delta + a^2 s^2) + \frac{2acDR}{s\rho^4} + \frac{a^2 scQ}{\rho^2}$$

and then

$$(\rho^2\vartheta')' = \frac{cD}{s^3\rho^2}(D + 2as^2 E) + \frac{a^2 scQ}{\rho^2}$$

(Definition of Δ and $R = \rho^2 E - aD$). Using the definition of D, this finally yields

$$\rho^2(\rho^2\vartheta')' = \frac{1}{s^4}(scD^2 - s^2 D\partial_\vartheta D) + a^2 scQ = -\frac{1}{2\vartheta'}\left(\frac{D^2}{s^2}\right)' + a^2 scQ\,.$$

Theorem 16.12 *Both sides of the equation*

$$\rho^4(\vartheta')^2 + a^2 c^2 Q + \frac{D^2}{s^2} = -\frac{\rho^4}{\Delta}(r')^2 - Qr^2 + \frac{R^2}{\Delta}$$

are constant along a geodesic γ.

Proof The equation corresponds to Theorem 16.10. We verify the constancy of the left-hand side. By Theorem 16.11, we have

$$\left(\rho^4(\vartheta')^2 + a^2 c^2 Q + \frac{D^2}{s^2}\right)' = 2\rho^2\vartheta'(\rho^2\vartheta')' + \partial_\vartheta(a^2 c^2 Q)\vartheta' + \left(\frac{D^2}{s^2}\right)'$$

$$= 2\vartheta'\left(-\frac{1}{2\vartheta'}\left(\frac{D^2}{s^2}\right)' + a^2 scQ\right) - 2a^2 scQ\vartheta' + \left(\frac{D^2}{s^2}\right)' = 0\,.$$

Definition 16.3 The Carter constant K is the conserved quantity

$$K = \rho^4(\vartheta')^2 + a^2 c^2 Q + \frac{D^2}{s^2} = -\frac{\rho^4}{\Delta}(r')^2 - Qr^2 + \frac{R^2}{\Delta}\,.$$

◆

The importance of this fourth conserved quantity is that we can now write down a system of differentials equations for the geodesic γ. It consists of the two equations from Theorem 16.9 and the two differential equations for K.

Theorem 16.13 *The four coordinate functions of the geodesic* γ *satisfy the following system of differential equations:*

$$\rho^2 t' = aD + (r^2 + a^2)R/\Delta$$

$$\rho^2 \varphi' = D/s^2 + aR/\Delta$$

$$\rho^4 (r')^2 = -\Delta(Qr^2 + K) + R^2$$

$$\rho^4 (\vartheta')^2 = K - a^2 c^2 Q - D^2/s^2 .$$

The geodesic can be determined from the given initial values of t, φ, r, ϑ and the signs of r' and ϑ'.

We now restrict ourselves to motion in the equatorial plane $\vartheta = \pi/2$. Then the fourth equation is omitted, and the Carter constant is reduced to

$$K = D^2 = (L - aE)^2$$

and the equations for t', r', φ' can be traced back to the physically interpretable quantities E (energy) and L (angular momentum). Elementary calculations give

$$t' = \frac{1}{\Delta}\left[\left(r^2 + a^2 + \frac{2\mu a^2}{r}\right)E - \frac{2\mu a}{r}L\right]$$

$$\varphi' = \frac{1}{\Delta}\left[\frac{2\mu a}{r}E + \left(1 - \frac{2\mu}{r}\right)L\right]$$

$$r^2(r')^2 = -\Delta + \frac{2\mu}{r}(L - aE)^2 + (r^2 + a^2)E^2 - L^2 .$$

We finally study this system of differential equations for a particle that starts from rest state $r' = \varphi' = \vartheta' = 0$ at $r = \infty$ in the equatorial plane. From these initial conditions, it follows that $E = 1$ and $L = 0$. The equations are then given by

$$t' = (r^2 + a^2 + 2\mu a^2/r)/\Delta$$

$$\varphi' = (2\mu a/r)/\Delta$$

$$r' = -\sqrt{2\mu(r^2 + a^2)/r^3} .$$

To interpret these equations, it should first be recalled that the independent variable with respect to which the derivative dash is taken, is the proper time of the particle (of the comoving observer). As the particle falls towards the black hole, the proper time increases, r decreases and φ increases. The right-hand side of the equations for φ' shows that the angular velocity φ' depends very much on r, and for $r \searrow r_+$ it tends to ∞. Since the expression for Δ has a pole at $r = r_+$, the improper integral of φ' from r_+ to some upper limit diverges. This means that the particle performs an infinite number of revolutions until the event horizon r_+ is reached. And because of the continuity of the expression for r', this happens in finite proper time.

A Glimpse of String Theory

17

17.1 Quantum Theory Versus Relativity Theory

It has always been a dream of physicists to describe all physical phenomena in the universe via a single, all-encompassing theory. Then all physical phenomena should be deductively derived from this theory. Unfortunately, general relativity is not suitable for this, as shown by the following explanations.

A central result of quantum theory, aimed at the microworld, is the Heisenberg uncertainty principle. This is the inequality

$$\Delta x \Delta p \geq ah.$$

In the context of an experiment, Δx is the maximum possible distance between two measured positions of a particle, Δp the corresponding quantity with respect to the momentum of the particle, h the Planck quantum of action, and a a fixed positive number. In particular, this means that the measurement results $\Delta x = 0$ and $\Delta p = 0$ are not possible. Such precise measurements of the position and momentum of a particle would not contradict the theory of relativity. So the Heisenberg uncertainty principle cannot be derived from the theory of relativity. The theory of relativity is therefore not a theory describing all physical phenomena.

A promising attempt to remedy this deficiency is to break away from the idea that elementary particles are point-like.

17.2 Elementary Particles as Strings

In classical physics, an elementary particle at time t is regarded as a point in three-dimensional space $\mathbb{R}^3$. In contrast, in string theory, an elementary particle at time t is an (extremely tiny) string in $\mathbb{R}^d$. Each type of elementary particle (Leptons, hadrons, quarks,...) has a specific form. The three-dimensional space $\mathbb{R}^3$ is too small to accommodate all of them. The natural number $d > 3$ is different in different versions of string theory. In the standard version, $d = 9$, and in M-theory, $d = 10$.

In the theory of relativity, a point particle moves along a curve in spacetime M, described by a function $\gamma : \mathbb{R} \to M$. At any point t of time, this has a future-pointing, time-like tangent vector $\gamma'(t)$ with

$$g(\gamma'(t), \gamma'(t)) = 1,$$

and which maps a differentiable function f to the number

$$\gamma'(t)f = (f \circ \gamma)'(t).$$

In string theory, the particle is a string of length s_l, that moves in spacetime M. The M-valued function γ describing this motion therefore depends on two variables, and in addition to the time $t \in \mathbb{R}$, we have also a variable $s \in [0, s_l]$, which describes the position of the string at time T, in the form of a parametric representation. This gives, in addition to the tangent vector γ', characterised by

$$\gamma' f = \frac{\partial}{\partial s}(f \circ \gamma),$$

the spatial tangent vector $\dot{\gamma}$, characterised by

$$\dot{\gamma} f = \frac{\partial}{\partial t}(f \circ \gamma).$$

In the time interval $t_1 \leq t \leq t_2$, the string sweeps an (extremely small) surface in spacetime with the parametric representation $\gamma(s, t)$ with $t \in [t_1, t_2]$ and $s \in [0, s_l]$. We are now interested in the size of this area.

In classical analysis, we know how to calculate the are of a curved smooth surface in space $\mathbb{R}^3$ in Cartesian coordinates x, y, z, given by a parametric representation $x(s, t), y(s, t), z(s, t)$: The function $\sqrt{EG - F^2}$ has to be integrated on the param-eter domain. E is the square of the length of the vector $(\frac{\partial x}{\partial s}, \frac{\partial y}{\partial s}, \frac{\partial z}{\partial s})$, G is the square of the length of the vector $(\frac{\partial x}{\partial t}, \frac{\partial y}{\partial t}, \frac{\partial z}{\partial t})$ and F is the dot product of these two vectors. If we formulated the length of the vectors and the scalar product with the tensor g, then the integrand in our context would amount to the square root of the expression

$$g(\gamma)(\dot{\gamma}, \dot{\gamma})g(\gamma)(\gamma', \gamma') - (g(\gamma)(\dot{\gamma}, \gamma'))^2.$$

However, it should be noted here that spacetime is not a Euclidean space, and vectors can also have a negative length, expressed via the tensor g. In fact, the expression formulated above is negative for all s and t. This can be justified as follows: For each pair (s, t) of numbers, the graph of the function on $\mathbb{R}$ given by

$$h(\lambda) = g(\gamma)(\gamma' + \lambda\dot{\gamma}, \gamma' + \lambda\dot{\gamma}) = g(\gamma)(\gamma', \gamma') + 2\lambda g(\gamma)(\gamma', \dot{\gamma}) + \lambda^2 g(\gamma)(\dot{\gamma}, \dot{\gamma})$$

thanks to $g(\gamma)(\dot{\gamma}, \dot{\gamma}) < 0$, is a parabola opening downwards. Since

$$h(0) = g(\gamma)(\gamma', \gamma') > 0$$

the function h then has two distinct zeros. The solution formula for the quadratic equation

$$0 = \frac{h(\lambda)}{g(\gamma)(\dot{\gamma}, \dot{\gamma})} = \lambda^2 + 2\frac{g(\gamma)(\gamma', \dot{\gamma})}{g(\gamma)(\dot{\gamma}, \dot{\gamma})}\lambda + \frac{g(\gamma)(\gamma', \gamma')}{g(\gamma)(\dot{\gamma}, \dot{\gamma})}$$

implies the inequality

$$\left(\frac{g(\gamma)(\gamma', \dot{\gamma})}{g(\gamma)(\dot{\gamma}, \dot{\gamma})}\right)^2 - \frac{g(\gamma)(\gamma', \gamma')}{g(\gamma)(\dot{\gamma}, \dot{\gamma})} > 0,$$

and so

$$(g(\gamma)(\gamma', \dot{\gamma}))^2 - g(\gamma)(\gamma', \gamma')g(\gamma)(\dot{\gamma}, \dot{\gamma}) > 0.$$

17.3   The Extremal Principle

In Sect. 10.1, we postulated that the integral

$$\int_{t_1}^{t_2} \sqrt{g(\gamma(t))(\gamma'(t), \gamma'(t))}\, dt$$

is maximal for $\gamma : [t_1, t_2] \to M$ with $\gamma(t_1) = P$ and $\gamma(t_2) = Q$, describing the force-free motion of a (point) particle in spacetime M from point P to point Q. String theory postulates that for the force-free motion of a particle described by $\gamma : [0, s_l] \times [t_1, t_2] \to M$, the area

$$\int_{t_1}^{t_2} \int_0^{s_l} \sqrt{(g(\gamma)(\dot{\gamma}, \gamma'))^2 - g(\gamma)(\dot{\gamma}, \dot{\gamma})g(\gamma)(\gamma', \gamma')}\, ds dt$$

that the string sweeps is minimal. In order to derive the properties of motion-describing function γ from this extremum principle, we have to apply the calculus of variations technique to the integral

$$\int_{t_1}^{t_2} \int_0^{s_l} L(s, t) ds dt$$

with the integrand

$$L(s, t) = \sqrt{(g(\gamma)(\dot{\gamma}, \gamma'))^2 - g(\gamma)(\dot{\gamma}, \dot{\gamma})g(\gamma)(\gamma', \gamma')}.$$

For an arbitrary function

$$\eta : [0, s_l] \times [t_1, t_2] \to M$$

we define the function

$$h(\varepsilon) = \int_{t_1}^{t_2} \int_0^{s_l} \sqrt{(a(s, t, \varepsilon))^2 - b(s, t, \varepsilon)c(s, t, \varepsilon)} \, ds dt$$

with

$$a(s, t, \varepsilon) = g(\gamma + \varepsilon\eta)(\dot{\gamma} + \varepsilon\dot{\eta}, \gamma' + \varepsilon\eta')$$
$$b(s, t, \varepsilon) = g(\gamma + \varepsilon\eta)(\dot{\gamma} + \varepsilon\dot{\eta}, \dot{\gamma} + \varepsilon\dot{\eta})$$
$$c(s, t, \varepsilon) = g(\gamma + \varepsilon\eta)(\gamma' + \varepsilon\eta', \gamma' + \varepsilon\eta').$$

The extremum principle implies $\frac{dh}{d\varepsilon}(0) = 0$. To obtain $\frac{dh}{d\varepsilon}(0)$, we need the partial derivative of $a^2 - bc$ with respect to ε at $\varepsilon = 0$. Thanks to the bilinearity of the metric g, we have

$$a = g(\dot{\gamma}, \gamma') + \varepsilon g(\dot{\gamma}, \eta') + \varepsilon g(\gamma', \dot{\eta}) + \varepsilon^2 g(\dot{\eta}, \eta')$$
$$b = g(\dot{\gamma}, \dot{\gamma}) + 2\varepsilon g(\dot{\gamma}, \dot{\eta}) + \varepsilon^2 g(\dot{\eta}, \dot{\eta})$$
$$c = g(\gamma', \gamma') + 2\varepsilon g(\gamma', \eta') + \varepsilon^2 g(\eta', \eta'),$$

and altogether

$$\begin{aligned}
a^2 - bc &= (g(\dot{\gamma}, \gamma'))^2 - g(\dot{\gamma}, \dot{\gamma})g(\gamma', \gamma') + 2\varepsilon g(\dot{\gamma}, \gamma')g(\dot{\gamma}, \eta') \\
&+ 2\varepsilon g(\dot{\gamma}, \gamma')g(\gamma', \dot{\eta}) - 2\varepsilon g(\dot{\gamma}, \dot{\gamma})g(\gamma', \eta') - 2\varepsilon g(\gamma', \gamma')g(\dot{\gamma}, \dot{\eta}) \\
&+ 2\varepsilon^2 g(\dot{\gamma}, \gamma')g(\dot{\eta}, \eta') + 2\varepsilon^2 g(\dot{\gamma}, \eta')g(\gamma', \dot{\eta}) \\
&- \varepsilon^2 g(\dot{\gamma}, \dot{\gamma})g(\eta', \eta') - 4\varepsilon^2 g(\dot{\gamma}, \dot{\eta})g(\gamma', \eta') - \varepsilon^2 g(\gamma', \gamma')g(\eta', \dot{\eta}) \\
&+ 2\varepsilon^3 g(\dot{\gamma}, \eta')g(\dot{\eta}, \eta') + 2\varepsilon^3 g(\gamma', \dot{\eta})g(\dot{\eta}, \eta') - 2\varepsilon^3 g(\dot{\gamma}, \dot{\eta})g(\eta', \eta') \\
&- 2\varepsilon^3 g(\eta', \eta')g(\dot{\eta}, \dot{\eta}) + \varepsilon^4 (g(\dot{\eta}, \eta'))^2 - \varepsilon^4 g(\dot{\eta}, \dot{\eta})g(\eta', \eta').
\end{aligned}$$

Fortunately, when calculating $\frac{d}{d\varepsilon}(a^2 - bc)$ at $\varepsilon = 0$, we need to consider the first six summands in the above expression for $a^2 - bc$, since the derivatives of the other summands become zero after insertion of $\varepsilon = 0$. For the first summand

$$(g(\dot{\gamma}, \gamma'))^2 = \left[\sum_{i,k} g_{ik}(\gamma + \varepsilon\eta)(\dot{\gamma})^i (\gamma')^k \right]^2$$

we have

$$\frac{\partial}{\partial\varepsilon} \left(g(\dot{\gamma}, \gamma') \right)^2 \bigg|_{\varepsilon=0} = 2g(\dot{\gamma}, \gamma') \sum_{i,j,k} \frac{\partial g_{ik}}{\partial x_j}(\gamma)(\dot{\gamma})^i (\gamma')^k \eta^j$$

and for the second summand

$$g(\dot{\gamma}, \dot{\gamma})g(\gamma', \gamma') = \left[\sum_{i,k} g_{ik}(\gamma + \varepsilon\eta)(\dot{\gamma})^i (\dot{\gamma})^k \right] \left[\sum_{i,k} g_{ik}(\gamma + \varepsilon\eta)(\gamma')^i (\gamma')^k \right]$$

which gives

$$\frac{\partial}{\partial \varepsilon} g(\dot{\gamma}, \dot{\gamma})g(\gamma', \gamma') \bigg|_{\varepsilon=0} = g(\gamma)(\dot{\gamma}, \dot{\gamma}) \sum_{i,j,k} \frac{\partial g_{ik}}{\partial x_j}(\gamma)(\gamma')^i (\gamma')^k \eta^j$$

$$+ g(\gamma)(\gamma', \gamma') \sum_{i,j,k} \frac{\partial g_{ik}}{\partial x_j}(\gamma)(\dot{\gamma})^i (\dot{\gamma})^k \eta^j.$$

For the summand number three to six, we have

$$\frac{\partial}{\partial \varepsilon} \varepsilon g(\dot{\gamma}, \gamma')g(\dot{\gamma}, \eta') \bigg|_{\varepsilon=0} = g(\dot{\gamma}, \gamma')g(\dot{\gamma}, \eta')$$

$$\frac{\partial}{\partial \varepsilon} \varepsilon g(\dot{\gamma}, \gamma')g(\gamma', \dot{\eta}) \bigg|_{\varepsilon=0} = g(\dot{\gamma}, \gamma')g(\gamma', \dot{\eta})$$

$$\frac{\partial}{\partial \varepsilon} \varepsilon g(\dot{\gamma}, \dot{\gamma})g(\gamma', \eta') \bigg|_{\varepsilon=0} = g(\dot{\gamma}, \dot{\gamma})g(\gamma', \eta')$$

$$\frac{\partial}{\partial \varepsilon} \varepsilon g(\gamma', \gamma')g(\dot{\gamma}, \dot{\eta}) \bigg|_{\varepsilon=0} = g(\gamma', \gamma')g(\dot{\gamma}, \dot{\eta}).$$

Summarising

$$\frac{\partial}{\partial \varepsilon}(a^2 - bc) \bigg|_{\varepsilon=0} = 2g(\dot{\gamma}, \gamma') \sum_{i,j,k} \frac{\partial g_{ik}}{\partial x_j}(\gamma)(\dot{\gamma})^i (\gamma')^k \eta^j$$

$$- g(\gamma)(\dot{\gamma}, \dot{\gamma}) \sum_{i,j,k} \frac{\partial g_{ik}}{\partial x_j}(\gamma)(\gamma')^i (\gamma')^k \eta^j$$

$$- g(\gamma)(\gamma', \gamma') \sum_{i,j,k} \frac{\partial g_{ik}}{\partial x_j}(\gamma)(\dot{\gamma})^i (\dot{\gamma})^k \eta^j$$

$$+ 2g(\gamma)(\dot{\gamma}, \gamma')g(\gamma)(\dot{\gamma}, \eta')$$
$$+ 2g(\gamma)(\dot{\gamma}, \gamma')g(\gamma)(\gamma', \dot{\eta})$$
$$- 2g(\gamma)(\dot{\gamma}, \dot{\gamma})g(\gamma)(\gamma', \eta')$$
$$- 2g(\gamma)(\gamma', \gamma')g(\gamma)(\dot{\gamma}, \dot{\eta}).$$

We can now adopt the viewpoint that for the metric coefficients we have $g_{ik} = 0$ for $i \neq k$, $g_{00} = 1$ and $g_{ii} = -1$ for $i > 0$. Then in particular the partial derivatives of g_{ik} are zero. With that we obtain

$$
0 = \frac{dh}{d\varepsilon}(0) = \int_{t_1}^{t_2} \int_0^{s_l} \frac{\partial}{\partial \varepsilon} \sqrt{a^2 - bc} \bigg|_{\varepsilon=0} ds\,dt = \int_{t_1}^{t_2} \int_0^{s_l} \frac{\frac{1}{2}\frac{\partial}{\partial \varepsilon}(a^2 - bc)}{\sqrt{a^2 - bc}} \bigg|_{\varepsilon=0} ds\,dt
$$

$$
= \int_{t_1}^{t_2} \int_0^{s_l} \frac{g(\dot{\gamma}, \gamma')g(\dot{\gamma}, \eta') + g(\dot{\gamma}, \gamma')g(\gamma', \dot{\eta}) - g(\dot{\gamma}, \dot{\gamma})g(\gamma', \eta') - g(\gamma', \gamma')g(\dot{\gamma}, \dot{\eta})}{\sqrt{a^2 - bc}} ds\,dt.
$$

This applies to all M-valued piecewise differentiable functions η on $[t_1, t_2] \times [0, s_l]$, in particular for those that have only an mth-component η^m and for which $\eta^m(t_1) = 0 = \eta^m(t_2)$. For such η, we have

$$
0 = \int_{t_1}^{t_2} \int_0^{s_l} \frac{g(\dot{\gamma}, \gamma')(\gamma^m)^{\cdot} - g(\dot{\gamma}, \dot{\gamma})(\gamma^m)'}{\sqrt{a^2 - bc}} (\eta^m)' ds\,dt
$$

$$
+ \int_0^{s_l} \int_{t_1}^{t_2} \frac{g(\dot{\gamma}, \gamma')(\gamma^m)' - g(\gamma', \gamma')(\gamma^m)^{\cdot}}{\sqrt{a^2 - bc}} (\eta^m)^{\cdot} dt\,ds.
$$

The integrals inside can be transformed by partial integration into

$$
\int_0^{s_l} \frac{g(\dot{\gamma}, \gamma')(\gamma^m)^{\cdot} - g(\dot{\gamma}, \dot{\gamma})(\gamma^m)'}{\sqrt{a^2 - bc}} (\eta^m)' ds =
$$

$$
- \int_0^{s_l} \frac{\partial}{\partial s} \frac{g(\dot{\gamma}, \gamma')(\gamma^m)' - g(\dot{\gamma}, \dot{\gamma})(\gamma^m)'}{\sqrt{a^2 - bc}} \eta^m ds
$$

and

$$
\int_{t_1}^{t_2} \frac{g(\dot{\gamma}, \gamma')(\gamma^m)' - g(\gamma', \gamma')(\gamma^m)^{\cdot}}{\sqrt{a^2 - bc}} (\eta^m)^{\cdot} dt =
$$

$$
- \int_{t_1}^{t_2} \frac{\partial}{\partial t} \frac{g(\dot{\gamma}, \gamma')(\gamma^m)' - g(\gamma', \gamma')(\gamma^m)^{\cdot}}{\sqrt{a^2 - bc}} \eta^m dt.
$$

Thus we obtain the equation

$$
\int_{t_1}^{t_2} \int_0^{s_l} \left(\frac{\partial}{\partial t} \frac{g(\dot{\gamma}, \gamma')(\gamma^m)' - g(\gamma', \gamma')(\gamma^m)^{\cdot}}{\sqrt{a^2 - bc}} \right.
$$

$$
\left. + \frac{\partial}{\partial s} \frac{g(\dot{\gamma}, \gamma')(\gamma^m)^{\cdot} - g(\dot{\gamma}, \dot{\gamma})(\gamma^m)'}{\sqrt{a^2 - bc}} \right) \eta^m ds\,dt = 0.
$$

Since this equation holds for all η chosen as described above, we conclude that in fact the integrand must be identically zero, that is

$$
\frac{\partial}{\partial t} \frac{g(\dot{\gamma}, \gamma')(\gamma^m)' - g(\gamma', \gamma')(\gamma^m)^{\cdot}}{\sqrt{(g(\dot{\gamma}, \gamma'))^2 - g(\dot{\gamma}.\dot{\gamma})g(\gamma', \gamma')}} + \frac{\partial}{\partial s} \frac{g(\dot{\gamma}, \gamma')(\gamma^m)^{\cdot} - g(\dot{\gamma}, \dot{\gamma})(\gamma^m)'}{\sqrt{(g(\dot{\gamma}, \gamma'))^2 - g(\dot{\gamma}, \dot{\gamma})g(\gamma', \gamma')}} = 0.
$$

The solution γ to the extremum problem formulated above satisfies these equations, the equations of force-free motion of a string.

References

1. ABRAHAM, R., MARSDEN, J.E., RATIU, T.: *Manifolds, Tensor Analysis, and Applications*, Springer, 1983.
2. BERRY, M.: *Kosmologie und Gravitation*, Teubner, 1990.
3. BISHOP, R.L., GOLDBERG, S.J.: *Tensor Analysis on Manifolds*, Macmillan, 1968.
4. CHANDRASEKHAR, S.: *The Mathematical Theory of Black Holes*, Oxford, 1998.
5. DIRSCHMID, H.J.: *Tensoren und Felder*, Springer, 1996.
6. FLIESSBACH, T.: *Allgemeine Relativit"atstheorie*, Bibl.Inst, 1990.
7. GOENNER, H.: *Einf"uhrung in die spezielle und allgemeine Relativit"atstheorie*, Spektrum, 1996.
8. HAWKING, S.W., ELLIS, G.F.R.: *The Large Scale Structure of Space-Time*, Cambridge, 1973.
9. D'INVERNO, R.: *Einführung in die Relativit"atstheorie*, VCH, 1995.
10. J"ANICH, K.: *Vektoranalysis*, Springer, 1992.
11. LANDAU, L.D., LIFSCHITZ, E.M.: *Lehrbuch der theoretischen Physik, Bd.VI, Hydrodynamik*, Akademie-Verlag, 1991.
12. LUMINET, J.- P.: *Schwarze L"ocher*, Vieweg, 1997.
13. v.MEYENN, K.: *Albert Einsteins Relativit"atstheorie*, Vieweg, 1990.
14. MICHOR, P.: *Differentialgeometrie*, Vorlesungsausarbeitung, Wien, 1983.
15. MIESNER, C.W., THORNE, K.S., WHEELER, J.A.: *Gravitation*, Freeman, 1973.
16. O'NEILL, B.: *Semi-Riemannian Geometry*, Academic Press, 1983.
17. O'NEILL, B.: *The Geometry of Kerr Black Holes*, Peters, 1995.
18. RUDER, H. UND M.: *Die Spezielle Relativit"atstheorie*, Vieweg, 1993.
19. SACHS, R.K., WU, H.: *General Relativity for Mathematicians*, Springer, 1977.
20. SCHMUTZER, E.: *Relativit"atstheorie–aktuell*, Teubner, 1979.
21. SCHR"ODER, U.E.: *Spezielle Relativit"atstheorie*, Harry Deutsch, 1981.
22. SEXL, R.U., URBANTKE, H.K.: *Gravitation und Kosmologie*, Bibl.Inst., 1983.
23. STEPHANI, H.: *Allgemeine Relativit"atstheorie*, Verl.d.Wiss., 1977.
24. STRAUMANN, N.: *General Relativity and Relativistic Astrophysics*, Springer, 1984.
25. TRIEBEL, H.: *Analysis und Mathematische Physik*, Teubner, 1981.
26. WARNER, F.W.: *Foundation of Differentiable Manifolds and Lie-Groups*, Springer, 1983.
27. WASSERMAN, R.H.: *Tensors and Manifolds*, Oxford, 1992.
28. WEYL, H.: *Raum–Zeit–Materie*, Springer, 1918.

© The Editor(s) (if applicable) and The Author(s), under exclusive license to Springer
Nature Switzerland AG 2023
R. Oloff, *The Geometry of Spacetime*, Graduate Texts in Physics,
https://doi.org/10.1007/978-3-031-16139-1

Index

A
Age of current cosmos, 241
Alternating, 63
Alternatisation, 65
Arc length, 38
Atlas, 1
  maximal, 4

B
Basis of the topology, 197
Bianchi identity
  first, 97
  second, 175
Big bang, 240
Binet differential equation, 157
Black hole, xii, 138, 223
Boundary of an open set in $\mathbb{R}^n$, 205
Boyer-Lindquist blocks, 244

C
Carter constant, 260
Charge, 52
Charge-current-form, 126
Charge density, 126
Chart, 1
Christoffel symbols, 86
  of a surface, 87
  of Boyer-Lindquist coordinates, 252
  of Schwarzschild coordinates, 93
Components, 27
Connection, 84
  Levi-Civita, 86
  Riemannian, 86

Connection forms, 109, 114, 256
Continuity equation, 119
Contraction, 30
Contraction in length, 117
Contraction of a pair, 31
Conversion
  field strength, xi
  of charge density and current density, 126
  of Christoffel symbols, 87
  of field strengths, 61
  of momentum, 59
  of relative energy, 59
  of relative speed, 56
  of tensor components, 28
Co-oriented bases, 42
Coordinate vector fields, 19
Coordinates
  Boyer-Lindquist, 243
  Kerr-star, 248
  Schwarzschild, 91
Cosmic time, 236
Cosmological constant, 133
Cosmological principle, 234
Cotangent space, 35
Covariant derivative
  for the standard metric, 136
  in $\mathbb{R}^3$ with constant curvature, 232
  in Einstein-de Sitter spacetime, 111
  in Robertson-Walker spacetime, 235
  in Schwarzschild spacetime, 93
  in the Kerr spacetime, 255
  of a tensor field, 171–173
  of a vector field, 86

Printed in the United States
by Baker & Taylor Publisher Services